ACTIVE ELECTRICAL DISTRIBUTION NETWORK

ACTIVE ELECTRICAL DISTRIBUTION NETWORK

Issues, Solution Techniques, and Applications

Edited by

SANJEEVIKUMAR PADMANABAN
Department of Electrical and Electronics Engineering, KPR Institute of Engineering and Technology, Tamilnadu, India

BASEEM KHAN
Department of Electrical and Computer Engineering, Faculty of Electrical Engineering, Institute of Technology, Hawassa University, Ethiopia

OM PRAKASH MAHELA
Power System Planning Division, Rajasthan Rajya Vidyut Prasaran Nigam Limited, Jaipur, India

HASSAN HAES ALHELOU
Tishreen University, Lattakia, Syria

S. RAJKUMAR
Department of Mechanical Engineering, Faculty of Manufacturing, Institute of Technology, Hawassa University, Ethiopia

ELSEVIER

ACADEMIC PRESS
An imprint of Elsevier

Academic Press is an imprint of Elsevier
125 London Wall, London EC2Y 5AS, United Kingdom
525 B Street, Suite 1650, San Diego, CA 92101, United States
50 Hampshire Street, 5th Floor, Cambridge, MA 02139, United States
The Boulevard, Langford Lane, Kidlington, Oxford OX5 1GB, United Kingdom

ISBN 978-0-323-85169-5

For information on all Academic Press publications
visit our website at https://www.elsevier.com/books-and-journals

Publisher: Charlotte Cockle
Acquisitions Editor: Lisa Reading
Editorial Project Manager: Michelle Fisher
Production Project Manager: Nirmala Arumugam
Cover Designer: Greg Harris

Typeset by STRAIVE, India

Contents

SECTION 4 Distribution system state estimation

SECTION 5 Micro phasor measurement units' placement

SECTION 6 Micro-grid optimization and integration

SECTION 7 Electric vehicle technology

SECTION 12 Demand side response techniques

SECTION 13 Smart metering

SECTION 14 Optimization of operational issues

Contributors

Dessalegn Bitew Aeggegn
School of Electrical and Computer Engineering, Debre Markos University, Debre Markos, Ethiopia

Takele Ferede Agajie
Department of Electrical and Electronics Engineering, University of Buea, Buea, Cameroon; Debre Markos University, Debre Markos; Hawassa University, Hawassa, Ethiopia

Ziad M. Ali
College of Engineering at Wadi Addawasir, Prince Sattam Bin Abdulaziz University, Wadi Addawasir, Saudi Arabia

Yayehyirad Ayalew Awoke
School of Electrical and Computer Engineering, Debre Markos University, Debre Markos, Ethiopia; Department of Electrical and Electronics Engineering, University of Buea, Buea, Cameroon

Mulusew Ayalew
Department of Electrical and Computer Engineering, Hawassa University, Hawassa, Ethiopia

I. Baranilingesan
Renewable Energy Lab, Department of Electrical and Electronics Engineering, KPR Institute of Engineering and Technology, Coimbatore, Tamil Nadu, India

R. Bharathwaaj
Department of Mechanical Engineering, KPR Institute of Engineering and Technology, Coimbatore, Tamil Nadu, India

Gajendra Singh Chawda
Department of Electrical Engineering, IIT Jodhpur, Jodhpur, India

V. Kumar Chinnaiyan
Renewable Energy Lab, Department of Electrical and Electronics Engineering, KPR Institute of Engineering and Technology, Coimbatore, Tamil Nadu, India

Emenike C. Ejiogu
Laboratory of Industrial Electronics, Power Devices and New Energy Systems; Africa Centre of Excellence for Sustainable Power and Energy Development, University of Nigeria, Nsukka, Nigeria

Mkpamdi Eke
Department of Mechanical Engineering, University of Nigeria, Nsukka, Nigeria

Ragab A. El-Sehiemy
Kafrelsheikh University, Kafrelsheikh, Egypt

Ahmad Esmaeili
Power Systems Operation and Planning Research Department, Niroo Research Institute, Tehran, Iran

Mesfin Fanuel
Debre Markos University, Debre Markos; Hawassa University, Hawassa, Ethiopia

Damot Gebregziabher
Department of Electrical and Computer Engineering, Hawassa University, Hawassa, Ethiopia

Josep M. Guerrero
The Villum Center for Research on Microgrids (AAU CROM), Aalborg University, Aalborg, Denmark

Issaias Gidey Hagos
Department of Electrical and Computer Engineering; Institute of Technology, Hawassa University, Hawassa, Ethiopia

Amoussou Isaac
Department of Electrical and Electronics Engineering, University of Buea, Buea, Cameroon

Salah Kamel
Department of Electrical Engineering, Aswan University, Aswan, Egypt

J. Karpagam
Renewable Energy Lab, Department of Electrical and Electronics Engineering, KPR Institute of Engineering and Technology, Coimbatore, Tamil Nadu, India

Alagar Karthick
Renewable Energy Lab, Department of Electrical and Electronics Engineering, KPR Institute of Engineering and Technology, Coimbatore, Tamil Nadu, India

Ola Khaddam
Electrical Power Engineering, Tishreen University, Latakia, Syria

Baseem Khan
Department of Electrical and Computer Engineering; Institute of Technology, Hawassa University, Hawassa, Ethiopia

Ashish Kumar
Department of Electrical Engineering, NIT Hamirpur, Hamirpur, India

Pankaj Kumar
Department of Electrical Engineering, RTU Kota, Kota, India

Ravita Lamba
Department of Electrical Engineering, Malaviya National Institute of Technology Jaipur, Jaipur, India

Chika Maduabuchi
Department of Mechanical Engineering, University of Nigeria, Nsukka, Nigeria

Om Prakash Mahela
Power System Planning Division, Rajasthan Rajya Vidyut Prasaran Nigam Ltd., Jaipur, Rajasthan, India

Ram Niwash Mahia
Department of Electrical Engineering, NIT Hamirpur, Hamirpur, India

Abbas Marini
Power Systems Operation and Planning Research Department, Niroo Research Institute, Tehran, Iran

V. Mohanavel
Centre for Materials Engineering and Regenerative Medicine, Bharath Institute of Higher Education and Research, Chennai, Tamil Nadu, India

Howard O. Njoku
Applied Renewable and Sustainable Energy Research Group, Department of Mechanical Engineering, University of Nigeria, Nsukka, Nigeria; Department of Mechanical Engineering Science, University of Johannesburg, Johannesburg, South Africa

Chigbogu Ozoegwu
Department of Mechanical Engineering, University of Nigeria, Nsukka, Nigeria

S. Rajkumar
Department of Mechanical Engineering, Faculty of Manufacturing, Institute of Technology, Hawassa University, Hawassa, Ethiopia

Abdelhady Ramadan
Department of Electrical Engineering, Aswan University, Aswan, Egypt

Manickam Ravichandran
Department of Mechanical Engineering, K. Ramakrishnan College of Engineering, Trichy, Tamil Nadu, India

T. Sakthi
Department of Mechanical Engineering, National Engineering College, Kovilpatti, Tamil Nadu, India

Morteza Shabanzadeh
Power Systems Operation and Planning Research Department, Niroo Research Institute, Tehran, Iran

Avdhesh Sharma
Department of Electrical Engineering, Faculty of Engineering and Architecture, Jai Narain Vyas University, Jodhpur, Rajasthan, India

Surendra Singh
Department of Electrical Engineering, Faculty of Engineering and Architecture, Jai Narain Vyas University, Jodhpur, Rajasthan, India

Vishwajeet Kumar Sinha
Department of Electrical Engineering, NIT Hamirpur, Hamirpur, India

Endeshaw Solomon
Department of Electrical and Computer Engineering; Institute of Technology, Hawassa University, Hawassa, Ethiopia

S. Vasanthaseelan
Department of Mechanical Engineering, KPR Institute of Engineering and Technology, Coimbatore, Tamil Nadu, India

SECTION 1

Introduction of electrical distribution network

CHAPTER 1

Electrical distribution network: An introduction

Baseem Khan
Department of Electrical and Computer Engineering, Hawassa University, Hawassa, Ethiopia

Contents

1 Introduction

Distribution carries energy from the transmission point and supplies it to users. It is the last step in the supply of electricity. Low-voltage substations (distribution substations) link to the transmission lines through transformers and lower the voltage of the transmission to a medium level, which ranges from 2 to 35 kV. Fig. 1 shows a circuit diagram of the electrical power system [1,2].

The primary goal of distribution is to supply energy to customers after obtaining large amounts of power through a transmission substation. Primary and secondary (customer) substations are the two main types of

Active Electrical Distribution Network
https://doi.org/10.1016/B978-0-323-85169-5.00016-2

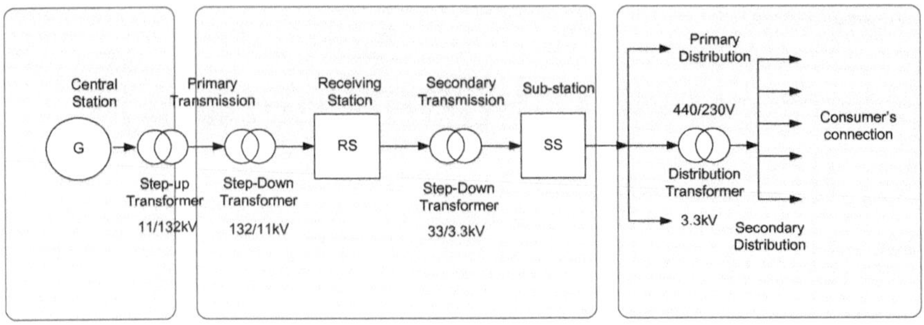

Fig. 1 Electrical power system.

distribution substations. The customer substation communicates with the low-voltage (LV) network, whereas the major substation serves as a load center. A customer substation is a distribution room that is often provided by the client. The transformer and a number of high-voltage switchgear panels can be accommodated in the distribution room to permit LV connection to the customer's incoming switchboard [3].

Transmission and distribution meet at a power substation, which has many functions. The transmission grid can be disconnected from the substation, and distribution lines can be terminated with circuit breakers and switches. Transformers convert high-voltage transmission lines (35 kV or more) to low-voltage main distribution lines. This is a medium-voltage line with a voltage range of 600–35,000 V. The transformer delivers electricity to the bus-bar, which can divide the power for distribution in many ways. The bus transports energy to distribution lines, which subsequently deliver it to customers [4].

The distribution network might take the shape of overhead wires or underground cables, depending on the geographical area. In metropolitan areas, cables are often employed, but in rural regions, overhead lines are used. In order to meet the required supply reliability, various network designs are conceivable. To ensure that the distribution network runs smoothly, protection, control, and monitoring equipment is installed.

2 History

When electricity began to be created at power stations in the 1880s, electric power distribution became necessary. Previously, power was typically generated onsite. The first power distribution systems built in European and American towns were with arc lighting (approximately 3000 V) alternating current (AC) or direct current (DC) and incandescent lighting (100 V) DC.

For street lighting, arc lighting is used while gas is replaced by incandescent lighting for household and commercial illumination [3].

As high voltage is utilized in arc lighting, a single generating station may power circuits that are up to 11-km long [4]. For transporting four times the power, the voltage is doubled with the same size conductor and power losses. In 1882, the Edison Pearl Street Station was built, but it had trouble reaching consumers more than a mile distant. This was because of the use of a 110 V LV system. This system required conductor lines (thick copper) and generating facilities within 2.4 km of the furthest consumer to prevent cables that were too large and costly.

3 Structure of power distribution network

A power system developed near or within a city or industrial zone is known as a distribution system. It receives electricity through transmission lines. A step-down transformer is used to reduce the high transmission voltage to low distribution voltage [5–8]. Generally, electricity distribution is performed at 11 kV. It can vary from 2.4 to 33 kV, depending upon the demand and type of customer. The structure of the electricity distribution network is depicted in Fig. 2.

3.1 Distribution system components

Generally, an electric distribution system consists of [9]:

- Low-voltage substation
- Feeder lines

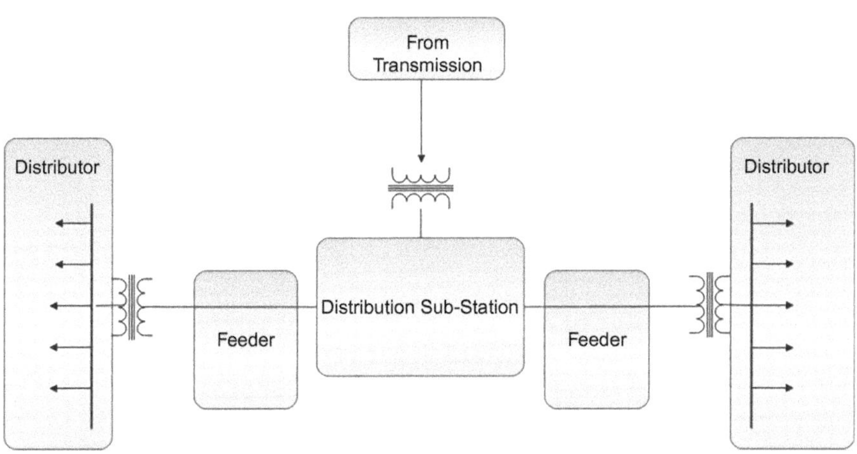

Fig. 2 Structure of a power distribution network.

- Transformers (distribution)
- Distributors
- Service mains

Switches, protection devices, measuring devices, and other components are included in a distribution system.

A. Feeder lines

The voltage, which is stepped down, is carried via feeder cables after the substation through the distribution transformers. No tappings are taken from the feeders in the majority of situations, ensuring that the current remains constant. The current-carrying capacity is the most essential aspect to consider while constructing a feeder (Fig. 3).

B. Transformer (distribution level)

A transformer at the distribution level, sometimes known as a service transformer, is the last transformer in the power network. Generally, it is a three-phase step-down transformer. It steps down the input supply to 400Y/230V. The phase voltage (the voltage between any phase and the neutral) is 230V while the line voltage is 400V.

C. Distributors

The transformer's output is passed through the distributor. The distributor's tappings are used to distribute electricity to users. In a distributor, the flow of current is not continuous because tappings are performed at different locations along its distance. As a result, the drop in voltage over the distance of a distributor is the most essential element to consider.

D. Service main

This is a tiny cable that runs from the nearest pole's distributor conductor to the customer's premises.

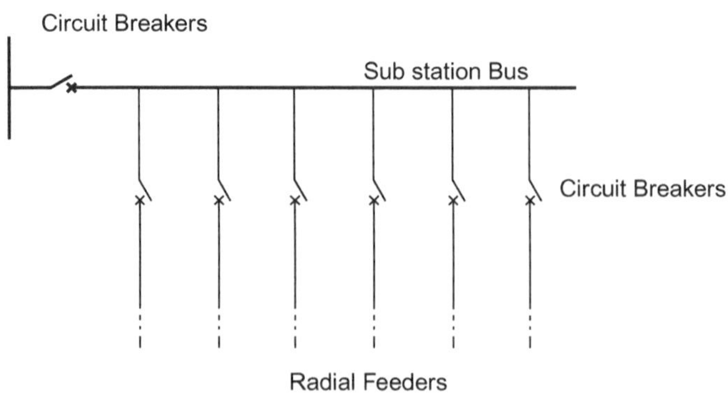

Fig. 3 Feeders.

4 Classification of distribution network

Distribution systems can be classified as follows [10].

4.1 Type of nature of current

Based on the type of the current, distribution systems may be divided into the following categories:

- Direct current (DC)
- Alternative current (AC)

4.1.1 DC distribution system

DC technology was used to build the first electrical distribution networks, which were built by Edison near the end of the 19th century. AC systems, on the other hand, proved to be considerably better than DC systems at the time, and AC systems were widely employed for power production, transmission, and distribution. The fact that electricity is nearly entirely produced, transferred, and supplied as AC is general information. Some industrial systems, however, necessitate the use of DC power. Electrical machines such as DC machines and industrial processes such as electrochemical processes demand DC power.

The DC supply from the substation might have the following forms:

- Two-wire
- Three-wire

A. Two-wire

This distribution system is made up of two wires (+ and −), as the name suggests. The positive (+) wire is the outgoing conductor while the negative conductor provides the return path. Between these two wires, loads are interconnected. The two-wire system is shown in Fig. 4.

B. Three-wire

As shown in Fig. 5, this is made up of two outside wires and a neutral conductor in the middle, which is earthed at the substation. The difference in voltage between the outer and neutral conductors is two times the voltage difference between the outer and neutral wires.

High-voltage loads (such as motors) are linked in parallel to the outside conductors while low-voltage loads are linked among the outer and the neutral.

4.1.2 AC distribution system

Electricity is basically generated, transmitted, and distributed in AC form. A distribution system normally starts at a substation from which a

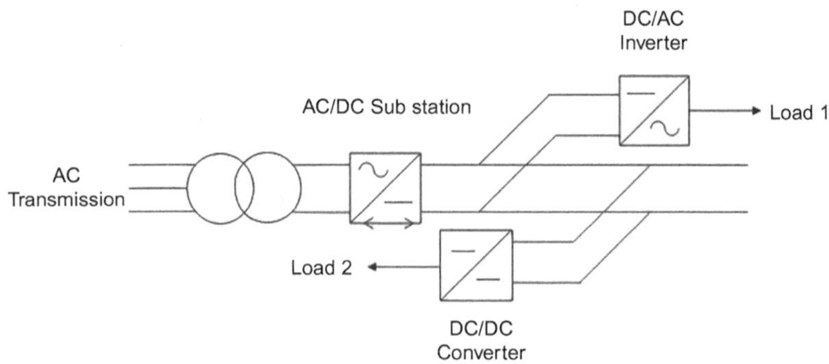

Fig. 4 Two-wire system for distribution.

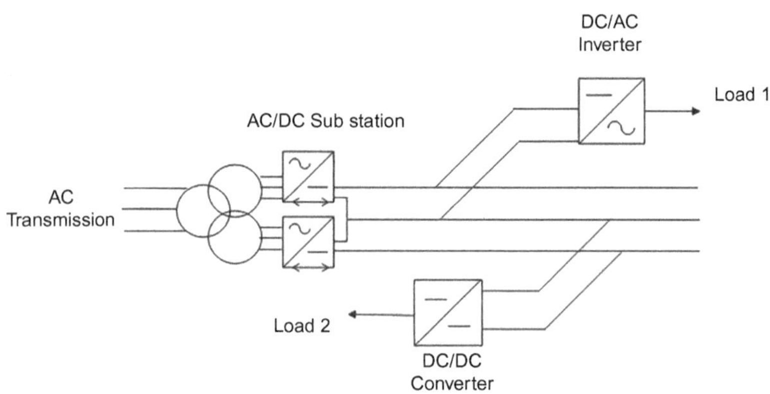

Fig. 5 Three-wire system for distribution.

transmission network delivers power. The distribution system may start at the generating station in some cases, such as loads available near the generating system. The terms primary and secondary distribution are used for large areas and industrial locations. The detailed description of these two types of distribution systems is as follows:

- Primary
- Secondary

A. Primary

This operates at slightly higher voltages than normal and can handle larger amounts of electricity than a regular low-voltage consumer. Because of the expense, primary distribution is done with a three-phase, three-conductor network.

The energy from the generator is sent to a substation at a high potential. Afterward, the potential is lowered to 11 kV using a step-down transformer. Fig. 1 shows the basic principle of a main distribution system.

B. Secondary

The rating of secondary AC distribution is 400/230 V. It is a three-phase, four-conductor network. Electricity is distributed to a number of distribution substations via the major distribution circuit. As illustrated in Fig. 1, the line potential is 400 V while the phase potential is 230 V. Domestic loads that are normally one-phase are linked to the neutral via any one phase. On the other hand, commercial loads, which are generally three-phase, are linked directly to the three-phase lines. The AC method is currently extensively utilized for distributing electricity due to its simplicity and cost effectiveness compared to the DC system.

4.2 Based on the types of connection

Based on the scheme of connection, distribution networks can be categorized into the following categories:

- Radial
- Ring main
- Interconnected

A. Radial

This system utilizes distinct feeder lines extending from a substation and serve the distributor conductors from just one end. A radial distribution system is depicted as a single line diagram in Fig. 6. This type of distribution is useful when the supply voltage is low and the substation is within the city. The advantages of this system are its simplicity and low initial cost.

It does, however, contain the shortcomings listed below.

(1) The distributor's end near the supply point will be severely loaded.
(2) All the customers are served by a lone distributor and feeder.

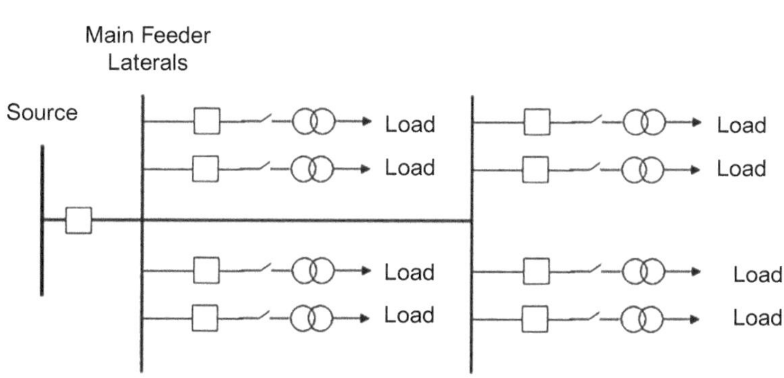

Fig. 6 Electricity distribution in a radial way.

(3) As a result, outages in any feeder or distributor lines cut power to customers on the fault's side.

(4) Considerable voltage deviations are faced by the consumers at the far end due to fluctuations in the load on the distributor.

Because of these drawbacks, this technique is only employed over small lengths. For further expansion of the radial network, more laterals and sublaterals are required.

B. Ring main

As the name indicates, in this system the distribution transformer's primary winding creates a ring. It is initiated from the bus-bars of the substation, loops over the service area, and then returns to the substation. Fig. 7 shows the ring main system.

Some of the advantages of the ring main system include:

(1) The distribution quality is good as there are fewer voltage deviations at the consumer end.

(2) The reliability of the system is increased compared to the radial system as each distributor is supplied by two feeder lines. Therefore, with an outage in any part of the feeder, the supply will be continuously available.

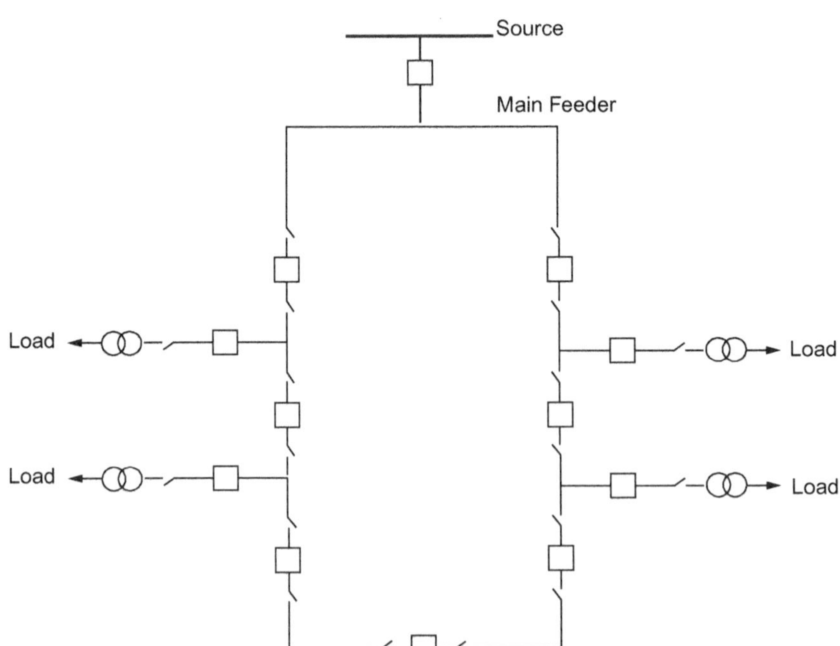

Fig. 7 Ring main system.

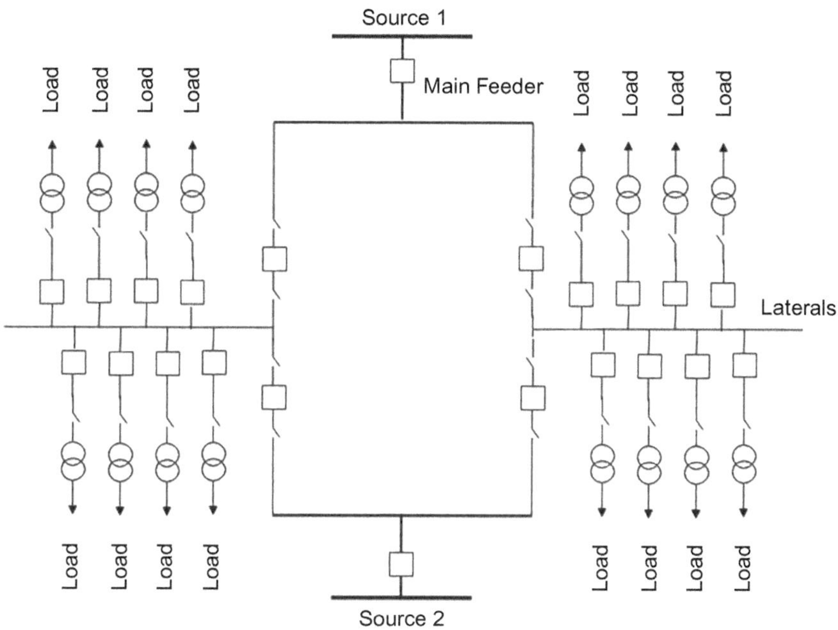

Fig. 8 Interconnected systems.

C. Interconnected network

The ring of the feeder of an interconnected system is electrified by two or more sources. Fig. 8 depicts a single line representation of an interconnected system.

The following are some of the benefits of an interconnected system:

(1) It improves service dependability.

(2) Any region served by one producing station can be fed by other generating stations during peak demand hours. This diminishes the system's backup power capacity while increasing its efficiency.

5 Features of ideal distribution system

It requires a lot of effort to keep energy distribution that fulfills the demands of numerous categories of customers. For adequate distribution of electricity, different criteria include:

- Proper voltage
- Energy availability on demand
- Reliable system

5.1 Proper voltage

Voltage deviation at the user's end must be as low as possible. It is a critical criterion. Load variation on the system typically causes variations in voltage. Low voltage results in revenue loss, inadequate lighting, and the risk of motor burnout. High voltage can permanently burn out bulbs and cause other equipment to fail.

5.2 Energy availability on demand

Consumers must be able to obtain power in any amount they demand at any time. Without notifying the electric utility, motors and lights can be started or stopped, and lights can be switched on or off. The distribution system must be able to satisfy customer load needs because electrical energy cannot be stored. This requires that operation staff research load trends on a regular basis to anticipate big load fluctuations that follow established schedules.

5.3 Reliable supply

The operation of modern industry is nearly entirely reliant on electric power. Electric power is used to light, heat, chill, and ventilate homes and office buildings. This necessitates dependable service. Electric power, like everything else created by humans, can never be completely trustworthy. However, there are a number of things that may be done to improve reliability:

- System interconnection.
- Automatic control system.
- Reserve facility incorporation.

6 Design requirements

A distribution network's voltage control is most likely the most significant aspect in providing good service to customers. Feeder and distributor design must be carefully considered for this reason.

6.1 Feeders

The current carrying capacity of a feeder is the most essential feature because the voltage drop is very negligible in it. This is the case because voltage-regulating technology at the substation can compensate for voltage drops in a feeder.

6.2 Distributors

When constructing a distributor, the voltage drop is taken into account. This is because consumers are serviced by a distributor, and potential fluctuations at the consumer's terminals are restricted by law (6% of rated value).

6.3 Earthing

It is critical that distribution systems are properly earthed so that high voltages do not occur on individual customer connections.

The points connected to the earth are the transformer secondary output, the load point with a local meter and protection fuse, the neutral conductor of the four-wire system, and the star point of the low voltage winding on the step-down transformer. Fig. 9 shows the protective multiple earth system.

The protective multiple earth (PME) system protects the metallic coverings and equipment from the supply. This system prevents dangerously high voltages that could jeopardize people's lives.

7 Conclusion

A brief overview of the electrical distribution system is presented in this chapter. The chapter discusses the background and history of the distribution system. The structure of the network and its types are also discussed in detail. Moreover, the features and design requirements of the electrical distribution network are also discussed.

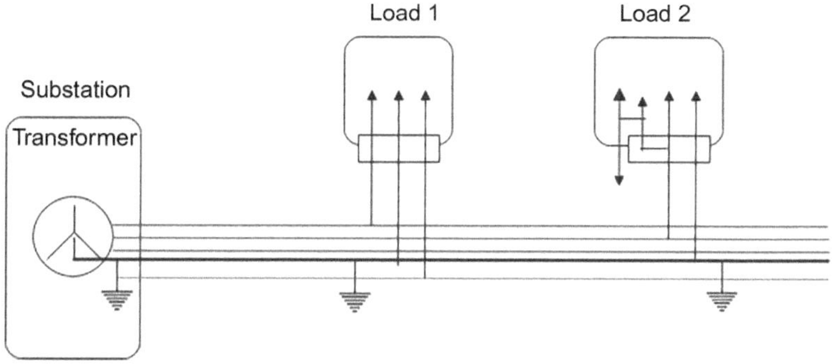

Fig. 9 Protective multiple earth system.

References

[1] T.A. Short, Chapter 1—Fundamentals of distribution systems gives an introduction on distribution network and its components; Chapter 8—Short circuit protection outlines the basic principle and calculation of distribution protection, in: Electric Power Distribution Handbook, CRC Press, Boca Raton, FL, 2004.
[2] J.J. Grainger, W.D. Stevenson Jr., Chapter 10—Symmetrical fault gives logical discussion and numerical examples on 3-phase symmetrical fault, in: Power System Analysis, McGraw-Hill, New York, 1994.
[3] N. Hadjsaïd, J.-C. Sabonnadi're, Electrical Distribution Networks, Wiley, 2013.
[4] E. Lakervi, E.J. Holmes, Electricity Distribution Network Design, second ed., IET, 2003.
[5] A. Arefi, F. Shahnia, G. Ledwich, Electric Distribution Network Management and Control, Springer Singapore, 2018.
[6] T.F. Agajie, B. Khan, J.M. Guerrero, O.P. Mahela, Reliability enhancement and voltage profile improvement of distribution network using optimal capacity allocation and placement of distributed energy resources, Comput. Electr. Eng. 93 (2021) 107295.
[7] T.F. Agajie, B. Khan, H.H. Alhelou, O.P. Mahela, Optimal expansion planning of distribution system using grid-based multi-objective harmony search algorithm, Comput. Electr. Eng. 87 (2020) 106823.
[8] B. Khan, H.H. Alhelou, F. Mebrahtu, A holistic analysis of distribution system reliability assessment methods with conventional and renewable energy sources, AIMS Energy 7 (4) (2019) 413–442.
[9] https://www.electricaleasy.com/2018/01/electric-power-distribution-system.html (Accessed 16 June 2021).
[10] https://electrical-engineering-portal.com/electrical-distribution-systems (Accessed 18 June 2021).

SECTION 2

Issues related to existing electrical distribution network

CHAPTER 2

Electrical distribution network: Existing problems

Baseem Khan[a] and Josep M. Guerrero[b]
[a]Department of Electrical and Computer Engineering, Hawassa University, Hawassa, Ethiopia
[b]The Villum Center for Research on Microgrids (AAU CROM), Aalborg University, Aalborg, Denmark

Contents

1 Introduction

Power is generated at the generating plant and transmitted at a higher voltage through transmission lines. Distribution substation transformers scale down the transmission system voltage to lower levels. The part of the power network between the distribution substation and the usage transformers is known as the primary distribution system. The primary distribution system is made up of circuits known as primary or distribution feeders that begin at the distribution substation's secondary bus. In big industrial or commercial applications, the distribution substation is generally the point of electric power supply [1].

Active Electrical Distribution Network
https://doi.org/10.1016/B978-0-323-85169-5.00017-4

Distribution networks provide a critical and often challenging role in ensuring that we, the customers, receive dependable and safe power. Compared to their transmission counterparts, they often have lower budgets and considerably broader service regions. As a result, network management is a major endeavor. Obviously, each network region and the difficulties it faces are distinct. However, it is thought that there are a number of common points that impact the majority of operators [2].

Customers must call to report an issue because the network, particularly at lower power levels, has little or no real-time monitoring. Although some automatic control actions are available, remote control is only available to a limited extent. At lower voltage levels, an "automated" reaction may consist of merely removing a load, generator, or network segment until an engineer arrives.

2 Problems associated with distribution system

The traditional distribution network faces various problems [2]. Some of these issues are presented in Fig. 1.

2.1 Inadequacy of the existing grid

The traditional sustainability of the large-scale power grid (generation and transmission) has already been deteriorating for more than a decade and is anticipated to continue to do so at an increasing rate [2].

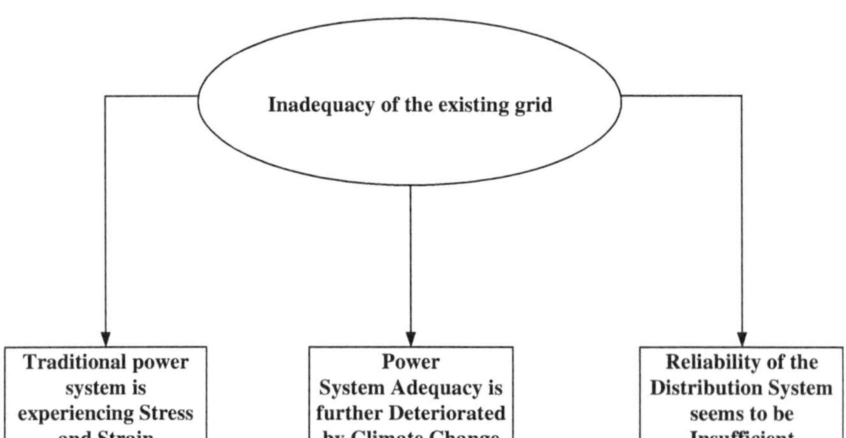

Fig. 1 Inadequacy issues in the existing grid.

A. The traditional power system is experiencing stress and strain

The large-scale power grid is just failing. The cost of depreciation surpasses the cost of fresh investment. Because of the following reasons, economic expansion is not catching up with network degradation [3]:

(1) The rate of continuous exponential rise in power consumption has slowed.

(2) Environmental and sustainability-related legislative and regulatory hurdles obstruct, if not outright impede, the siting, building, and operation of new grid infrastructure.

(3) In a more competitive economy, the costs of new generating, transmission, and distribution plants is rising, and any major new construction means increased rates for customers.

(4) Utilities stand a substantial danger of not recouping all their expenses, much alone receiving a sufficient return on new infrastructure investments.

As a consequence, the traditional large energy grid is starting to show its age, and it is progressively being maintained in "run to fail" mode. As a result, significant disruptions have almost quadrupled each 5 years over the last 20 years.

B. Power system adequacy is further deteriorated by climate change

Climate change puts grid adequacy at jeopardy. Apart from the contentious relationship between climate change and the use of carbon-based fuels in utilities, the frequency, length, and intensity of weather events have been progressively increasing. Even though the ultimate causes are debatable, the negative impact on grid dependability is obvious [2].

In the meantime, when it comes to climate change, society is increasing opposed to the use of carbon-based fuels, believing that they are partly responsible for climate change. Moreover, the use of "dirty" or hazardous fuels pollutes our air, land, and water more and more. "Perception is reality," as the old adage goes. If users start believing this or accepting it, you are impacted by it. It not only restricts the deployment of new power generation, but also the operation of current generation [4].

C. The reliability of the distribution system seems to be insufficient

Distribution network breakdowns generate more customer support disruptions than large electric grid breakdowns. Although the electric grid's capability is deteriorating, flawless large electric utility dependability does not imply perfect distribution grid reliability. However, as in the digital world, it will become extremely relevant.

2.2 Physical and cyber security issues of the conventional grid

The power network architecture is vulnerable, as if there weren't already sufficient risks to system sufficiency and dependability. Although data protection has received a lot of attention in the business, it is not the sole or even the most significant issue. Information security may ultimately pose as awesome a danger to power system reliability as it does to utility company and consumer data protection, as the power system is becoming more virtual and automated. But even more alarming, the grid is incredibly vulnerable in terms of physical security [2,5].

There isn't a vital component in the network (e.g., production, transmitting, or distributing) that can't be accessed by almost anyone, generally without even being observed or screened by the company or any other agency. Even if they don't reach or cut through the barrier, they may certainly get close enough to deliver a punch and surely close enough to fire a rifle or rocket-propelled grenade. Direct assaults on the network, including the main electric grid and energy distribution systems, are now on the increase. Identifying and reacting to direct assaults as well as predicting and avoiding them will become especially critical.

One of the most serious cybersecurity concerns of the day may be the grid's physical vulnerability. Even so, the web is reliant on consistent energy. There can be no web without power. The only way to assure a dependable web in the present state of things, with the electric grid's deteriorating reliability and worrying physical insecurity, is to discover solutions to traditional network connectivity.

2.3 Traditional vs future grid

Customers are looking for options to traditional grid service as costs go up, dependability falls, and sustainability has become more essential. They come up with a growing number of methods to spend less money with their dominant cost-plus monopolistic utilities. Companies often seek sustainability and energy economy, which are the most straightforward and reliable options [2,6].

With demand response and crucial peak pricing efforts, utilities have helped customers become more aware of their energy consumption and alternatives. Those were all designed to alter consumer habits to keep the traditional system alive. Consumers, on the other hand, are not driven to modify their habits to help their electric company stay afloat. However, people are much more concerned with more practical and cost-effective methods to get, control, and consume electricity with their own life quality, company productivity, and investment goals.

Consumers are prepared to give up only so much in terms of comfort to reduce consumption, even during periods when they need electricity the most. They generally seek to restrict consumer usage at peak hours when providers are looking to buy electricity. Consumers are discovering new options on their side of the system, including distributed generation (DG) units such as traditional emergency power, solar rooftop panels, combined heat and power (CHP) plants, or even microgrids. With the introduction of electric vehicles, battery banks, smart homes and buildings, and other technologies, even the way customers use and manage their electric energy is changing dramatically.

2.4 Integration of distributed generation, energy storage with management systems

The traditional power system is supplied by company-owned generating facilities. These are deployed from a centralized spot to transfer electricity through power lines that transport it to load centers, where it is distributed to users via utility meters. Distributed generation, storage, and administration are turning this centralized, one-way approach on its head. At the grid's distribution edges, there are currently hundreds of thousands of energy generation units on the consumer sides of the meters [2,7].

Self-generation is seen in about half of all major enterprises. Between now and 2030, it is anticipated that nearly all new generation capacity in the electric grid will be customer-owned and operated, accounting for 30% of total capacity.

The use of distributed energy is increasing at an exponential rate. Consider the rapidly expanding market for distributed solar PV production. Every 3–4 min, a new rooftop solar array is installed.

2.5 Incorporation of novel disruptive enabling techniques

This is a sort of innovation that fundamentally alters the way consumers, businesses, and organizations operate. Disruptive technology washes away the institutions or practices it substitutes because it has plainly superior features.

A fast expanding variety of new power, electrical, information, and telecommunications technologies, gadgets, and applications underpins the advances mentioned above. For the same reasons that Moore's Law holds true for integrated circuits, their performance vs. cost improves rapidly. That is, their prices are constantly decreasing while their worth is steadily increasing. This hastens the demise of traditional energy production, transmission, distribution, sales, administration, and usage methods, which have all hit the limit of their usefulness while continuing to rise in price [2,8].

Technological advances are nonetheless changing the power system from the boundaries inside, but they are also addressing the problems they provide as well as taking advantage of totally novel opportunities. Electricity companies should keep on top of who they are, what they are doing, and how they might effectively use existing resource scheduling to run and govern their growing, difficult, and complex operations.

2.6 Enhance complexities

The cost-plus monopolistic business model, which is fairly simple, vertically integrated, centrally monitored, and regulated, will be on its way out. Inside its replacement, a novel, decentralized paradigm with more elements is gaining traction. Power companies are unlikely to control and supervise these from a centralized location. Consumers will still own and run them autonomously to their own advantage. They'll be tiny, irregular, and perhaps unpredictable. These would not be run solely to preserve the old grid's dependability and efficacy [2,9].

By adding, the ever-increasing number of electric cars that constitute roaming electricity utilization are also equivalent to a household's demand. The electric energy industry defies conventional centralized supervision and control. In the presence of probabilistic rather than deterministic factors, new techniques of detecting, assessing, and operating electric distribution networks will be required. A really smart distribution grid, not simply so-called "smart" meters and time-of-use pricing systems, will be required to accommodate more autonomous moving parts.

The "grid edge" is the term used to describe this area. The much more significant and far-reaching changes in the electricity company sector in the near future will occur at the grid's distribution edges, rather than in the bulk power grid. The majority, if not all, of these changes will occur on the consumer side of the devices. This presents utilities (distribution) with significant difficulties as well as wonderful possibilities to provide new and improved goods and services to their customers and localities.

Not only is complexity linked to planning, operations, and management, but it's also linked to data management, or "big data," data that must be perceived, evaluated, and acted upon in real time.

2.7 Revamping of generation cost and revenue computation

Generation costs, both variable and constant, are continuously increasing. Transmission and distribution costs are also increasing. Doing business is becoming more expensive. Utility earnings from electricity services, on the other hand, are falling owing to savings, efficacy, microgrids, and market competitiveness. The usual method of raising prices to achieve financial sustainability with a profit simply encourages consumers to explore other options [2,10].

Energy sector companies typically earn the bulk of their income from payments for electricity consumption, which is a variable amount, while most of their expenditures are related to capability, which is a set amount that does not reduce when energy demand decreases. Alternatively, in the presence of a grid edge, their expenses do not fall as often as their earnings.

Energy is delivered in varying volumes from moment to moment, and the price of delivering it varies according to the time duration. It is made up of the result of several types of fuel-burning power plants. Consumers, on the other hand, are charged in a single amount or in advance. Commodities, fuel, appliances, and most other items are not purchased in this manner.

Conventional price and cost determination techniques aren't adequate. As well, the main goal couldn't be to ensure revenue recovery and a fair return. This type of assurance is not available to any other entity. Companies should charge a premium that is commensurate with both the perceived worth of their products and services in contrast to rival options. Quite often the cost will be higher than the cost, and other times it may be lower.

A successful business's brilliance is earning greater profits, despite the high danger associated. Companies will be expected to deliver products and services based on what consumers would spend in the electricity industry, even while considering how to preserve the company's current economic health via best organizational processes.

2.8 Digital reevaluation

Utilities, like any other corporation, should change into completely digital companies in order to remain competitive. This only enables the continuous operation of business while maintaining the flexibility, personalization, and availability that consumers have come to expect from other organizations [2,11].

The creation of a website for energy utility enterprises is a transition into the digital business. It entails becoming able to transact virtually in the same way that you would in reality, if not faster. It entails a significant presence on social media as well as being larger in cyberspace than in the real world. New rivals will emerge as digital businesses, free of the constraints of having to mitigate an existing, nonvirtual corporate culture and structure.

A contemporary, smart power grid will be a digital grid in addition to the company being virtual. The fields of information technology and operations technology must grow and combine. In reality, the contemporary distribution system will bring the power system and the Internet of Things together. To reiterate, not only would the web be impossible without power, but so would a modern, intelligent grid.

2.9 Information and communication technologies

The widespread presence of high speed, two-way communication channels is the foundation for a digital company. It refers to high-speed Internet. Without it, becoming a digital company or having a contemporary smart power network will be difficult. All equipment, applications, information, and actions will be unable to integrate and interoperate until they are all immediately connected to the web [12].

The grid edge's diversity and complexity will necessitate the surveillance, analytics, and automating of the distribution network as well as grid edge elements. Closed corporate divisions of equipment, telecommunications networks, datasets, and software will not be able to do this. It necessitates three-dimensional interconnection. Each piece of equipment, software, and communication method should communicate with others in a consistent way. This is independent of manufacturer standards. It all has to run within the same system. That framework, of course, must be broadband Internet.

It is feasible that a distribution system will be able to accomplish whatever it requires on the open network in the future if it is commonly accessible. If it isn't already, it should be the beginning for a utility's technological system in order to support a good strategy. To reiterate, not only will the web be unfeasible without power, but so will a sophisticated, smart network.

2.10 Requirement and development of trained manpower

There are three aspects to the manpower problem. For starters, the power sector's long-term consistency has resulted in job longevity that is unmatched in many other industries. Therefore, companies are dealing with an elderly staff that is quickly dwindling. The outgoing employees have years of organizational experience and knowledge. As a result, companies are faced with the task of attracting, hiring, retraining, and keeping new staff.

At the very same moment, companies will require totally new skills and experiences to cope with the new power system elements and complexity. This necessitates the hiring of new types of personnel as well as the emergence of new communication and information technology (ICT) skills [13].

3 Impact on electrical power utilities

Innovative, advanced techniques will dramatically enhance the overall performance cost. There will also be innovative and disruptive techniques and business practices from time to time that constitute great leaps rather than constant though exponential progress. To cope with a rapidly evolving and mostly unpredictable future, companies will need extraordinary flexibility and creativity. The best approach to decrease uncertainty and risk is to accept and influence the future rather than resisting it. The much more effective distribution network will be the one that initiates radical transformation rather than the one that overlooks or fights it [14].

To address the problems outlined above and the many more that will develop as well as capitalize on the possibilities that they provide, they must be acknowledged, recognized, and embraced. Power companies cannot thrive by preserving the past in a back room while attempting to exclude recruits.

4 Conclusion

At the other end of the line, the distribution network links the transmission system with its customers. Due to the direct connectivity of the distribution system with customers, so many challenges are faced by this system. The inadequacy of the system, physical and cyber security, the integration of distributed resources, disruptive enabling techniques, and information and communication technologies are creating the recent challenges facing the future electrical distribution sector.

References

[1] T. Gonen, Electric Power Distribution Engineering, third ed., CRC Press Taylor and Francis Group, January 2014.
[2] Top 10 Challenges for Electric Distribution Utilities, https://www.linkedin.com/pulse/top-10-challenges-electric-distribution-utilities-steven-collier/
[3] Smoothing the Rocky Road to Distribution Automation, Case Study, Motorola, 2010.
[4] J. Casazza, F. Delea, Understanding Electric Power Systems, John Wiley & Sons, Inc, Hoboken, NJ, 2003.
[5] M. Bollen, Y. Beyer, E. Styvactakis, J. Trhulj, R. Vailati, W. Friedl, A European benchmarking of voltage quality regulation, in: Proceedings of International Conference on Harmonics and Quality of Power, ICHQP, 2012, pp. 45–52.
[6] M. Juamperez, G. Yang, S.B. Kjær, Voltage regulation in LV grids by coordinated volt-var control strategies, J. Mod. Power Syst. Clean Energy 2 (2014) 319–328.
[7] S. Borenstein, M. Jaske, A. Rosenfeld, Dynamic Pricing, Advanced Metering, and Demand Response in Electricity Markets, Center for the Study of Energy Markets, 2002.
[8] T.F. Agajie, B. Khan, J.M. Guerrero, O.p. Mahela, Reliability enhancement and voltage profile improvement of distribution network using optimal capacity allocation and placement of distributed energy resources, Comput. Electr. Eng. 93 (2021) 107295.
[9] T.F. Agajie, B. Khan, H.H. Alhelou, O.P. Mahela, Optimal expansion planning of distribution system using grid-based multi-objective harmony search algorithm, Comput. Electr. Eng. 87 (2020) 106823.
[10] O.P. Mahela, V.S. Bhati, G. Ahmad, B. Khan, P. Sanjeevikumar, A.R. Garg, R. Mahla, A protection scheme for distribution utility grid with wind energy penetration, Comput. Electr. Eng. 94 (2021) 107324.
[11] T. Molla, B. Khan, B. Moges, H.H. Alhelou, R. Zamani, P. Siano, Integrated optimization of smart home appliances with cost-effective energy management system, CSEE J. Power Energy Syst. 5 (2) (2019) 249–258.
[12] O.P. Mahela, J. Sharma, B. Kumar, B. Khan, H.H. Alhelou, An algorithm for the protection of distribution feeder using Stockwell and Hilbert transforms supported features, CSEE J. Power Energy Syst. (2021), https://doi.org/10.17775/CSEEJPES.2020.00170.
[13] B. Simachew, B. Khan, J.M. Guerrero, S. Padmanaban, O.P. Mahela, H.H. Alhelou, Capacity-based optimization using whale optimization technique of a power distribution network, Eng. Rep. (2021), e12455.
[14] M. Tesfaye, B. Khan, O.P. Mahela, H.H. Alhelou, N. Gupta, M. Khosravy, T. Senjyu, J.M. Guerrero, Analysing integration issues of the microgrid system with utility grid network, Int. J. Emerg. Electr. Power Syst. 22 (1) (2021) 113–127.

SECTION 3

Smart and active distribution of electrical energy

CHAPTER 3

Active electric distribution network: Smart and active

Baseem Khan[a] **and Ziad M. Ali**[b]
[a]Department of Electrical and Computer Engineering, Hawassa University, Hawassa, Ethiopia
[b]College of Engineering at Wadi Addawasir, Prince Sattam Bin Abdulaziz University, Wadi Addawasir, Saudi Arabia

Contents

1 Introduction

Electricity distribution systems are responsible for transferring electricity between both the power grid and end consumers and also performing voltage conversions. This component of the power grid is also in charge of integrating dispersed generating power and keeping voltage levels and power quality in prescribed limits. As a result, this part of the electricity system has the greatest impact on consumer perceptions of power quality [1]. The system has lately taken on a more active role, and it is now in charge of incorporating power generation from distributed units deployed throughout the system.

Furthermore, additional advancements are predicted, such as automated voltage regulation, significant integration of distributed energy sources, storage devices, electric cars, and, especially, bidirectional power flow.

The evolution of electric distribution networks is also described as a "smart grid." While the names are interchangeable, active distribution networks (ADNs) are concerned with the incorporation, deployment, and

Active Electrical Distribution Network
https://doi.org/10.1016/B978-0-323-85169-5.00019-8

monitoring of large-scale deployments of distributed generators with energy storage systems in distribution system (DS). The phrase "smart grid" is frequently attributed to a broader area that includes bulk production, transmitting, and the use of improved measurement techniques enabling earlier fault identification and self-healing capacities to reduce interruptions.

2 Various components of ADN

An active distribution network (ADN) has different elements that can be classified as follows.

2.1 Components with existing system

- Substations
- Distribution feeder
- Transformers
- Cabling
- Metering and control equipment.

2.2 Added components with ADN

- Distributed generators
- Microgrid
- Demand side management (DSM) techniques
- Electricity markets
- Smart metering
- Variable electricity pricing
- Heat pumps
- Electric vehicles
- Information and communication technologies (ICT)

The distribution power station needs to connect the primary transmission line to the subtransmission system, converting high tension to medium tension voltage levels. These voltage levels are used on the main distribution system, in which many transformers are linked together to transform medium-tension to low-tension voltage levels. It is appropriate to meet the domestic and business category customers' demands in the supply system at the secondary level. As per the needed topology and operating current and voltage, this system is linked with a set of four-wire, three-phase overhead and subterranean cables.

Most of the electric factors such as current, voltage, and maximum power are monitored regularly in the power station. It has become frequent to also observe the distribution transformers, distribution feeder lines, and terminating nodes to facilitate power failure identification and track the quality of power on the network.

Furthermore, modifications to the energy metering system are necessary to effectively implement the electricity regulatory frameworks. It enhanced flexibility and lowered the prices of microcontrollers and information and communication technologies. These technologies have enabled the deployment of automated metering systems that measure and record power and energy usage in a dynamic manner. Fig. 1 shows the transition of the power industry.

2.2.1 Distributed generation

Distributed generation (DG) generates electricity on a local level, whether using traditional or renewable energy methods. Numerous factors contribute to its mass acceptance: DG improves delivery performance by shortening the length and power conversions among consumption and production; it can also delay grid improvements by boosting locally accessible energy capacity. However, the ability of large DGs to effectually incorporate green energy sources is strongly motivated by their numerous environmental and technical advantages [2].

A. Photovoltaic solar

The photovoltaic solar (PV) technique turns light into power directly. The solar cell, which is constructed of semiconductor materials, is the individual photovoltaic element. Photovoltaic technology deployments are categorized as:

(1) Large scale (utility)

(2) Small scale (commercial/residential)

These classifications are based on their power production capacity. The former are centralized and have significant power production capacity (>1 MWp). Moreover, they are frequently linked to a utility network. The second category has a lower output of electrical power (typically ranges from 20 kWp to 1 MWp). They are also dispersed, and are directly linked to a low-voltage system.

B. Microgas-based turbine system

These are tiny power producers that operate on the freestanding joule cycle [3]. They typically range in size from 15 to 300 kWe. Microgas

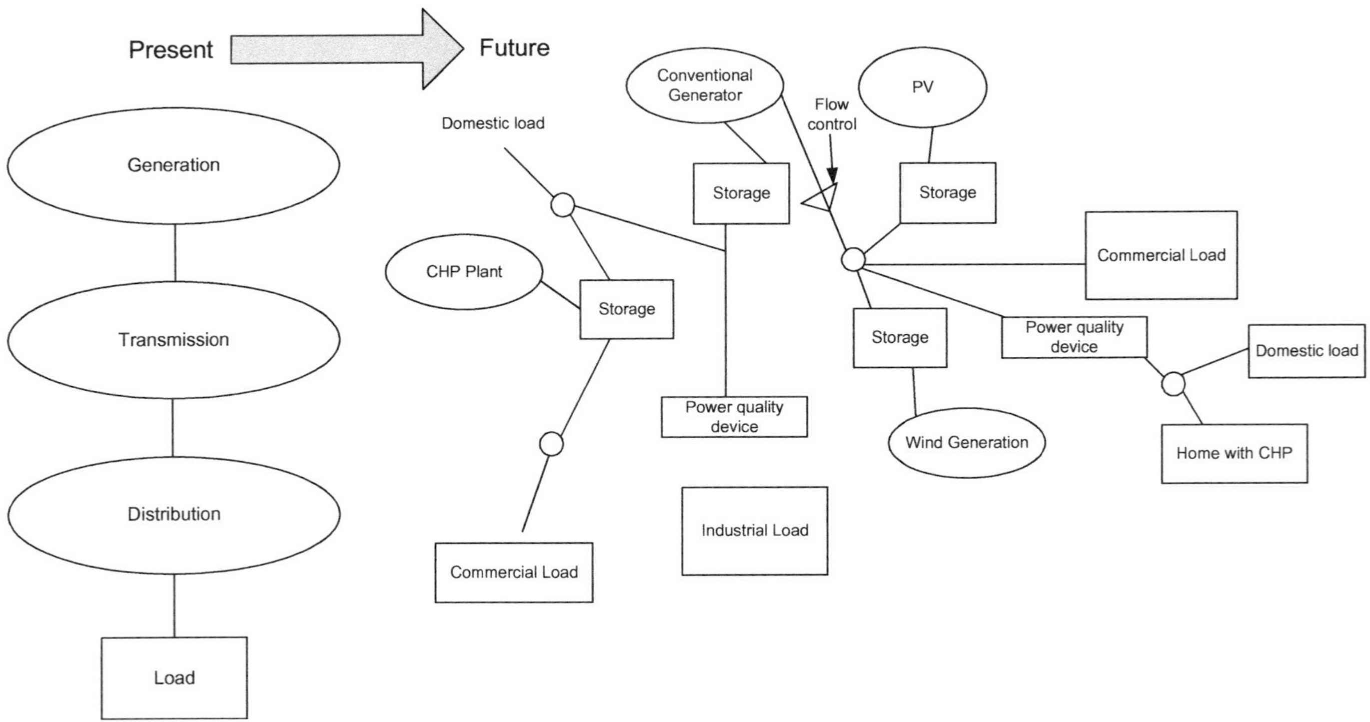

Fig. 1 Transition of the power industry.

generators are widely employed as emergency energy production systems, and have an overall system efficacy around 22%–30%. However, many types are used in a combined heat and power (CHP) system that simultaneously produce heat and electricity, thus enhancing efficacy to more than 80%. This property makes them especially appealing for applications that require both power and heat, such as residential areas, hospitals, and business buildings. Microgas turbines, in general, have a number of benefits such as great dependability, fast operation, and little maintenance [4].

C. Wind

Distributed wind energy sources can be placed at domestic, business, industrial, or community sites. The capacities of such systems can vary from 5 kW to multimegawatts for different locations at a production plant or linked to a regional supply grid.

The following characteristics are used to classify the distributed wind energy system:

- Wind turbines (WTs) erected at the load point for supplying onsite power requirements or auxiliary support for system operation of a distribution system are known as proximity to end-use wind turbines.
- WTs linked to the consumer end of the meter and connected straight to the low-voltage supply network or off-grid in a distant area are referred to as point of interconnection.

Households, schools, farms, ranches, companies, cities, villages, and rural places all benefit from distributed wind energy systems, which provide reliable electricity generation.

D. Fuel cells

These have a remarkable ability to turn a large percentage of the chemical energy in a fuel into electricity. Fuel cell power facilities are approximately twice as efficient as traditional power plants, even without cogeneration, with efficiencies nearing 60%. Fuel cells contribute considerably to a cleaner environment by emitting significantly fewer pollutants and producing mostly hot water and tiny quantities of carbon dioxide as byproducts. Fuel cells can also be installed at or near load centers because of their modular design, saving money on transmission network expansion. The commercial usefulness of fuel cells is progressively being recognized by the industry.

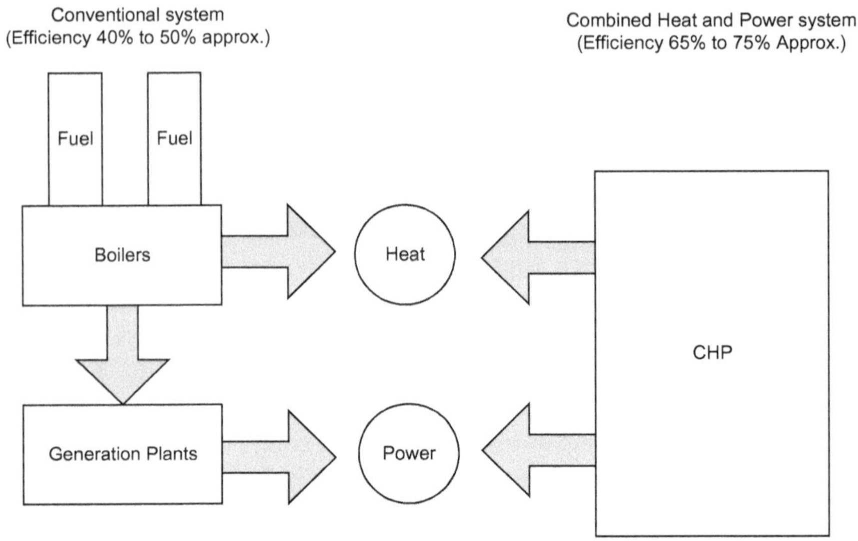

Fig. 2 Combined heat and power plant.

E. CHP

Cogeneration (combined heat and power) is a term used to describe the process of combining energy and heat generation. It is the process of concurrent generation in which working heat and electrical energy are generated by using the same production process. Different from the conventional energy production process, combined heat and power plants can be placed near energy users. Fig. 2 shows the process of a CHP plant.

F. Energy storage

These systems enable power to be saved and delivered in bright and breezy seasons. Although it may appear to be a simple idea, energy storage may be accomplished in a variety of ways.

- Compressed air
- Pumped-storage hydroelectricity
- Flywheel
- Lithium-ion battery
- Liquid air
- Pumped heat electrical storage

2.2.2 Microgrid

This is a local energy power grid with monitoring and control capabilities allowing it to function independently of the regular grid.

It is a network of interrelated demands and power sources in a distributed manner that works like a unified controlled system, connected to the utility grid through circuit breakers. Different energy sources produce electricity within their mentioned technical constraint limits. It can be worked as grid-linked and islanded modes of operation based on the operating requirements. Fig. 3 shows a schematic diagram of the microgrid system.

2.2.3 DSM techniques

These techniques are related to the adjustments made to the network's energy demand with respect to the information sent by the independent system operators, such as load shedding of the nonessential loads at the peak energy consumption periods. Customers can, though, use recently established variable pricing rates to deliberately adjust their power consumption patterns in reaction to market changes by moving heavy demand burdens to low-cost hours.

A. Demand-side management strategies

DSM techniques are used to improve the various load characteristics such as load duration curve, load factor, and demand factor along with minimizing the losses and overall cost. Basic DSM techniques consist of load shifting, peak clipping, valley filling, etc. For that purpose, numerous optimization methods such mathematical, metaheuristic, and artificial intelligent are used.

In a DSM strategy, voltage is maintained in the distribution by reducing the losses and load. This action is performed by utilizing a home energy management system with an intelligent controller, which responds to the varying energy rates on an hourly basis.

2.2.4 Electricity markets

To keep the load and supply balanced at all times, different power sources must be deployed based on demand variations. Different energy generators have exclusive structures of generation costs, as these generators operate under specific operating scenarios. Utility companies determine which energy generator to put in the system at any given moment depending on a number of factors, including operating costs, energy demand forecasts, weather conditions, and reservoir capacity. Different options available to effectively charge customers for energy use are extremely changeable in nature.

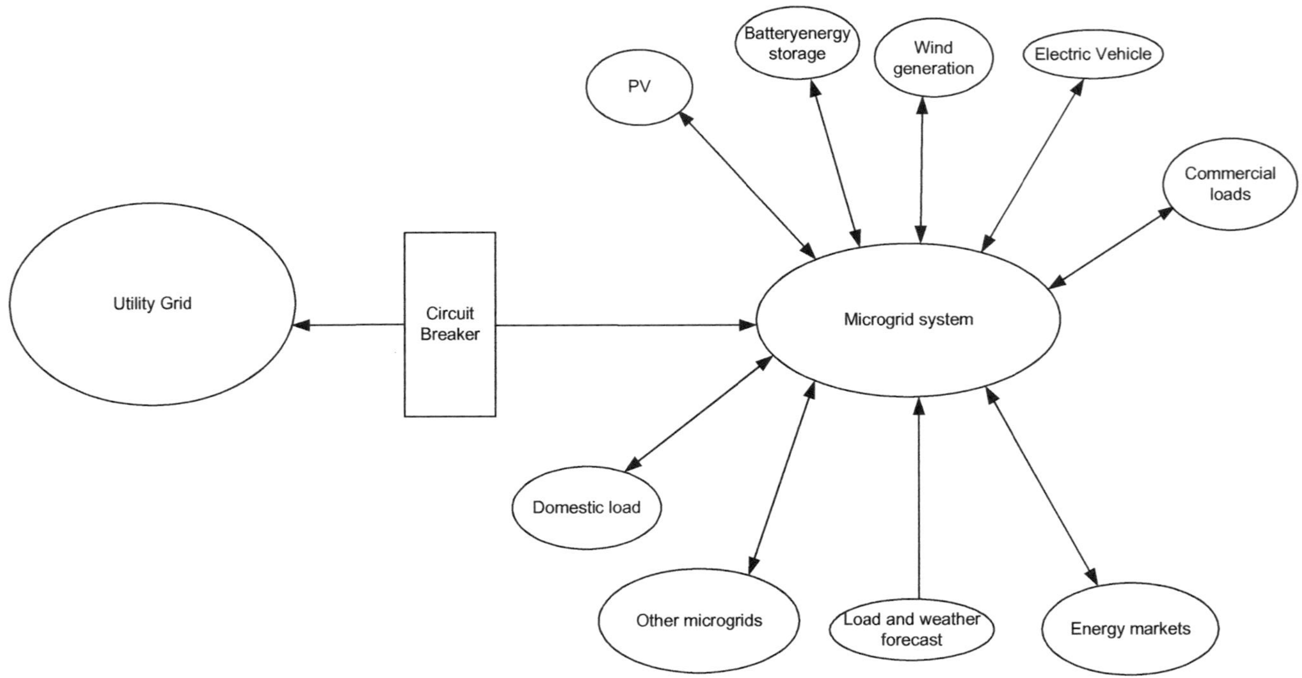

Fig. 3 Microgrid system.

2.2.5 Smart metering

These are devices that monitor and report energy usage for invoicing. Often, these intelligent meters have two-way communication capabilities, allowing them to send information about energy usage and generation (for example, to deployed PV systems) back to the utility, and also other parameters (such as quality, supply, and so on) that are used to monitor the network.

2.2.6 Variable electricity pricing

These methods are used to incentivize customers to move their electricity usage from peak to off-peak hours, therefore stabilizing the demand on the grid. This is done through the use of dynamic pricing techniques that are more closely aligned with the true cost of power [5].

These techniques are as follows:

1. Monthly variable pricing
2. Time-of-use (TOU)
3. Critical peak pricing (CPP)
4. Critical peak rebate (CPR)
5. Real-time pricing (RTP)

2.2.7 Heat pumps

These pumps are used to transfer thermal energy from a low-temperature node to a high-temperature node through heating and compressing the refrigerant to the boiling point of the lower temperature. To transfer the thermal energy from the refrigerant material to the thermal sink, a heat exchanger is used. The ground-supplied heat pump is a heat pump that converts one unit of electricity into three thermal energy units. As a result, its expansion makes it easier to reduce not just fossil fuel-dependent heating. The direct heating used in the heat pumps enhances the overall efficacy of the system.

2.2.8 Electric vehicles

In traditional cars, high energy-density fuels (mostly petroleum) are burned to provide the necessary force for propulsion. An electric motor provides the needed propulsion in an electric vehicle, and the required energy is obtained through the battery system [6].

Different types of available electric vehicles (EV) are:

1. EV based on a battery.
2. EV based on an extended range battery system.
3. Hybrid EV.
4. Plug-in hybrid EV.

The battery systems normally used in EV technology are:
1. Lithium ion (Li-ion).
2. Metal hydride (NiMH).

2.2.9 ICT for ADN

ICT is a term frequently used to describe the elements brought together via telecommunication services, Internet technology, and audio-video systems [7].

Basically, an ADN uses two-way communication with a real-time control system to integrate the renewable energy-based distributed generation sources to increase system dependability and operation efficacy. A distribution management system is used to integrate the distributed energy into the whole energy market to control the load and voltage.

The ADN requires a wide area communication system to transfer the data with control instructions between the onsite smart devices or master and distribution points. The following are the features of the communication network used in ADN:
1. Because a DN has so many substations, switches, transformers, and compensation devices, a high number of communication terminals is necessary.
2. The communication network in the ADN uses a mix of the communication main channel and the cell branch communication network because the distance between communication nodes is quite small.
3. The data flow in the ADN is significantly lower than in a transmission substation.

Based on the above characteristics, the types of communication techniques in the ADN are:
1. Power line carrier (PLC).
2. Fiberoptic communications.
3. Wireless public network communication.
4. Micropower wireless technology.
5. Fieldbus technology.

3 Active distribution network operation and control

The supply network is developed for working as a 3-Φ network with perfect balance. This indicates that each phase's voltage is the same magnitude with 120° phase displacement. Unbalanced demands, high capacity DG integration, and different network-related components create disturbances in the system, necessitating the placement of elements (active and passive) for operating the network near its operating limits.

There will be power quality issues when the operating limits are not met. Voltage is the important parameter that should be kept within the limit. Generally, the allowable voltage deviation in a distribution network is 10% [8,9]. Various voltage control methods are often employed. At the distribution substation, the primary voltage is regulated while the feeder linked to the customers uses tap changers of transformers to maintain the voltage level. Manually (for seasonal load fluctuations) or automatically, these tap changers can be adjusted by utilizing automatic voltage regulators.

For example, a high level of distribution generation penetration and the rising load of electric cars forces automatic voltage regulators to change over many times every day, rather than just a few times as they were built for, in systems needed to handle voltage operation utilizing automatic voltage regulators (AVRs). As a result, several of these systems' operational lifetimes have been significantly reduced [10]. As a result, in huge distribution systems with higher integration of distributed generation and electric vehicles, an automatic voltage regulator with capacitor banks can't provide adequate performance and voltage deviation is sometimes beyond the permissible ranges.

However, the similar technology that caused these annoyances can also assist in alleviating them. Several writers have looked into the possibility of using the ability of photovoltaic system inverters to supply VARs (voltage ampere reactive) for the necessary assistance to keep the system operating within its limits [11–13]. As a result, future active distribution networks must employ these characteristics to preserve network stability while also allowing for greater integration of the distributed generation. At the load point, emerging techniques could help sustain voltage levels by managing the load response to system conditions. Fig. 4 shows the controlling of the active distribution network.

4 Problems associated with ADN

ADN is very useful for implementation of a smart grid as well as to improve the operating efficiency of a system. But there are several issues that should be addressed for the adequate implementation of the ADN in the existing system:

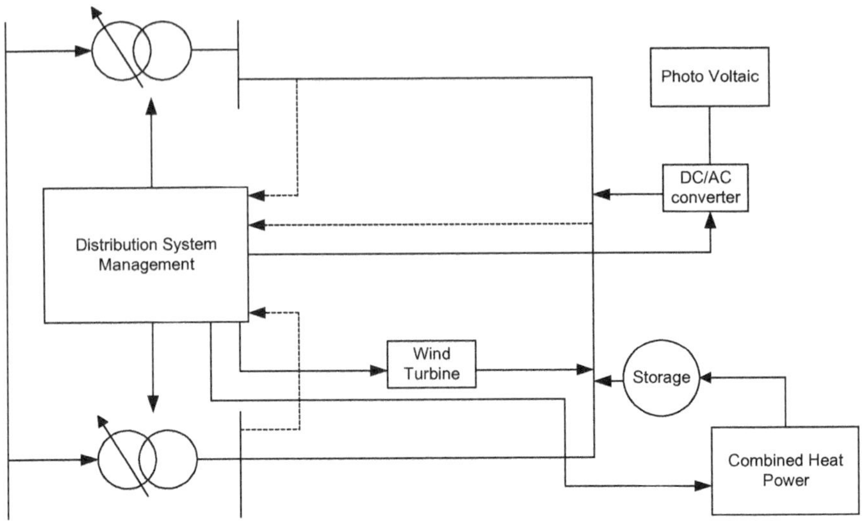

Fig. 4 Controlling of the active distribution network.

- Flexibility.
- Controllability.
- Adequate monitoring of the system state (state estimation).
- Placement of the monitoring devices.
- Microgrid optimization.
- Integration of EV technology.
- Power quality issues.
- Reactive power compensation problem.
- Integration of adequate demand response techniques.

Most of the issues discussed here are further explored and addressed in the next chapters of this book.

5 Conclusion

This chapter provided a brief overview of the active distribution network. Different components of the ADN such as distributed generations, microgrids, DSM techniques, electricity markets, smart metering, variable electricity pricing, heat pumps, electric vehicles, and information and communication technologies were discussed in this chapter. Further, the operation and control of the ADN is also presented. Different issues associated with the ADN are also discussed.

References

[1] J. Casazza, F. Delea, Understanding Electric Power Systems, John Wiley & Sons, Inc, Hoboken, NJ, October 2003.

[2] W.S. Tan, M.Y. Hassan, M.S. Majid, H.A. Rahman, Optimal distributed renewable generation planning: a review of different approaches, Renew. Sust. Energ. Rev. 18 (2013) 626–645.

[3] H.I.H. Paul Saravanamuttoo, G.F.C. Rogers, H. Cohen, Straznicky., Gas Turbine Theory, sixth ed., Pearson Education Canada, 2008.

[4] Resource Dynamics Corporation, Assessment of distributed generation technology applications, Technical Report February, 2001.

[5] Energy Markets Inspectorate, The Swedish electricity and natural gas markets 2014. Technical report, Energy Markets Inspectorate, Eskilstuna, Sweden, 2014.

[6] L. Guzzella, A. Sciarretta, Vehicle Propulsion Systems—Introduction to Modeling and Optimization, Springer, 2013.

[7] T. Yang, Chapter 10—ICT technologies standards and protocols for active distribution network, in: Q. Yang, T. Yang, W. Li (Eds.), Smart Power Distribution Systems, Academic Press, 2019, pp. 205–230.

[8] M. Bollen, Y. Beyer, E. Styvactakis, J. Trhulj, R. Vailati, W. Friedl, A European benchmarking of voltage quality regulation, in: Proceedings of International Conference on Harmonics and Quality of Power, ICHQP, 2012, pp. 45–52.

[9] American National Standards Institute, ANSI C84.1-2011: Electric Power Systems and Equipment Voltage Ratings (60 Hertz), 2011.

[10] J. McDonald, Solar Power Impacts Power Electronics in the Smart Grid, 2013. https://www.electronicdesign.com/technologies/power-electronics-systems/article/21196254/solar-power-impacts-power-electronics-in-the-smart-grid. (Accessed 15 January 2021).

[11] N. Rakibuzzaman Shah, R.C.B. Mithulananthan, V.K. Ramachandaramurthy, A review of key power system stability challenges for large-scale PV integration, Renew. Sust. Energ. Rev. 41 (2015) 1423–1436.

[12] C. Bornholm, A. Constantin, R.D. Lazar, S.B. Kjær, Voltage Control in Low Voltage Networks By (December), 2012.

[13] A.G. Marinopoulos, F. Papandrea, M. Reza, S. Norrga, F. Spertino, R. Napoli, Grid integration aspects of large solar PV installations: LVRT capability and reactive power/voltage support requirements, in: 2011 IEEE Trondheim PowerTech, June 2011, pp. 1–8.

SECTION 4

Distribution system state estimation

CHAPTER 4

Smart distribution system state estimation

Endeshaw Solomon and Issaias Gidey Hagos
Department of Electrical and Computer Engineering, Hawassa University, Hawassa, Ethiopia

Contents

1 Introduction

Distribution system state estimation (DSSE) is the method of finding the state variable of a system by using some known measured data at a definite place within the system. DSSE is typically a mathematical method that uses measurements of data for computing state variables. However, state

Active Electrical Distribution Network
https://doi.org/10.1016/B978-0-323-85169-5.00015-0

estimation (SE) is an intense and commonly utilized idea in the electrical grid but isn't widely applicable in DSSE and may show the way toward future research. A distribution system (DS) is planned to transfer power from the substation to the load. SE is the procedure of conveying significance to a new system variable based on conformation and measurements attained from the system. A variety of methods have been used to achieve an SE solution. New technological progress in distribution system automation improves checking the system state to enhance the controllability of the system. Using distributed generation (DG), microgrids and real-time measurements lead to the development of a state estimator at the distribution operation. Therefore, advanced measurement and monitoring devices such as a flexible alternate current (AC) transmission system (FACTS), a phasor measurement unit, and an intelligent electronic device are used to attain dynamic real-time measurements. Fig. 1 shows a flow chart of the conventional state estimator.

2 Distribution system state estimation

This is utilized to get the representation of the distribution system status at a given time constant. The state of the system is the minimum set of variables

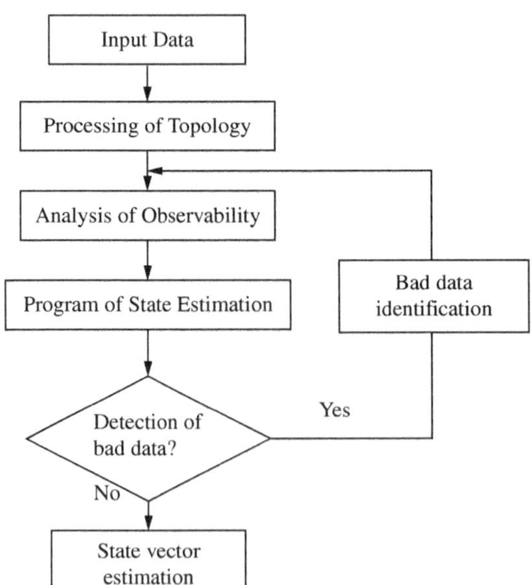

Fig. 1 Conventional state estimator.

that completely defines the power system by using transmission line topology and line parameters. SE is a way of shaping the state of the system using network measurement by minimizing definite parameters [1–5].

The main objectives of state estimation are:

- Detecting of measured bad data.
- Minimizing small errors.
- Topology error detection.
- Stipulation of estimates for the unmonitored part of the system.
- Estimating distribution system parameters using redundancy in measurements.

3 Techniques and application of DSSE

3.1 Conventional DSSE

Fig. 2 shown below is an overview of the major process information flow of SE. First, the topology processor proves that the network parameters are correct, guaranteeing the distribution network model is exact and up to date. An observability study creates adequate measurement data of the existing state estimation. Observability could be rapidly decided by testing the empty space of the Jacobian matrix [6]. Lastly, bad data are processed to detect and

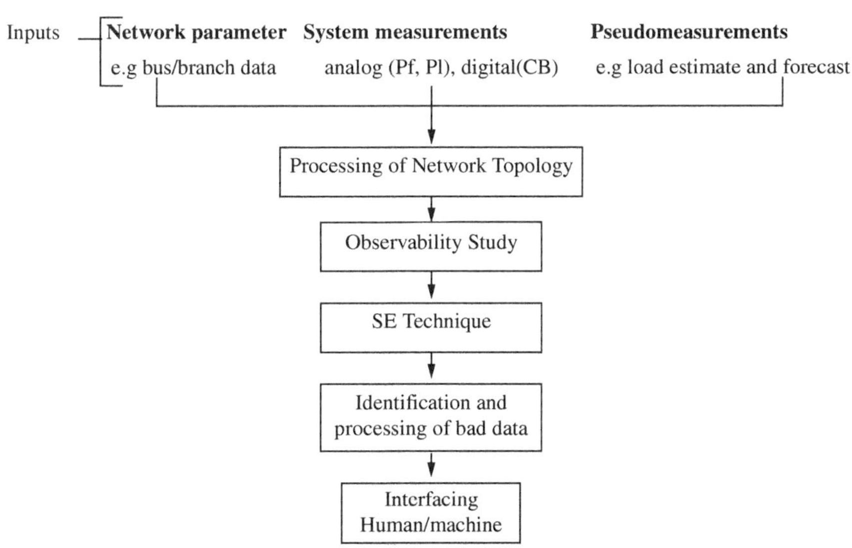

Fig. 2 State estimation overview function.

eliminate the information influenced by gross errors and noises, such as measurement or communication failure [7–11].

3.2 Challenges associated with distribution system state estimation

DSSE presents many challenges because distribution network characteristics are different from transmission network ones, including [12–20]:

- **Construction**: Many distribution systems have radial construction while transmission networks are interconnected with a high resistance-to-reactance ratio.
- **Redundancy**: In a distribution system for economic and technical feasibility, the number of measurement points is less compared to the transmission grid. Therefore, the system is underdetermined.
- **Measurement types**: Measurements or pseudomeasurements of power/current injections are now used as input data in the distribution system, although the actual measurements of voltage and power are only used on rare occasions.
- **Scale and complexity**: DSs are varied and have a huge number of components.
- **Phase imbalance**: The conventional SE method assumes that the distribution system is a balanced system.

4 Measurement model

The measurement model used in DSSE can be described as follows [21–32]:

$$z = h(x) + e \tag{1}$$

where, $z = (m \times 1)$ measurement vector,

$x = (n \times 1)$ is the true state variable vector,

$h(.)$ = a $(m \times 1)$ nonlinear vector function,

$e = (m \times 1)$ measurement error vector, m = number of measurements and n = number of state variables.

The standard options of the state variable are the bus voltage magnitude and the phase angle. However, the data to be measured are the real and reactive power flow, the node injections, and the magnitude of voltage.

4.1 Weighted least square (WLS)

Residual (r) is the difference between the measurement (z) and model value ($h(x)$).

$$r = z - h(x) \tag{2}$$

In WLS,

$$J(x) = \sum_{i=1}^{M} W_i[z_i - h_i(x)]^2 \tag{3}$$

where w_i is the weight to the ith measurement and M is the total number of available measurements

$$J(x) = [z - h(x)]^T W[z - h(x)] \tag{4}$$

5 DSSE problem formulation

The distribution system SE problem formulation differs from the standard state estimate because of the major distinction between transmission networks and distribution systems. The key distinction is that in DSSE, the measurement function is modeled. As a result, this chapter examines two important issue formulations for estimating the state of a distribution system:

5.1 Voltage-based DSSE

In this example, the magnitude and angle of the bus voltage in the transmission system are utilized as state variables [33]. This usual method has also been engaged in DSSE.

5.2 Branch current-based SE (BCSE)

In this situation, a prominent set of branch currents are utilized as state variables, which provide a significant path in the distribution system SE formulation.

6 Types of state estimation

There are three main categories of state estimation based on the time variant and time invariant of the measurement and the static dynamic model of the power system state.

(i) Dynamic
(ii) Static
(iii) Tracking

6.1 Dynamic

This also utilizes the current states and the previous estimates of the states. Due to economic dispatching, security evaluation, and other activities that may be performed in addition to the state estimate, system operators have more time to make decisions.

6.2 Static

In this case, the data processing technique is utilized for metered data and other information is processed into a vector estimator. The measured data is time-invariant.

6.3 Tracking

The methods in this approach rely on the simple inclusion of a static state estimate methodology. They provide current existing results of the system state to renew their estimated results.

7 DSSE algorithm

DSSE is a numerical approach that uses a series of measurements from the power system to calculate the best estimate of node voltage and phase angle for each node. Although transmission system SE methods are superior in terms of their straights, they can't be used in DS because it has different features as compared to the transmission system. Table 1 shows the characteristics of a distribution and transmission network.

Many techniques that require minimum input data are implemented for DSSE because it is not common or nonvariable to get sufficient measurement in DS. Table 2 shows the DSSE algorithm comparison results.

Table 1 Characteristics of a distribution and transmission network.

Distribution system	Transmission system
Radial topology	Meshed topology
Power flows in both direction	Power flows only in one direction
Lines and loads are variable	Loads and transmission lines are constant
Observability isn't accomplished	Observable
High resistance-to-reactance ratio	Low resistance-to-reactance ratio

Table 2 DSSE algorithm comparison results.

Type of estimation	Pros	Disadvantages
Bus voltage (Polar form)	Estimating of amplitude and angle of voltage directly More appropriate for interconnected system	Nonlinear measurement functions Computational loads are high
Rectangular node voltage	Many measurement functions are linear Computational speed is high Appropriate to interconnected network	Measurement of current magnitude has different management
Polar branch current	Easy managing of measurements of current magnitude	More nonlinear measurement functions Computation burdens are high Used for only interconnected networks
Rectangular branch current	Measurement functions are linear Very sparse system matrix Computational speed is high	For radial meshed networks only

8 Error classification

The observability study determines if the quantity of measurements supplied is sufficient to predict the state vector. The state estimator estimates the voltage magnitude and phase angle at each bus using real-time data that include both active and reactive power. Errors can be categorized into three different classifications according to the severity of errors introduced into the telemetered measurement.

8.1 Extreme error

This is the error in which the difference between the measured and actual value exceeds 20σ, where σ is the measurement's standard deviation.

8.2 Gross error (bad data)

When the absolute magnitude of the difference between the measured and real value is between 5σ and 20σ, it is said to be an error.

8.3 Normal error

This error takes place when the absolute difference between the measured and actual values is smaller than 5σ. Prefiltering the measurements eliminates a large amount of erroneous data. Performing a simple check on the incoming raw measurement data prevents further errors.

9 Detection and identification of bad data

SE senses and recognizes errors in measurement and tries to reduce it. Measurements may comprise errors because of the erroneousness of meters or communication media. Therefore, the estimator filters such kinds of expected errors if measurements are suitably redundant. First, bad data processing is achieved by computing the estimates several times. Every time one measurement is detached from the measured data, it is a less effective technique and impractical as well. Therefore, the weighted least-square performance index test $J(x)$ is used to identify the bad data. To validate the measures impacted by the main error, the standardized residual recognition test is utilized. Bad data can be recognized by normalized residuals.

The capacity to recognize and classify faulty data and system configuration errors is determined by a set of measures. The detection of incorrect data is challenging because measurement redundancy is poor and load models are ambiguous. In the weighted least square method, it is introduced after the estimation process. When utilizing load profile data to detect faults in feeder line flow measurements, the commonly used 3σ poor data identification minimum is inadequate.

10 Observability of network

Different types of measurements in state estimation are:

(i) **Analog measurement**: bus voltage, active and reactive power injection.

(ii) **Logic measurements**: the status of the switch and circuit breaker.

(iii) **Pseudo measurement**: forecasted bus loads and generators and passive node zero injections.

The state estimator processes the above measurements to predict the state. The system is visible if the measurement state is sufficient and purely well distributed for a precise state estimate. Extra measurements should be placed intentionally for the case of unobservability such as the inclusion of pseudomeasurements.

11 Conclusion

This chapter provides a detailed analysis of different methods related to DSSE in power systems. DSSE includes the forecasting capability in which the future state of the system should be predicted. It is based on current variables that have been measured. Its capacity to forecast events is critical for power system security and emergency controls. Due to the unobservable nature of distribution systems, such as three-phase unbalanced systems, radial topology, restricted real-time measurement, high resistance-to-reaction ratio, and others, power system state estimation methods cannot be directly applied to DSSE. DSSE research has been performed in past years and is being unceasingly developed to satisfy the ever-evolving distribution system. In recent years, research works on various standards of distribution system state estimation methodologies have been introduced, taking specific improvements and characteristics of distribution networks into account. Many existing methods give an effective process for constructing online network models. New methods and accurate data sources to eliminate the flat race for extensive applications of DSSE are required. In spite of the high interest of researchers in the past two decades, DSSE hasn't been mostly developed in distribution systems to date. In this paper, an overview of the critical aspects of DSSE is presented. Active studies such as DSSE problem formulation, DSSE measurement model, bad data recognition identification, system topology, network observability, and DSSE algorithms are reviewed.

References

[1] J.F. Dopazo, State estimation for power systems: detection and identification of gross measurement errors, in: Proceeding of the 8th IEEE PICA Conference, June 1973, 1973.
[2] W.-H. Liu, F. Wu, S.-M. Lun, Estimation of parameter errors from measurement residuals in state estimation [power systems], Power Syst. IEEE Trans. 7 (1) (1992) 81–89.
[3] I. Costa, J. Leao, Identification of topology errors in power system state estimation, Power Syst. IEEE Trans. 8 (4) (1993) 1531–1538.
[4] W.-H. Liu, S.-L. Lim, Parameter error identification and estimation in power system state estimation, Power Syst. IEEE Trans. 10 (1) (1995) 200–209.
[5] O. Alsac, N. Vempati, B. Stott, A. Monticelli, Generalized state estimation, Power Syst. IEEE Trans. 13 (3) (1998) 1069–1075.

[6] E. Castillo, A. Conejo, R. Pruneda, C. Solares, State estimation observability based on the null space of the measurement Jacobian matrix, Power Syst. IEEE Trans. 20 (3) (2005) 1656–1658.
[7] X. Nian-de, W. Shi-ying, Y. Er-keng, A new approach for detection and identification of multiple bad data in power system state estimation, Power Appar. Syst. IEEE Trans. 101 (2) (1982) 454–462.
[8] A. Monticelli, A. Garcia, Reliable bad data processing for real-time state estimation, Power Appar. Syst. IEEE Trans. 102 (5) (1983) 1126–1139.
[9] H.-J. Koglin, T. Neisius, G. Beiler, K. Schmitt, Bad data detection and identification, Int. J. Electr. Power Energy Syst. 12 (2) (1990) 94–103.
[10] J. Chen, A. Abur, Placement of PMUs to enable bad data detection in state estimation, Power Syst. IEEE Trans. 21 (4) (2006) 1608–1615.
[11] D. Singh, R. Misra, V. Singh, R. Pandey, Bad data pre-filter for state estimation, Int. J. Electr. Power Energy Syst. 32 (10) (2010) 1165–1174.
[12] H. Liao, J.V. Milanovic, Pathway to cost-efficient state estimation of future distribution networks, in: Proceeding of IEEE PES General Meeting, Boston, MA, USA, 2016, pp. 1–5.
[13] J. Peppanen, M.J. Reno, R.J. Broderick, S. Grijalva, Distribution system model calibration with big data from AMI and PV inverters, IEEE Trans. Smart Grid 7 (5) (2016) 2497–2506.
[14] P.A. Pegoraro, A. Meloni, L. Atzori, P. Castello, S. Sulis, Adaptive PMU-based distribution system state estimation exploiting the cloud based IoT paradigm, Proc. IEEE Int. Instrum. Meas. Technol. Conf. (2016) 1–6.
[15] F. Magnago, L. Zhang, R. Nagarkar, Three phase distribution system state estimation utilizing common information model, in: Proc. IEEE PowerTech, Eindhoven, The Netherlands, 2015, pp. 1–6.
[16] F. Castanedo, A review of data fusion techniques, Sci. World J. 2013 (2013), 704504.
[17] L. Xie, D.H. Choi, S. Kar, H.V. Poor, Fully distributed state estimation for wide-area monitoring systems, IEEE Trans. Smart Grid 3 (3) (2012) 1154–1169.
[18] Cisco Systems, Inc., Smart grid reference architecture, vol. 1, using information and communication services to support a smarter grid, in: International Business Machines Corporation, Edison Company, San Jose, CA, 2016. [Online]. Available: www.pointview.com/data/files/1/636/2181.pdf.
[19] A. Gomez-Exposito, A. Abur, A. de la Villa Jaen, C. Gomez-Quiles, A multilevel state estimation paradigm for smart grids, Proc. IEEE 99 (6) (2011) 952–976.
[20] S. Deshmukh, B. Natarajan, A. Pahwa, State estimation and voltage/VAR control in distribution network with intermittent measurements, IEEE Trans. Smart Grid 5 (1) (2013) 200–209.
[21] F. Meng, D. Haughton, B. Chowdhury, M.L. Crow, G.T. Heydt, Distributed generation and storage optimal control with state estimation, IEEE Trans. Smart Grid 4 (4) (2013) 2266–2273.
[22] A. Zegers, H. Brunner, TSO-DSO interaction: an overview of current interaction between transmission and distribution system operators and an assessment of their cooperation in smart grids, in: ISGAN, Seoul, South Korea, Discussion paper Annex 6 Power T D Syst. Task 5, Sep. 2014.
[23] E. Handschin, et al., Bad data analysis for power system state estimation, IEEE Trans. Power Syst. 94 (2) (1975) 329–337.
[24] J. Souza, A.L. Da Silva, A.A. de Silva, Online topology determination and bad data suppression in power system operation using artificial neural networks, IEEE Trans. Power Syst. 13 (3) (1998) 796–803.
[25] D.C. Filho, M. Brown, J.C.S. de Souza, Forecasting-aided state estimation part I: panorama, IEEE Trans. Power Syst. 24 (4) (2009) 1667–1677.

[26] M.B. Do Coutto Filho, J.C. Stacchini de Souza, R.S. Freund, Forecasting-aided state estimation—part II: implementation, IEEE Trans. Power Syst. 24 (4) (2009) 1678–1685, https://doi.org/10.1109/TPWRS.2009.2030297.
[27] Y. Gu, et al., Bad data detection method for smart grids based on distributed state estimation, in: 2013 IEEE International Conference On Communications (ICC), 2013.
[28] Y. Liu, P. Ning, M.K. Reiter, False data injection attacks against state estimation in electric power grids, ACM Trans. Inf. Syst. Secur. 14 (1) (2011), https://doi.org/10.1145/1952982.1952995. Article 13.
[29] A. Monticelli, F.F. Wu, Network observability: identification of observable islands and measurement placement, IEEE Trans. Power Appar. Syst. PAS-104 (5) (1985) 1035–1041, https://doi.org/10.1109/TPAS.1985.323453.
[30] B. Gou, A. Abur, A direct numerical method for observability analysis, IEEE Trans. Power Syst. 15 (2) (2000) 625–630.
[31] B. Gou, A. Abur, An improved measurement placement algorithm for network observability, IEEE Trans. Power Syst. 16 (4) (2001) 819–824.
[32] A. Jain, et al., Network observability: a solution technique using neural networks, in: TENCON 2003. Conference on Convergent Technologies for the Asia-Pacific Region, 2003.
[33] A. Monticelli, State Estimation in Electric Power Systems: A Generalized Approach, Springer Science & Business Media, 1999.

SECTION 5

Micro phasor measurement units' placement

CHAPTER 5

Microphasor measurement unit placement

Mulusew Ayalew and Issaias Gidey Hagos
Department of Electrical and Computer Engineering, Hawassa University, Hawassa, Ethiopia

Contents

Active Electrical Distribution Network
https://doi.org/10.1016/B978-0-323-85169-5.00011-3

1 Introduction

A phasor measuring unit (PMU) is a power system device used to monitor synchronized current and voltage in a power system. PMU synchronization is achieved by sampling the voltage and current waveforms at the same sample rate [1]. A PMU is one of the most important measuring devices in power system monitoring, control, and protection because it can find current synchronized readings [2]. PMUs have an intriguing use in state estimation (SE), which is an important function in energy management systems (EMS) because it provides voltage phasors at all network nodes. Due to the high cost, until now PMUs were used in transmission networks and thus the electric distribution networks are enormous in number [3].

The fitting of PMUs in every power system bus is neither cost-effective nor essential. However, valuable and appropriate PMU sites with a minimum number of accessible PMUs must be established. The different PMU site methods are reviewed in detail in the following subsection. The Electric Power Research Institute (EPRI) considers two main factors for the optimal placement of PMUs. This is based on characteristics of the power system and the applications of PMUs [1].

Scholars have advanced topological or arithmetical methods to decide the best placement of PMUs to reduce the price of fitting in the power system. These approaches achieve an optimization procedure, taking into consideration all power system network buses.

2 Integration of micro-PMU (μPMU) into the grid

Integrating a microphasor measurement unit (μPMU) into the power system network plays a vital role in power system synchronization, as shown in Fig. 1.To collect real-time data, a large number of μPMUs are linked to a central GPS server. All μPMUs that are subject to time signal accuracy constraints utilize their ESP 8266 WiFi device to collect time evidence from a main GPS server at the power distribution substation.

3 Literature review

Different scholars have done work on μPMU placement by applying their own different methods. Some of them are reviewed next.

According to the authors in [4], the PMU location can be found through theoretical and graphical methods. These methods were studied to compute

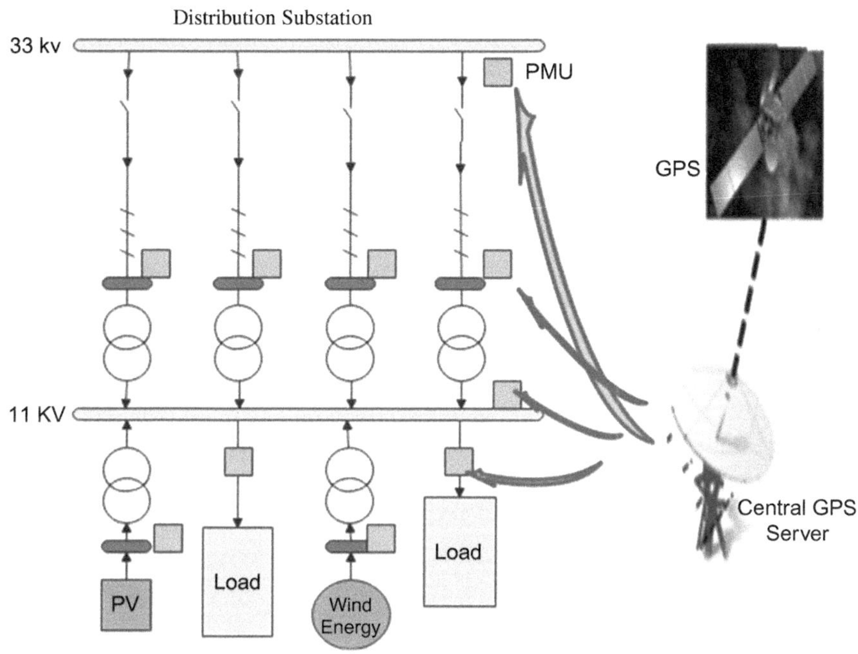

Fig. 1 Installation of PMUs into the power system.

the observability of the system, so as to determine the sufficient number of PMUs in the electrical network. This procedure applies spanning tree illustrations of the power system to catch the optimal PMU placement.

The authors in [5] proposed a fast investigation technique for electrical power system topological observability. In this paper, a linear power system model and an incident matrix were investigated for the state estimator. The target of the work is to reduce the number of PMUS while selecting the minimum number of PMUS involved in the network. A Tabu search optimization algorithm was proposed to solve the optimization problem.

The authors in [6] proposed the synchronized optimal placement of PMUs to monitor the power system network current and phasors. The authors also reviewed previous research about PMU placement.

In [7], the authors introduced the idea of sensitivity-constrained best PMU placements for the whole power system parameters of sensitivity and observability. In this paper, a simulated annealing procedure was proposed for PMU optimal placement and for full system observability. Also, the objective functions were introduced to increase the observability and measurement sensitivity.

The authors in [8] presented a site program to increase electrical system observability. In this work, the problem formulation was analyzed for PMU optimal placement to monitor in real-time power system measurements. The branch and bound were involved for optimal placement.

The authors in [9] proposed PMU placement based on two steps; the selection of the minimum number of PMUs was analyzed in the first step, and the second stage is suggested to check if the outcome of PMU location in step one leads to a complete ranked Jacobian measurement.

The authors in [10] proposed the binary optimization of a particle swarm (BPSO) metaheuristic algorithm for the best PMU placements. In this paper, the PMU best location problem formulations were done.

The author in [11] proposes the behavior of the PMUs in wide area measurement system (WAMS) based on experimental analysis for the best PMU locations. Also in this paper, the connected power positive sequence voltage and currents are connected to the PMUs for optimal location.

Numerous review papers on μPMUs have mostly regarded the monitoring, diagnostic, and control application areas of electrical power distribution networks [12–14].

The authors in [15] reviewed μPMUs for emerging active distribution networks. They also discussed state awareness and event detection.

The aim of this research is to deliver a wide-ranging review of different μPMU location approaches for diverse application areas.

4 Microphasor measurement units

μPMU equipment delivers phasor information (both magnitude and phase angle) in actual time. Also, the information delivered by PMUs is precise and allows the power system engineer to measure during blackout/failures. The features of PMU technology are that they focus on measurements that are time-framed power system parameters.

4.1 Phasor diagram of a power system

The power system state variables such as voltage magnitude and phase angles need to be known [16]. The basic needs of these values are active and reactive power flow calculations. Synchronized phasor measurements are capable of measuring power system variables from a far distance. By using a phasor measurement unit, an electrical grid can have a time-synchronized voltage magnitude and phase angle [16].

Synchronized phasor measurements refer to the precise time-synchronized measurements of the magnitude and phase angle of the sine waves found in the electric grid. Subsequently, the phasor measurement units are integrated with GPS; each phasor measurement is timely synchronized.

4.2 Phasor and its representation

Let Eq. (1) be a pure sinusoidal component and given as

$$X(t) = \left(\sqrt{2}X\right) * \sin(\omega t + \varphi) \tag{1}$$

Where
$\omega =$ is the angular frequency in radian per second.
$\varphi =$ is the phase angle in the radians.
Expressing Eq. (1) again is given in Eq. (2) below

$$X(t) = \mathrm{Re}\left[Xe^{j(\omega t + \varphi)}\right] = \mathrm{Re}\left[\left\{e^{j(\omega t)}\right\} Xe^{j\varphi}\right] \tag{2}$$

Eq. (3) is the phasor representation of a complex number X^*

$$X^* = Xe^{j\varphi} = X\left[\cos\varphi + j\sin\varphi\right] \tag{3}$$

4.3 Schematic diagram of μPMU

μPMU equipment delivers phasor data, and its general block diagram is given in Fig. 2. The PMU measures the analog voltage and current

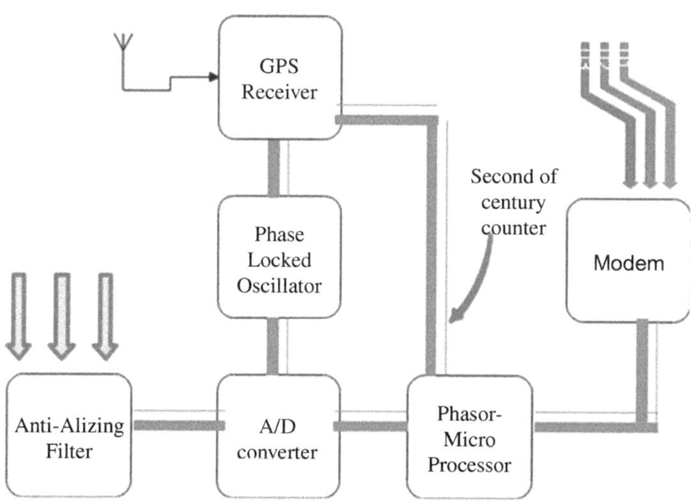

Fig. 2 Block diagram of the μPMU.

inputs from the secondary sides of the voltage and current transformers, respectively. The measured analog voltage and current measurements are then sent to the analog-to-digital converters.

4.4 Advantages of phasor measurement units

- Real-time operation applications.
- Wide area situational awareness.
- Frequency stability monitoring.
- Oscillation monitoring for damage control.
- Voltage monitoring.
- Alarm.
- Manage big data.
- Capturing various types of data for analysis.
- Potential backup to system or supervisory control and data acquisition (SCADA) system.
- Provide data to improve accuracy.
- Operations planning.
- Wide area controls.

5 Optimal placement methods of μPMU

Optimal PMU placement (OPP) can be defined as the minimum number of PMU placements in a given power system network. OPP has the ability to measure all timely synchronized data of a power system. OPP is a combinatorial problem. Let's K represents power system substation or amount of network buses and N is quantity of PMUs placed in a power system network.

To select the minimum quantity of PMUs, the combinations for a bus K are given by:

$$\text{Combinations size for bus}\,(K) = \sum_{i=1}^{k}\binom{K}{i} = \sum_{i=1}^{k}\left(\frac{K!}{(K-i)! * i!}\right) \quad (4)$$

The number of combinations given in Eq. (4) is very large, even for small systems. Let's consider a 14-bus test system; its number of possible combinations becomes 16,000. For a grid-connected power system with hundreds of buses, it is difficult to check all arrangements of optimal results, so the least optimal results have been established in a diverse way. Iterative search algorithms are the best solution for the minimum number of PMU optimal placement problems. In this paper, Tabu search, simulated annealing and genetic algorithms and particle swarm optimization methods are reviewed.

In Section 5.1, different iterative algorithms are reviewed for optimal PMU placements. In almost all methods, the basic aim is system observability. The observability of the topology constraint is conceivable to perceive the topology of the power network. The observability of power system parameters is different if the power system is under different contingencies.

The authors in [17–19] reviewed and classified works on PMU location in power networks. In those papers, the arrangements were based on seeing the application of PMUs. The application areas considered in this paper were observability of state estimation and monitoring of power system voltage stability.

5.1 Methods for observability

The power system is observable if power flow across the system circuits can be determined from the given power system network topology and its accessible measurements [20].

The author in [1] proposed power system partitioning using a metaheuristic algorithm and a spanning tree. The two numerical and topological observability approaches were considered. In this paper, optimal PMU locations were selected to minimize the installation cost.

In [4], PMU locations based on incomplete observability were proposed by using theoretical and graphical methods. In this paper, the spanning tree graphics methods was proposed to find the PMU optimal placement. For PMU placement problems and the communication system, a simulated annealing algorithm was proposed as a solution. The new and basic finding of this work considers the systematic method of the PMU optimal location in the power system network.

The authors in [5] proposed power system parameter observability based on the linearized power system state estimator model and the augmented incidence matrix of the power system network. This paper focuses on reducing the number of PMU installations by considering full network parameter observability and avoiding sufficient redundancy. A Tabu search algorithm was used to solve the optimal PMU placement problems.

The authors in [6,7] proposed optimal PMU placements based on the sensitivity constraint of the measurement parameters and the most appreciated data of a power system network were considered. Also, a simulated annealing algorithm was proposed for optimal PMU placement and observabilities of measurement variables. In addition, the author calculates the sensitivities of buses and PMUs are placed based on the sensitivity orders of distribution network buses.

The authors in [8] discussed improving the performance of voltage control in a power system by monitoring the real-time voltages at all buses of a power system network. For the full observability of voltages, a problem formulation was done to select the minimum number of PMUs. The author proposed the branch and bound optimization algorithm to solve the optimal PMU placement problem.

The authors in [9] proposed that optimal PMU placement was based on two steps: the selection of the minimum numbers of PMUs was analyzed in the first step, and then the location results of the PMU are checked.

The author in [10] proposed the BPSO metaheuristic algorithm for the best PMU placements. In this paper, the PMU optimal placement problem formulations were done. Also, the full observability minimizing measurement redundancy and selecting and installing the minimum number of PMUs were discussed in detail.

The author in [11] proposed the behavior of PMUs in WAMS based on an experimental analysis for the best locations of PMUs. Also, the connected power positive sequence voltage and currents are connected to the PMUs for optimal location.

5.2 State estimation methods

The concept of state estimation is applied for the best locations of PMUs. The nonlinear state estimation equations are introduced in Eq. (5) below.

$$z = h(x) + \in \tag{5}$$

Where

z ($z \in Rm$): Measurement vector.

x ($x \in Rn$): State vector.

$\in$ ($E \in Rm$): Measurement vector error.

h (h: $Rm \rightarrow Rm$): Relationship between measurement vector and state vector.

The Newton-Raphson method is used to solve Eq. (5). The state variables and the measurement variables (i.e., current and voltage) are given by PMUs; this is shown in Eq. (6) below.

$$z = Hx + \in \tag{6}$$

Where H ($H \in R^m\ X^n$) is the matrix of the "state" of the system.

The author in [21] proposed computing the quality of phasor estimations. In this paper, two procedures were introduced: the first is to enable

PMUs to report the measurement variables of the power system, and the second is to determine the quality of the estimated phasor.

The author in [22] presented the observability of variables based on optimal placements of PMUs. In this paper, a graph theorem analysis and simulated annealing algorithm were investigated to solve the PMU optimal placement problem.

The author in [23] proposed a genetic algorithm to resolve PMU unit optimal location problems. In addition, a graph theory procedure and genetic algorithm were involved for PMU optimal placement.

The author in [24] proposed integrating and the use of PMUs in power system protection control and monitoring. In this paper, binary integer linear programming (BILP) was investigated to solve PMU optimal locations and to obtain the observabilities of the intended state variables. The binary decision variables (0,1) are used to determine whether to install a PMU at each bus while providing full system observability and determining the minimum number of PMUs to reduce the investment costs.

5.3 Programming based methods

The author in [25] proposed an integer linear programming algorithm for optimal PMU placement. In this paper, the optimal placement problems were solved by integer linear programming and its advantage based on minimizing computing time was discussed in detail. The proposed algorithm was easier, efficient, and can be used in practice for optimal placement.

In [26,27], the authors presented integer programming for the PMU location. In this paper, the problem formulations were based on measurements such as injections and power flows. The integer programming is intended to obtain current and voltage measurements by selecting the minimum number of PMUs.

5.4 Applications of genetic algorithms for PMU placement

The authors in [23] discussed optimal PMU placement based on two conflicting ideas. The first is minimizing the number of PMUs and the second is the redundant measurement maximization. In this paper, a genetic algorithm (GA) was proposed to solve the PMU optimal placement problems.

The authors in [28] proposed a GA for the best PMU location problem of a power system. In this paper, an immunity GA was incorporated to increase the efficiency of the GA. This method improves the immobilization of degenerative phenomena during the evolution of the GA.

In [29], the authors selected the minimum number of PMUs and an appropriate site for the applications of failure in the power network. A GA was used to solve the optimization problems. This method was more economical and can be implemented in an interconnected power system. In this paper, two methods were introduced: the first is to determine the optimal number of PMUs and the second is the appropriate placement in a given bus of a power system network.

The authors in [30,31] proposed using GA to solve the best PMU location problems. In this paper, the applications of PMUs were investigated in an active distribution network. The PMU measurements were also used to measure the effects of the power flow while integrating distributed generators with an active distribution network. In addition to GA, the Monte Carlo method was used for PMU optimal placements.

5.5 Simulated annealing method

The Simulated Annealing (SA) method was introduced by Metropolis in 1953. It is an algorithm that is capable of resolving optimization problems [19]. SA solves a problem by combining the physical annealing process with solid matters.

SA methods for the PMU best location in an electrical power system are given in [32]. The SA flow chart is given in Fig. 3.

The author in [32] proposed that the optimal placement problems of PMU and remote terminal units (RTU) were resolved by using a hybrid genetic algorithm and simulated annealing (HGS). The proposed methods were used for multiarea power system state estimation. In this paper, a power system controller was investigated by using PMU and RTU.

5.6 Minimum spanning tree method

The minimum spanning tree method (MST) has three main steps, as investigated by the author in [33]:

(1) Spanning tree (N) generation: The generation of a spanning tree algorithm is illustrated as shown in Fig. 4.

As shown in the flow chart above, for N number of buses of a power system, the algorithm is also executed N times. The algorithm finds the optimal location of the PMUs when full observability is achieved.

(2) Alternative MST pattern searches: The minimum number of PMU locations gained in step one is processed again as follows: each PMU location is replaced by the node connected with its

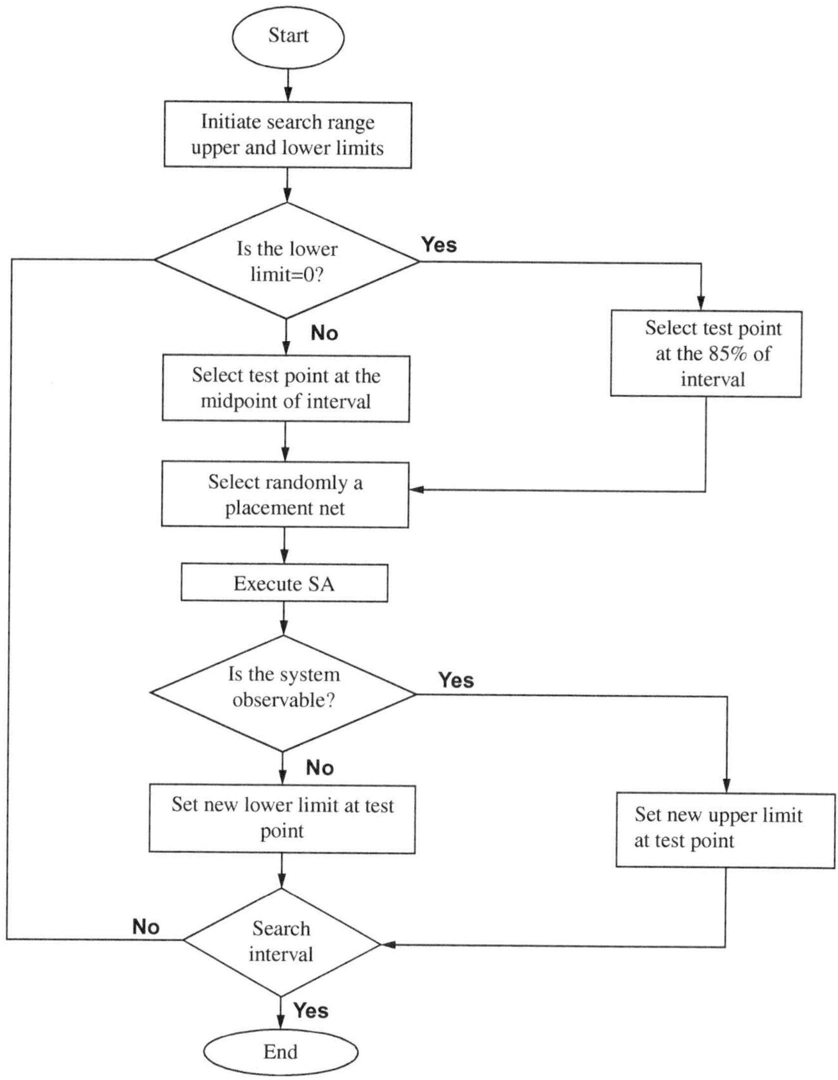

Fig. 3 Flow chart of the SA method.

corresponding bus. The full observability location/nodes of PMUs are illustrated as shown in Fig. 5 below.

(3) Reducing PMU number in case of pure transit nodes: This is the verification step when the PMU location gives full observability. If there is no temporary bus, the spinning tree procedures can end up at step two. Step three illustrates increasing the convergence of existing PMU locations and to set a selected bus, as shown in Fig. 6.

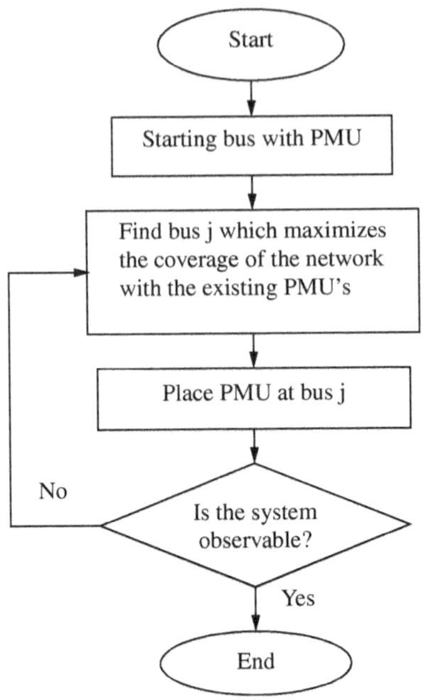

Fig. 4 Flow chart of the MST method.

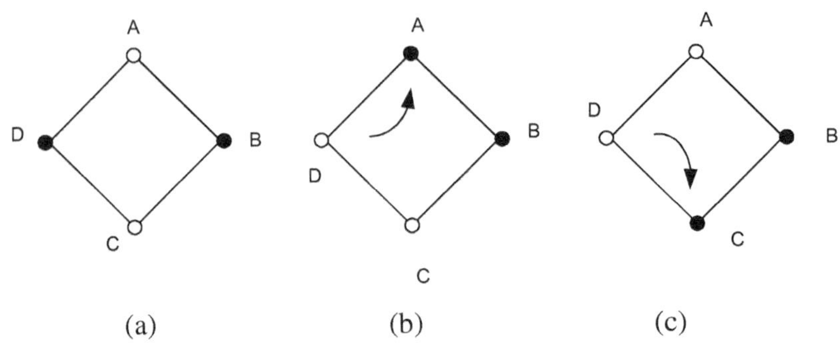

Fig. 5 Search of alternative placement sets.

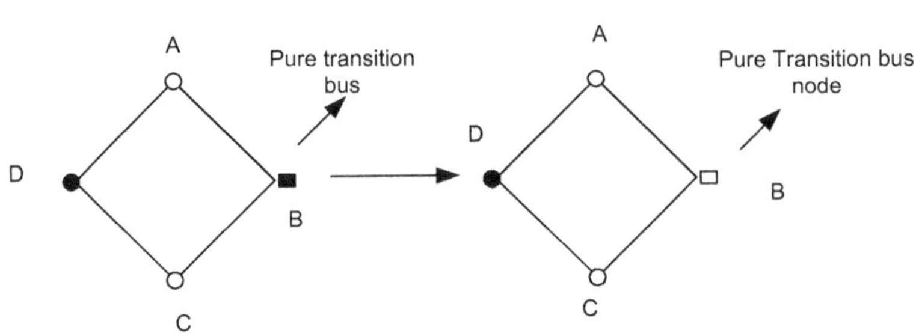

Fig. 6 Pure transition bus/node filtering.

The author in [34] proposed a solution for optimal PMU location to enhance the whole observability of the power system network. Also, the integer linear programming was investigated to divide the power system networks into subnetworks of a power system. The main objective of this research was to enhance the installation and investment cost of the PMUs. The objective was achieved using integer linear programming and by placing the PMUs on the subnetworks of the power system.

5.7 Particle swarm optimization method

PSO is an inspired movement of a group of birds or fish. When the particle swarms flow through their hunt space, the location of each particle is arranged based on its skill and the neighbors. Initially, PSO was advanced as an optimization problem [35]. Also, a discrete binary kind of PSO (BPSO) was advanced to resolve distinct optimization problems [36].

The author in [37] presented a discrete binary version of PSO (BPSO) and applied it to increase particle diversity. The initial locations of the PMUs are reassigned again so the possibilities of the particles getting trapped in local minima can be minimized. BPSO has the ability to filter out similar locations of PMUs at each iteration.

The author in [11] proposed BPSO to solve optimal PMU location problems. In this paper, power flow measurements, zero injection bus, and measurement redundancy at normal power flow were considered as constraint equations.

The authors in [38,39] proposed an improved PSO method of optimal PMU placement. The core aim of this paper was the feasibility of the optimal PMU placement solution.

5.8 Methods for voltage stability

The authors of [40] presented a voltage stability load index (VSLI) approach for optimal PMU placement. An echo state network (ESN) was used to evaluate the VSLI in this research. The process was carried out considering an islanding mode of operation of a power system. The key role of this research was to monitor voltage stability by placing the PMU at its optimal location to enhance the observability of the state variables.

The author in [41] discussed the optimal PMU location for different power system application areas. In this paper, the PMUs are intended to monitor the dynamics of the power system dynamics of generators, transmission lines, and load characteristics. A detailed discussion of measuring synchronized measurement values of voltage and current by using PMUs was presented.

The author in [42] proposed using the optimal PMU location to monitor voltage collapses of a power system. A voltage collapse is a critical problem in power system contingency analysis. For security assessment, the normal operation and the postcontingency timely synchronized power system phasor measurements can be measured by using PMUs.

6 Application of μPMU

In this section, different μPMU application areas are presented. The authors in [12,13] discussed the diverse applications areas of μPMUs for distribution network measurement as well as to monitor and control applications. The growth of μPMU trends for a power system network are discussed in the works. In [15], the author reviewed in detail μPMU for active distribution networks. Also in [14,43,44], the authors wrote a detailed review on μPMU so as to use it at the distribution level. The author in [45] discussed μPMU applications. In this paper, the author gives a detailed summary for the μPMU of the power system monitoring, diagnostics, planning, and operation and control applications.

Some of the diagnostic and monitoring applications areas of μPMU on distribution networks are:

- Incident recognition.
- High impedance fault recognition.
- Fault location.
- State estimation.
- Reverse power flow recognition.
- Power quality monitoring.
- Phase identification.

μPMU control applications on a distribution network include:

- Volt-Amps Reactive (VAR) optimization.
- Adaptive protection.
- Distribution network reconfiguration.
- Microgrid coordination.
- Flexible alternating current transmission system (FACTS) controller.

6.1 State estimation

This is computing the amplitude and phase angle of the voltage of a power system network. Measuring those values at the steady state conditions are better at the transmission line than at the distribution system. Due to the complexity, the calculation of the state estimation of a distribution network is more complex than a transmission line.

6.2 Fault recognition and site identification

Protective devices of power systems are rated based on the magnitude and duration of the current. Detecting and identifying the rated current and the fault currents are very difficult issues of power system protection. When the fault occurs, the circuit breaker and fuses can isolate a particular location at a power system. The typical method of fault detection is to physically look for damaged equipment and identify the faulted line section by inspection. This is a time-consuming process, and it is especially unreliable for underground cables. There are a number of algorithms to locate the fault, but they are limited in accessing accurate fault measurement. The magnitude of the fault impedance is used to locate the PMU at its optimal location.

6.3 Voltage stability monitoring

A synchronized phasor measurement has a great role in voltage stability monitoring issues. In transmission systems, the transmission line load can be monitored by using PMU measurements. There are no distinct transmission corridors in distribution networks. The reactive power is calculated by monitoring the dynamic voltages of the power system.

6.4 Reverse power flow

Reverse power flow is an important effect in a distribution network. In a given power system network, reverse power flow is used for the coordination of protective devices. Thus, due to the bidirectional protections, circuits can be protected from back power flow. Therefore, integrating PMU at an optimal location in a distribution network provides full information about the system and is more economical than conventional voltage and current transformers.

6.5 Monitoring of renewable generation in the distribution system

Monitoring distribution system network needs more information on line flows, existing load values, and required generation powers. Connecting large distributed generation (DG) will damage the measuring meter. Thus, it is important that the PMU's measured data are helpful for distribution system planning and operation.

6.6 Microgrid coordination

Due to the advancement of using distributed energy resources, PMUs have an adverse effect on the coordination of distributed energy resources. Thus, PMU measurement uses renewable energy resources to control and synchronize the distributed energy resource into an isolated or grid-connected power supply.

6.7 FACTS controllers

The author in [46] stated that flexible alternating current transmission systems (FACTS) have the capacity to enhance the transmission line and distribution network capability. Thus, PMUs can be used for large interconnected power systems to measure the synchronization parameters. FACTS is a compensation device based on the measurements of voltage, current, and powers of an interconnected system. As discussed in this paper, accurate reading can be done by using PMU units.

The author in [47] introduced using FACTS controllers to compensate for the nonlinear damping interarea oscillations of power systems. The compensation was achieved by controlling the bus angle and the frequencies of the oscillations were effectively minimized. Also, the nonlinearity dynamics of the power system were controlled by the PMU, and effective results were investigated.

7 Conclusion

μPMUs are used at the power system for measuring the voltage and current phasors for the monitoring, diagnostic, and control areas of the electrical power system. Due to the high installation cost of μPMU, in this chapter, a number of methods were reviewed for the best locations of μPMUs for full system observability of measurements in electrical power system networks.

References

[1] T.T.E. Lopes, Estimadores de estado em sistemas de energia elétrica incluindo restrições fuzzy (Ph.D. thesis), Universidade Federal do Rio de Janeiro, Rio de Janeiro, 2008. [Online]. Available at: http://www.pee.ufrj.br/teses/textocompleto/2008063002.pdf.

[2] M.U. Usman, M.O. Faruque, Applications of synchrophasor technologies in power systems, J. Mod. Power Syst. Clean Energy 7 (2019) 211–226.

[3] Power Standards Lab. n.d. Introduction to microPMU. Available online www.powerstandards.com$/wpcontent/uploads/dlm_uploads/2017/10/$Introduction-to-microPMU.pdf (Accessed 22 June 2019).

[4] R. Sodhi, S.C. Srivastava, Optimal PMU placement to ensure observability of power system, in: Presented at Fifteenth Natl. Power Syst. Conf. (NPSC), IIT Bombay, December, 2008.
[5] R.F. Nuqui, A.G. Phadke, L. Fellow, Phasor measurement unit placement techniques for complete and incomplete observability, IEEE Trans. Power Deliv. 20 (4) (2005) 2381–2388.
[6] J. Peng, Y. Sun, H. Wang, Optimal PMU placement for full network observability using Tabu search algorithm, Electr. Power Energy Syst. 28 (4) (2006) 223–231.
[7] R. Emami, A. Abur, Robust measurement design by placing synchronized phasor measurements on network branches, IEEE Trans. Power Syst. 25 (1) (2010) 38–43.
[8] Z. Hong-Shan, L. Ying, M. Zeng-qiang, Y. Lei, Sensitivity constrained PMU placement for complete observability of power systems, in: Transm. Distrib. Conf. Exhib. Asia Pacific, 2005 IEEE/PES, 2005, pp. 1–5.
[9] B. Mohammadi-Ivatloo, S.H. Hosseini, Optimal PMU placement for power system observability considering secondary voltage control, in: Electr. Comput. Eng. 2008. CCECE 2008. Can. Conf, 2008, pp. 365–368.
[10] R. Sodhi, S.C. Srivastava, S.N. Singh, Optimal PMU placement method for complete topological and numerical observability of power system, Electr. Power Syst. Res. 80 (9) (2010) 1154–1159.
[11] A. Ahmadi, Y. Alinejad-Beromi, M. Moradi, Optimal PMU placement for power system observability using binary particle swarm optimization and considering measurement redundancy, Expert Syst. Appl. 38 (6) (2011) 7263–7269.
[12] A. Von Meier, D. Culler, A. McEachern, R. Arghandeh, Micro-Synchrophasorsfor Distribution Systems, ISGT, Washington, DC, 2014, pp. 1–5.
[13] R. Arghandeh, Micro-synchrophasors for power distribution monitoring, a technology review, Syst. Control arXiv (2016). arXiv:1605.02813.
[14] A. Von Meier, E. Stewart, A. McEachern, M. Andersen, L. Mehrmanesh, Precision micro-synchrophasors for distribution systems: a summary of applications, IEEE Trans. Smart Grid 8 (2017) 2926–2936.
[15] Y. Liu, L. Wu, J. Li, D-PMU based applications for emerging active distribution systems: a review, Electr. Power Syst. Res. 179 (2020), 106063.
[16] K.J. Makasa, G.K. Venayagamoorthy, On-line voltage stability load index estimation based on PMU measurements, in: Power Energy Soc. Gen. Meet. 2011 IEEE, 2011, pp. 1–6.
[17] P.S. Sreenivasa Reddy, S.P. Chowdhury, S. Chowdhury, PMU placement—a comparative survey and review, in: presented at 10th IET International Conference on Developments in Power System Protection (DPSP 2010). Managing the Change, 2010, p. P51.
[18] W. Yuill, A. Edwards, S. Chowdhury, S.P. Chowdhury, Optimal PMU placement: a comprehensive literature review, in: Presented at 2011 IEEE Power and Energy Society General Meeting, 2011, pp. 1–8.
[19] N.M. Manousakis, G.N. Korres, P.S. Georgilakis, Optimal placement of phasor measurement units: a literature review, in: Presented at 2011 16th International Conference on Intelligent System Applications to Power Systems, 2011, pp. 1–6.
[20] A. VonMeier, R. Arghandeh, Chapter34-every moment counts: synchro phasors for distribution networks with variable resources, in: L.E. Jones (Ed.), Renewable Energy Integration, Academic Press, Boston, UK, 2014, pp. 429–438.
[21] M.A. Donolo, V.A. Centeno, A fast quality assessment algorithm for phasor measurements, IEEE Trans. Power Deli. 20 (4) (2005) 2407–2413.
[22] T.L. Baldwin, L. Mili, M.B. Jr, R. Adapa, Power system observability with minimal phasor measurement placement, IEEE Trans. Power Syst. 8 (2) (1993) 707–715.

[23] B. Milosevic, M. Begovic, Nondominated sorting genetic algorithm for optimal phasor measurement placement, IEEE Trans. Power Syst. 18 (1) (2003) 69–75.
[24] N.H. Abbasy, M.I. Hanafy, A unified approach for the optimal PMU location for power system state estimation, IEEE Trans. Power Systems 24 (2) (2009) 806–813.
[25] B. Gou, Optimal placement of PMUs by integer linear programming, Power Syst. IEEE Trans. 23 (3) (2008) 1525–1526.
[26] B. Gou, Generalized integer linear programming formulation for optimal PMU placement, Power Syst. IEEE Trans. 23 (3) (2008) 1099–1104.
[27] B. Xu, A. Abur, Observability analysis and measurement placement for systems with PMUs, Power Syst. Conf. Expo. 2 (2004) 943–946.
[28] F. Aminifar, C. Lucas, A. Khodaei, M. Fotuhi-firuzabad, Optimal placement of phasor measurement units using immunity genetic algorithm, IEEE Trans. Power Deliv. 24 (3) (2009) 1014–1020.
[29] S. Geramian, H.A. Abyane, K. Mazlumi, Determination of optimal PMU placement for fault location using genetic algorithm, in: Harmon. Qual. Power, 2008. ICHQP 2008. 13th Int. Conf, 2008, pp. 1–5.
[30] M. Campos, M. Arias, Ubicación óptima de unidades de medición fasorial aplicando swarm intelligence, 2010, [Online]. Available at: http://www.labplan.ufsc.br/congressos/Induscon2010/fscommand/web/docs/T1100.pdf.
[31] J. Liu, F. Ponci, A. Monti, C. Muscas, P.A. Pegoraro, S. Sulis, Optimal meter placement for robust measurement systems in active distribution grids, IEEE Trans. Instrum. Meas. 63 (5) (2014) 1096–1105.
[32] T. Kerdchuen, W. Ongsakul, Optimal placement of PMU and RTU by hybrid genetic algorithm and simulated annealing for multiarea power system state estimation, in: GMSARN International Conference on Sustainable Development: Issues and Prospects for GMS, 12-14 Nov. 2008, 2008.
[33] M. Farsadi, H. Golahmadi, H. Shojaei, Phasor measurement unit (PMU) allocation in power system with different algorithms, in: Proceedings of International Conference on Electrical and Electronics Engineering, Gamm, 2009, pp. 396–400.
[34] R. Sodhi, S.C. Srivastava, Optimal PMU placement to ensure observability of power system, in: Fifteenth National Power Systems Conference (NPSC), IIT Bombay, December 2008, 2008.
[35] J. Kennedy, R. Eberhart, Particle swarm optimization, in: Proc. 1995 IEEE Int. Conf. Neural Networks. Part 1 (of 6), vol. 4, 1995, pp. 1942–1948.
[36] J. Kennedy, R.C. Eberhart, A discrete binary version of the particle swarm algorithm, in: 1997 IEEE International Conference on Systems, Man, and Cybernetics. Computational Cybernetics and Simulation, vol. 5, 1997, pp. 4–8.
[37] Z.H. Rather, C. Liu, Z. Chen, P. Thogersen, Optimal PMU Placement by improved particle swarm optimization, in: 2013 IEEE Innovative Smart Grid Technologies-Asia (ISGT Asia), 2013, pp. 1–6.
[38] Y. del Valle, G.K. Venayagamoorthy, S. Mohagheghi, J.-C. Hernandez, R.G. Harley, Particle swarm optimization: basic concepts, variants and applications in power systems, IEEE Trans. Evol. Comput. 12 (2) (2008) 171–195.
[39] S. Chakrabarti, G.K. Venayagamoorthy, E. Kyriakides, PMU placement for power system observability using binary particle swarm optimization, in: Power Engineering Conference, 2008. AUPEC'08. Australasian Universities, 2008, pp. 1–5.
[40] S. Das, D. Ghosh, T. Ghose, D.K. Mohanta, Simulation of wide area measurement system with optimal phasor measurement unit location, in: presented at 2014 International Conference on Signal Processing and Integrated Networks (SPIN), 2014, pp. 226–230.
[41] A. Tiwari, V. Ajjarapu, Event identification and contingency assessment for voltage stability via PMU, in: Power Symposium, 2007. NAPS '07. 39th North American, 2007, pp. 413–420.

[42] D. Ruisheng, S. Kai, V. Vittal, J.O.K. Robert, R.R. Michael, B. Navin, S. Dwayne, K.S. Sanjoy, Decision tree-based online voltage security assessment using PMU measurements, IEEE Trans. Power Syst. 24 (2) (2009) 459–466.
[43] A. Jain, S. Bhullar, Micro-Phasor Measurement Units (μPMU) and its Applications in Smart Distribution Systems; ISGW: Compendium of Technical Papers, Springer, Singapore, 2018, pp. 81–92.
[44] A. Von Meier, R. Arghandeh, Diagnostic applications for micro-synchrophasor measurements, in: Proceedings of the Innovative Smart Grid Technologies Conference (ISGT), Washington, DC, USA, vols. 22–24, April 2014, pp. 1–5.
[45] R.K. Varma, S. Auddy, Migation of sub synchronous resonance by SVC using PMU-acquired remote generator speed, in: Power India Conference, IEEE 2006 conference, New Delhi, 2006, 2006, pp. 1–8.
[46] M. Zarghami, L.C. Mariesa, S. Jagannathan, Nonlinear control of FACTS controllers for damping interarea oscillations in power systems, IEEE Trans. Power Deliv. 25 (4) (2010) 3113–3121.
[47] C.S. Yu, C.W. Liu, S.L. Yu, J.A. Jiang, A New PMU-based fault location algorithm for series compensated lines, IEEE Trans. Power Deliv. 17 (1) (2002) 33–46.

SECTION 6

Micro-grid optimization and integration

CHAPTER 6

Impact of battery energy storage system integration on microgrid reliability improvement

Takele Ferede Agajie[a], Baseem Khan[b], Amoussou Isaac[a], and Yayehyirad Ayalew Awoke[c]
[a]Department of Electrical and Electronics Engineering, University of Buea, Buea, Cameroon
[b]Department of Electrical and Computer Engineering, Hawassa University, Hawassa, Ethiopia
[c]School of Electrical and Computer Engineering, Debre Markos University, Debre Markos, Ethiopia

Contents

1 Introduction

Sustainable and continuous electric power supply is a key component of developing a nation. The power of the nation can be taken from a national grid or from local microgrids. Nowadays, a microgrid is preferred to the national main grid. A microgrid contains renewable energy sources at the distributed level as well as, energy storage systems and controllable demand. Microgrid operation benefits consumers and electric companies through enhancing generation efficacy and reliability, minimizing losses and environmental impacts, and providing more cost-efficient electricity. A microgrid is typically a network of demands with renewable energy generators (DER) that is integrated into the main grid but with the ability to operate reliably in autonomous mode.

The ability of the entire system to produce and provide electrical power to end consumers is measured by microgrid system reliability (MSR). The adequacy of power production and supply by a microgrid system for a

Active Electrical Distribution Network
https://doi.org/10.1016/B978-0-323-85169-5.00023-X

particular combination of distributed generation (DG) units in the system as well as the subsystems contained in a DG unit are determined by this capacity. As a result, the most essential element for system designers to consider is dependability. The ability of the system to function or supply an adequate and desired amount of electric power to the associated load over an extended period of time is critical to any project's success.

Many scholars provide a definition as well as an analysis of the reliability of a microgrid, considering various techniques and methods. The potential that a system, product, or service functions well for a certain amount of time under continuous operating circumstances was outlined by the author in [1]. In [2], the researcher proposed a correlation between size, location, and the reliability of wind energy production. In [3], the research was carried out by adding storage in the microgrid modeling, which affect the reliability of the system. In [4,5], the researchers proposed the method of discrete wind speed analyses of the reliability of microgrids. In [6,7], the authors carried out a reliability analysis of grid -connected microgrids consisting of wind turbine energy production. In [8], the researcher investigated the estimated capacity as well as the charging and discharging behavior of a storage system on microgrid reliability. In [9,10], a reliability analysis of solar energy microgrid systems in a standalone mode was done. Further studies in [11,12], on microgrid reliability analysis development include an analysis of energy cost economic and reliability indices. In [13], a revised capacity outage probability table was used in the reliability analysis of a hybrid microgrid. [14] described the parallel operation of wind and solar energy sources that improved the reliability of hybrid energy microgrids. Also, a hybrid technique for a reliability evaluation of a hybrid renewable energy microgrid consisting of a wind, solar, and battery storage system was proposed. Further studies in [15,16] used a Monte Carlo simulation technique as well as the effect of load and generation on the reliability of the microgrid. Additionally, a similar analysis was carried out in [15] in which solid oxide fuel cell (SOFC) was added to the microgrid and the reliability of the system was analyzed. In [17], the author investigated the relationship between energy production charge and reliability analysis of a hybrid photovoltaic and wind microgrid. In [18], the researcher proposed a hybrid technique to investigate the effects of weather and battery storage on the availability of a hybrid renewable energy microgrid. In [19], the minimum path method was applied to explore the effects of operation mode and load types on the reliability of renewable energy microgrids. In [20], the reliability of a renewable energy microgrid considering the customer interruptions cost was considered.

In the same vein, studies on microgrid reliability include the development of a load model, which increased the degree to which microgrids satisfy load efficiently [21]. Ref. [22] describes how capacity outage probability is used to calculate the loss of load expectation, the energy index of reliability, renewable energy penetration, and conventional energy penetration reliability indices for microgrids. Almost all the studies on microgrid control looked at either photovoltaic or wind as a stand-alone power source, or both with or without batteries. In this paper, the optimal placing and sizing of a battery energy storage system (BESS) on an alternating current (AC) microgrid system is studied, and it explains the effects on the reliability indices in different scenarios.

This paper is organized as follows: Section 2 is problem formulation, Section 3 is a proposed methodology for the reliability assessment of a microgrid, Section 4 includes results and discussions, and Section 5 is the conclusion.

2 Problem formulation

Nowadays, radical increases in demands on the power system sector (microgrid) lead to mismatches between the output power and demand. Due to these mismatches, overloading (i.e., demand is greater than output power) and system component failure occur on the microgrid. For such reasons, the microgrid was unreliable for power delivery to the connected loads. Reliable microgrids with renewable energy resources are preferable due to greenhouse emissions and the fuel price [23–27].

3 Proposed methodology

The objective of this paper is to study the effect of reliability due to the operation of a microgrid integrating with BESS. To grasp the objective, to calculate the optimal placing and size of the BESS on a microgrid is performed by utilizing a weighted sum harmony search algorithm [28]. Reliability indices are calculated after the above objective functions are performed. Then, a BESS is integrated on a microgrid and all the aforementioned calculations are executed again.

A weighted sum harmony search algorithm (HSA) is an optimization algorithmic technique obtained from searching for perfect harmonious sound in music [29]. The weighted sum HSA is simple to program, gives a quicker convergence, and is efficient in searching for an optimal solution among other heuristic algorithms. The procedural steps of the proposed weighted sum HSA for the optimal placing and sizing of BESSs on a typical microgrid are shown in the flow chart in Fig. 1.

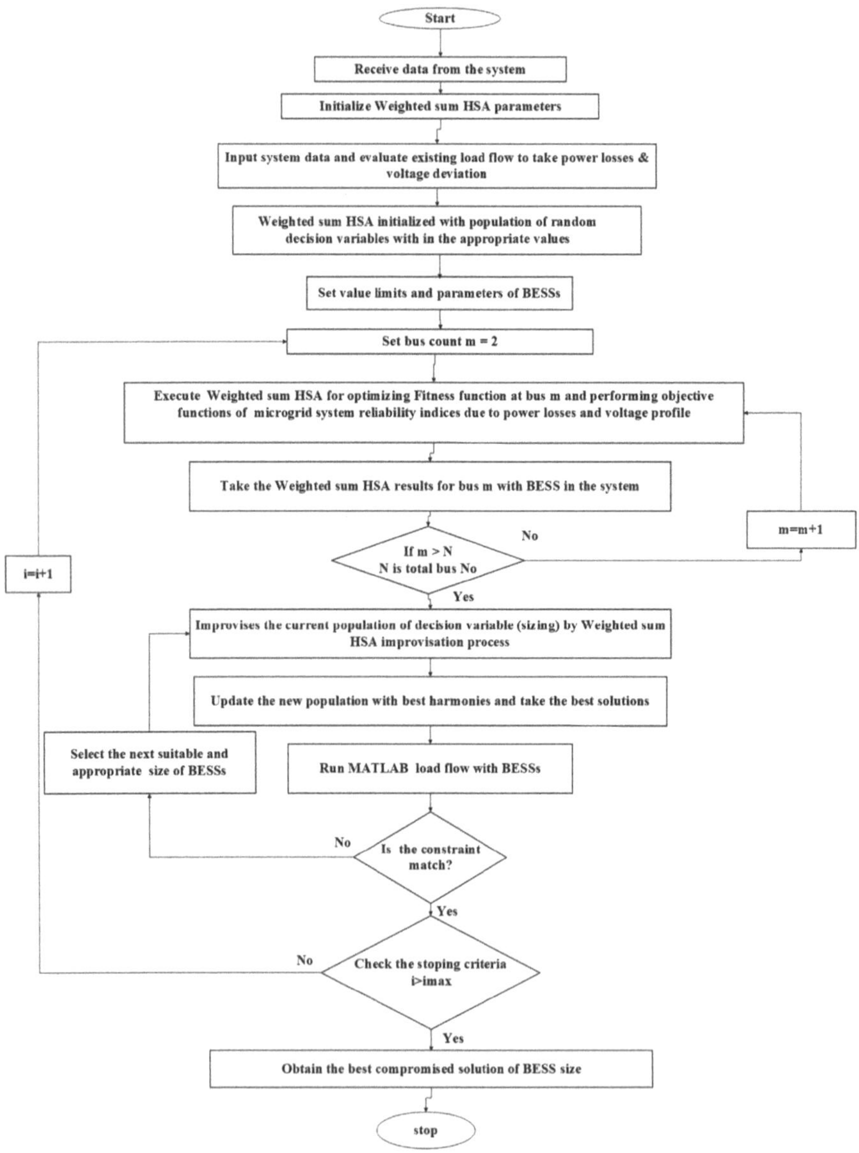

Fig. 1 Proposed weighted sum HSA flow chart for optimal placing and sizing of BESS on a typical microgrid.

Calculating the reliability indices on any given power system should be done by the following formulas. Unavailability is computed by the following formula:

$$U = 1 - A \tag{1}$$

Where U is unavailability and A is availability.

$$A = \frac{MTTF}{MTBF} \text{ and } U = \frac{MTTR}{MTBF} \tag{2}$$

where MTTR is mean time to repair, MTTF is mean time to failure, and MTBF is mean time between failures.

Where MTBF is calculated as:

$$MTBF = MTTR + MTTF \tag{3}$$

Also, calculating the failure rate (λ) and repair rates (μ), which are used to calculate the A and U because:

$$\lambda = \frac{1}{MTTF} \text{ and } \mu = \frac{1}{MTTR} \tag{4}$$

These equations give:

$$A = \frac{\mu}{\lambda + \mu} \text{ and } U = \frac{\lambda}{\lambda + \mu} \tag{5}$$

For the system that has more than two generation sources, the components are connected in parallel. Therefore, their equivalent unavailability is as follows:

$$U_{\text{sys}} = U_1 U_2 \tag{6}$$

Therefore,

$$A_{\text{sys}} = 1 - U_{\text{sys}} \tag{7}$$

In more detail:

$$A_{\text{sys}} = 1 - U_1 U_2 \tag{8}$$

$$A_{\text{sys}} = 1 - \frac{MTTR_1}{MTTF_1 + MTTR_1} \times \frac{MTTR_2}{MTTF_2 + MTTR_2} \tag{9}$$

Further,

$$U_{\text{sys}} = \frac{\lambda_1 \lambda_2}{(\lambda_1 + \mu_1) \times (\lambda_2 + \mu_2)} \tag{10}$$

$$A_{sys} = 1 - \frac{\lambda_1 \lambda_2}{(\lambda_1 + \mu_1) \times (\lambda_2 + \mu_2)} \tag{11}$$

To compute the whole repair and failure rates:

$$\mu_{sys} = \mu_1 + \mu_2 \tag{12}$$

$$U_{sys} = \frac{\lambda_1 \lambda_2}{(\lambda_1 + \mu_1) \times (\lambda_2 + \mu_2)} \tag{13}$$

$$\frac{\lambda_{sys}}{\lambda_{sys} + \mu_{sys}} = \frac{\lambda_1 \lambda_2}{(\lambda_1 + \mu_1) \times (\lambda_2 + \mu_2)} \tag{14}$$

$$\lambda_{sys} = \frac{\mu_{sys} \lambda_1 \lambda_2}{\mu_1 \mu_2 + \lambda_1 \mu_2 + \lambda_2 \mu_1} \tag{15}$$

which yields to:

$$\lambda_{sys} = \frac{(\mu_1 + \mu_2) \lambda_1 \lambda_2}{\mu_1 \mu_2 + \lambda_1 \mu_2 + \lambda_2 \mu_1} \tag{16}$$

Moreover,

$$MTTF_{sys} = \frac{1}{\lambda_{sys}} \text{ and } MTTR_{sys} = \frac{1}{\mu_{sys}} \tag{17}$$

Calculating the reliability indices is done as follows:

$$SAIDI = \frac{\text{Total duration of all interruptions}}{\text{Total number of customers connected}} \tag{18}$$

$$SAIDI = \frac{\sum_{i=1}^{n} r_i N_i}{N_T} \tag{19}$$

Where SAIDI is the system average interruption duration index

$$SAIFI = \frac{\text{Total number of all interruptions}}{\text{Total number of customers connected}} \tag{20}$$

$$SAIFI = \frac{\sum_{i=1}^{n} N_i}{N_T} = \frac{\sum_{i=1}^{n} \lambda_{LP} N_{LP}}{N_T} \tag{21}$$

Where SAIFI is the system average interruption frequency index

$$CAIDI = \frac{\text{Total duration of all interruptions}}{\text{Total number of all interruptions}} \tag{22}$$

$$CAIDI = \frac{\sum_{i=1}^{n} r_i N_i}{N_T} = \frac{SAIDI}{SAIFI} \tag{23}$$

Where CAIDI is the customer average interruption duration index.

The above formulas from Eq. (1)–(23) are utilized for computing analytical reliability factors [30].

4 Result and discussion

In this paper, we have studied two scenarios: (1) a microgrid without a BESS, and (2) a microgrid with a BESS. By using these two scenarios of reliability, indices are calculated and tabulated in these sections. Table 1 presents the load profile of the existing microgrid at a typical day.

The above shown in Fig. 2 and tabulated in Table 2 are the existing or base case of the microgrid system. The indention in Fig. 3 is the load profile and energy production of each generation unit of the system. The tabulated values of the reliability indices of the microgrid are not reliable to deliver the connected load for a long period without failure. Therefore, the system needs a mechanism to improve the reliability of the given typical microgrid. In this paper, the proposed solution is the optimal place and size of the battery energy storage system on the existing microgrid by utilizing the weighted sum harmony search algorithm optimization technique. Table 3 presents the existing case reliability indices of the microgrid.

The microgrid is integrated with BESS in optimal sizing and location utilizing weighted sum harmony search algorithm optimization approaches, as shown in Fig. 4, and the load profile and profile of each generation unit in the system are also displayed in Fig. 5. The microgrid's dependability indices were improved as a result of the proper integration of the battery energy storage technology. In Table 4, the dependability indices are listed.

According to Table 4, the values of the reliability indices of the microgrid are improved and are within the limit of the standard. The overall system comparison is shown in Table 5.

Table 1 Load profile of the existing microgrid at a typical day.

Time (h)	0:00–1:00	1:00–2:00	2:00–3:00	3:00–4:00	4:00–5:00	5:00–6:00	6:00–7:00	7:00–8:00	8:00–9:00	9:00–10:00	10:00–11:00	11:00–12:00
Load (MW)	218.9	200.6	200.6	200.6	207.6	208.6	219.6	228.6	230.3	250.3	250.3	240.3
Time (h)	12:00–13:00	13:00–14:00	14:00–15:00	15:00–16:00	16:00–17:00	17:00–18:00	18:00–19:00	19:00–20:00	20:00–21:00	21:00–22:00	22:00–23:00	23:00–00:00
Load (MW)	240.3	237.3	232.3	237.3	255.3	257.3	257.3	250.3	228.7	219.4	210.3	208.6

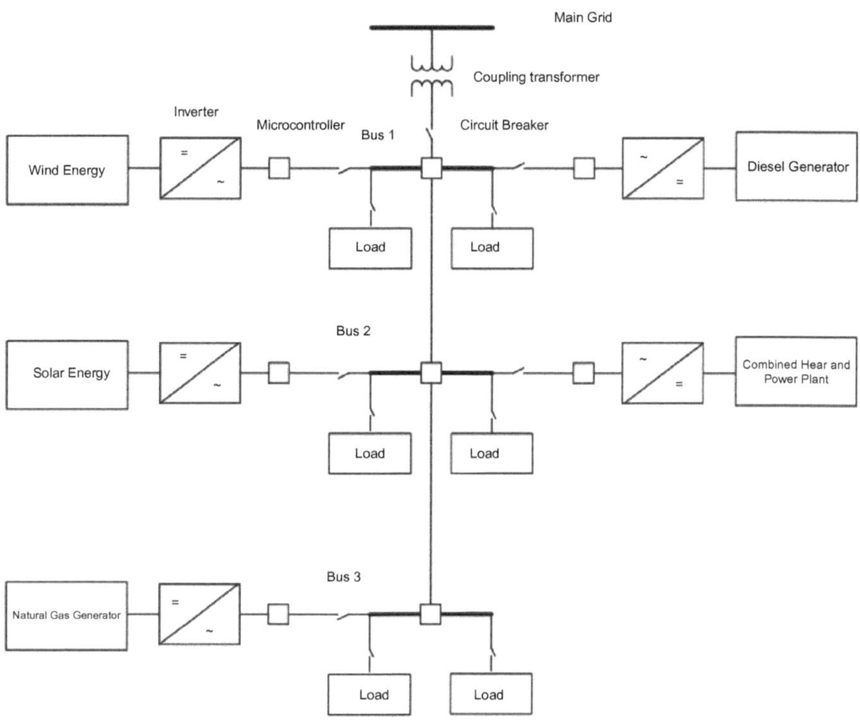

Fig. 2 One-line diagram of existing components of a microgrid.

Table 2 Characteristics of bus power generation units of the microgrid.

Bus unit no	Max. capacity	Min. capacity	Min. up time (h)	Min. down time (h)
1	113	78	3	2
2	130	49.6	4	2
3	40	21.75	5	4

The IEEE reliability test system (RTS) data are utilized in this work [31]. Further, the techniques developed in references [28,29,32] are utilized to optimally size and place the BESS on the microgrid. This work utilized DigSilent.

In this study, the BESS is utilized as a generator with adequate charging and discharging rates and enough storage to supply demand. The main goal of this study was to investigate the effect of optimally integrated BESS on a microgrid's operation cost. The presence of BESS enhances the reliability of the microgrid system as various reliability factors are available within the IEEE standard limits.

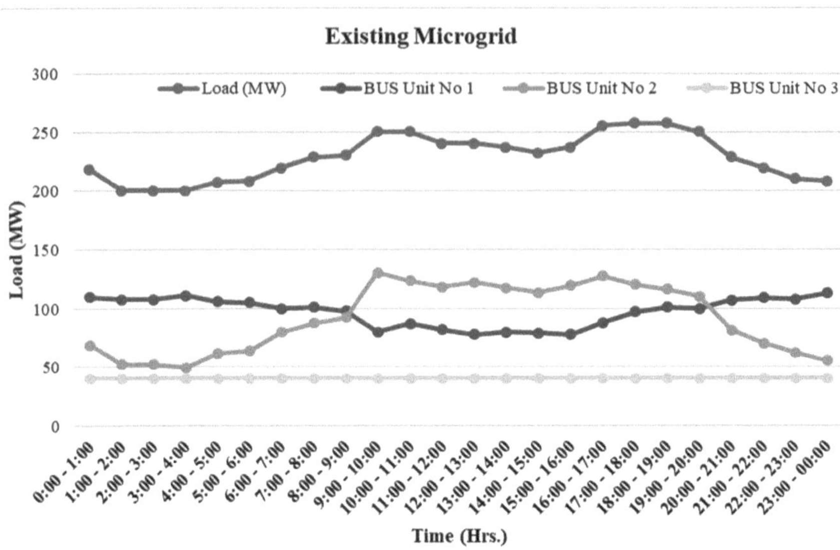

Fig. 3 Existing case bus power generation unit and load profile.

Table 3 Existing case reliability indices of microgrid.

Indices	Units	Value
ASAI	%	98.89924
ASUI	%	1.100760
SAIFI	Inter./customer. Year	13.859046
SAIDI	h/customer. Year	14.708952
CAIDI	h/customer interruption	1.001200

5 Conclusion

This chapter discusses the simulation techniques and reliability of the microgrid. A microgrid is considered a future grid due to the demand in power supply as well as its capability of integrating with renewable energy sources. The effective methods for the improvement of reliability using various techniques are proposed, that is, the integration of the BESS system. This work utilized a weighted HAS for optimal sizing (12.25 MW) and placement (Bus 3) of BESS. The reliability index calculation was performed and studied for the reliability

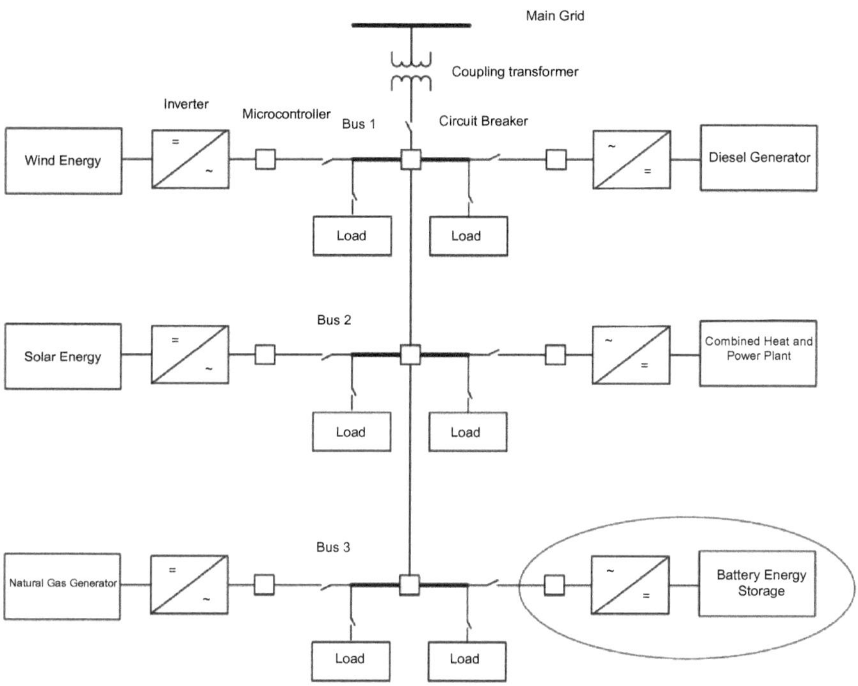

Fig. 4 Proposed one-line diagram of microgrid components with integration of BESS.

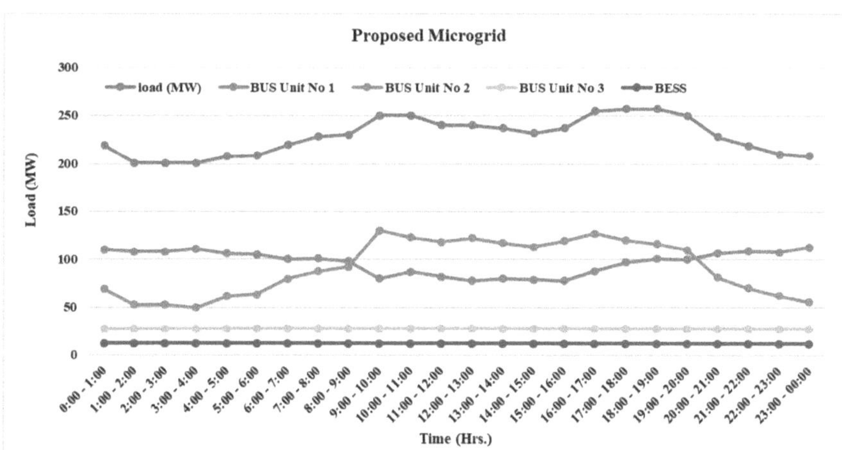

Fig. 5 Proposed microgrid bus power generation unit and load profile with BESS.

Table 4 Reliability indices of microgrid with integration of BESS.

Indices	Units	Value
ASAI	%	99.99124
ASUI	%	0.008760
SAIFI	Inter./customer. Year	0.659046
SAIDI	h/customer. Year	0.808952
CAIDI	h/customer interruption	1.190011

Table 5 Overall reliability index comparison in the two scenarios.

Indices	Existing case	Microgrid with BESS	Improvement (%)
ASAI	98.89924	99.99124	1.092096
ASUI	1.100760	0.008760	−99.20419
SAIFI	13.859046	0.672846	−95.145078
SAIDI	14.708952	0.809982	−94.493272
CAIDI	1.001200	1.190011	15.866324

evaluation. The reliability indices such as ASAI, ASUI, SAIFI, SAIDI, and CAIDI are computed, which show enhanced values, that is, 1.092096%, −99.20419%, −95.145078%, −94.493272%, and 15.866324%, respectively.

References

[1] A. Abdulkarim, S.M. Abdelkader, D.J. Morrow, Model for optimum design of standalone hybrid renewable energy microgrids, J. Fundam. Appl. Sci. 9 (2) (2017) 1074–1101.

[2] X. Liu, S. Islam, Wind-diesel-battery hybrid generation system reliability analysis on site and size factors, in: 2006 International Conference on Electrical and Computer Engineering, Dhaka, Bangladesh, Dec. 2006, pp. 229–232, https://doi.org/10.1109/ICECE.2006.355332.

[3] F.A. Bhuiyan, A. Yazdani, Reliability assessment of a wind-power system with integrated energy storage, IET Renew. Power Gener. 4 (3) (2010) 211–220.

[4] X. Liu, S. Islam, Reliability evaluation of a wind-diesel hybrid power system with battery bank using discrete wind speed frame analysis, in: 2006 International Conference on Probabilistic Methods Applied to Power Systems, Stockholm, Sweden, Jun. 2006, pp. 1–7, https://doi.org/10.1109/PMAPS.2006.360226.

[5] X. Liu, S. Islam, A.A. Chowdhury, D.O. Koval, Reliability evaluation of a wind-diesel-battery hybrid power system, in: 2008 IEEE/IAS Industrial and Commercial Power Systems Technical Conference, Clearwater Beach, FL, USA, May 2008, pp. 1–8, https://doi.org/10.1109/ICPS.2008.4606290.

[6] K. Cho, J. Park, T. Oh, J. Choi, S. Hong, M. Shahidehpour, Probabilistic reliability evaluation of composite power systems including wind turbine generators, J. Int.

Counc. Electr. Eng. 1 (3) (2011) 274–280, https://doi.org/10.5370/JICEE.2011.1.3.274.
[7] J. Park, et al., Reliability evaluation of interconnected power systems including wind turbine generators, Int. J. Innov. Comput. Inf. Control 8 (8) (2012) 5797–5808.
[8] Z.Y. Gao, P. Wang, Reliability evaluation of power systems with WTGs and energy storage, in: 2010 Conference Proceedings IPEC, 2010, pp. 654–659.
[9] J. Park, W. Liang, J. Choi, A.A. El-Keib, M. Shahidehpour, R. Billinton, A probabilistic reliability evaluation of a power system including solar/photovoltaic cell generator, in: 2009 IEEE Power & Energy Society General Meeting, 2009, pp. 1–6.
[10] J. Choi, J. Park, M. Shahidehpour, R. Billinton, Assessment of CO2 reduction by renewable energy generators, in: 2010 Innovative Smart Grid Technologies (ISGT), Gaithersburg, MD, USA, Jan. 2010, pp. 1–5, https://doi.org/10.1109/ISGT.2010.5434742.
[11] J. Park, K. Cho, T. Oh, J. Choi, A.A. El-Keib, Evaluation of economic impacts of wind turbine and solar cell generators on power system considering CO_2 emission, in: 2010 IEEE 11th International Conference on Probabilistic Methods Applied to Power Systems, 2010, pp. 413–418.
[12] M. Aien, A. Biglari, M. Rashidinejad, Probabilistic reliability evaluation of hybrid wind-photovoltaic power systems, in: 2013 21st Iranian Conference on Electrical Engineering (ICEE), 2013, pp. 1–6.
[13] B. Kekezoglu, O. Arikan, A. Erduman, E. Isen, A. Durusu, A. Bozkurt, Reliability analysis of hybrid energy systems: case study of Davutpasa campus, in: Eurocon 2013, 2013, pp. 1141–1144.
[14] Z. Liu, W. Liu, G. Su, H. Yang, G. Hu, Wind-solar micro grid reliability evaluation based on sequential Monte Carlo, in: 2016 International Conference on Probabilistic Methods Applied to Power Systems (PMAPS), 2016, pp. 1–6.
[15] M.M. Ghahderijani, S.M. Barakati, S. Tavakoli, Reliability evaluation of stand-alone hybrid microgrid using Sequential Monte Carlo Simulation, in: 2012 Second Iranian Conference on Renewable Energy and Distributed Generation, 2012, pp. 33–38.
[16] A.A. Recalde, T.K. Saha, M. Mosadeghy, Reliability evaluation with wind turbines and photovoltaic panels, in: 2014 IEEE PES Transmission & Distribution Conference and Exposition-Latin America (PES T&D-LA), 2014, pp. 1–5.
[17] M.M. Ghahderijani, S.M. Barakati, A. Jamshidi, Application of stochastic simulation method in reliability assessment of a PV-wind-diesel-SOFC hybrid microgrid, Int. J. Eng. Technol. 4 (5) (2012) 586.
[18] A. Abdulkarim, S.M. Abdelkader, D.J. Morrow, A.J. Falade, A.U. Lawan, H.R. Iswadi, Effect of weather and the hybrid energy storage on the availability of standalone microgrid, Int. J. Renew. Energy Res. 6 (1) (2016) 189–198.
[19] G. Li, Z. Bie, B. Hua, X. Wang, Reliability evaluation of distribution systems including micro-grids considering demand response and energy storage, in: 2012 47th International Universities Power Engineering Conference (UPEC), 2012, pp. 1–6.
[20] I.-S. Bae, J.-O. Kim, Reliability evaluation of customers in a microgrid, IEEE Trans. Power Syst. 23 (3) (2008) 1416–1422.
[21] P. Zhou, et al., Reliability evaluation of grid-connected micro-grid considering demand response, in: 2016 International Conference on Probabilistic Methods Applied to Power Systems (PMAPS), Beijing, Oct. 2016, pp. 1–5, https://doi.org/10.1109/PMAPS.2016.7764105.
[22] N. Nikmehr, S. Najafi-Ravadanegh, Reliability evaluation in multi-microgrids under probabilistic optimum operation using heuristic algorithm, in: 2015 Smart Grid Conference (SGC), Tehran, Iran, Dec. 2015, pp. 92–98, https://doi.org/10.1109/SGC.2015.7857416.
[23] A. Abdelsamad, D. Lubkeman, Reliability analysis for a hybrid microgrid based on chronological Monte Carlo simulation with Markov switching modeling, in: 2019

IEEE Power & Energy Society Innovative Smart Grid Technologies Conference (ISGT), 2019, pp. 1–5.
[24] T. Adefarati, R.C. Bansal, J.J. Justo, Reliability and economic evaluation of a microgrid power system, Energy Procedia 142 (2017) 43–48.
[25] M. Barani, V.V. Vadlamudi, P.E. Heegaard, Reliability analysis of cyber-physical microgrids: study of grid-connected microgrids with communication-based control systems, IET Gener. Transm. Distrib. 15 (4) (2021) 645–663.
[26] T. Adefarati, R.C. Bansal, Reliability, economic and environmental analysis of a microgrid system in the presence of renewable energy resources, Appl. Energy 236 (2019) 1089–1114.
[27] S.A. Arefifar, A.-R.M. Yasser, T.H. El-Fouly, Optimum microgrid design for enhancing reliability and supply-security, IEEE Trans. Smart Grid 4 (3) (2013) 1567–1575.
[28] H. Kanchev, B. Francois, V. Lazarov, Unit commitment by dynamic programming for microgrid operational planning optimization and emission reduction, in: International Aegean Conference on Electrical Machines and Power Electronics and Electromotion, Joint Conference, 2011, pp. 502–507.
[29] A. Assad, Recent advances in harmony search algorithm, Int. J. Comput. Eng. Technol. 10 (1) (2019) 157–165.
[30] T.F. Agajie, B. Khan, J.M. Guerrero, O.p. Mahela, Reliability enhancement and voltage profile improvement of distribution network using optimal capacity allocation and placement of distributed energy resources, Comput. Electr. Eng. 93 (2021), 107295.
[31] H.N. Soloklo, M.M. Farsangi, Multi-objective weighted sum approach model reduction by Routh-Pade approximation using harmony search, Turk. J. Electr. Eng. Comput. Sci. 21 (Suppl. 2) (2013) 2283–2293.
[32] C. Barrows, A. Bloom, A. Ehlen, J. Ikäheimo, J. Jorgenson, D. Krishnamurthy, J. Lau, et al., The IEEE reliability test system: a proposed 2019 update, IEEE Trans. Power Syst. 35 (1) (2019) 119–127.

SECTION 7

Electric vehicle technology

CHAPTER 7.1

Electric vehicle technology

Ola Khaddam
Electrical Power Engineering, Tishreen University, Latakia, Syria

Contents

1 Introduction

Fuel-based conventional vehicles consume fossil fuel and emit greenhouse gases (GHGs). These vehicles have an internal combustion engine where fossil fuel is burned to generate energy to move the vehicle. The energy efficiency of these vehicles is poor with a value between 15% and 18%. The environmental harm of such a process is devastating due to the huge number of these vehicles as they form the majority of transportation systems around the world. By 2030, there will be 2.08 billion vehicles on the roads compared to 70 million vehicles after World War II. This means that fuel consumption will increase drastically, leading to the depletion of energy resources [1]. The US Transportation Department and the European Commissioner for Climate Action stated that the average amount of emissions

Active Electrical Distribution Network
https://doi.org/10.1016/B978-0-323-85169-5.00007-1

from vehicles is 158.7 g/km. The US Environmental Protection Agency stated that the annual GHGs per passenger vehicle is 4.8 tons of CO_2. Fuel burning increases the levels of hydrocarbons, carbon dioxide, and nitrogen oxides in the atmosphere and makes fresh air poisonous. In addition, the use of petrol has reached a peak value. Therefore, in the next few decades, it is expected that we will face shortages of all petrol and gas resources. Because of this, there has been an orientation worldwide toward ultimate solutions for these issues, starting with vehicle emission standards and ending with mobility sustainability. Therefore, the world is in urgent need. The electrification of the transportation sector is a promising procedure to deal with these issues [2]. Hybrid electric vehicles (HEVs) were introduced first as vehicles that used less fossil fuel, therefore emitting fewer GHGs compared to conventional vehicles. They have an energy efficiency between 60% and 70%. which is much better than conventional vehicles. HEV technology presents a good alternative that can mitigate emissions and fuel consumption, but it can't be considered as the ultimate solution for the mentioned issues [3]. An HEV is a combination of gasoline and electrical energy, so it can be thought of as an intermediate stage between conventional internal combustion engine vehicles and fully electric vehicles (EVs); HEVs can be categorized as electric vehicles [4]. On the other hand, EV technology is a more environmental and economically friendly solution as it depends fully on clean electrical energy. In addition, EV energy efficiency is much higher than HEVs and fuel-based vehicles, with a value of 90%. Furthermore, EVs can be connected to smart grids as distributed energy storage devices to supply surplus energy to the grid during peak demand or for compensating renewable energy generation inconstancy. This process can have a positive effect on the quality and reliability of the grid and reduce the effects of renewable generation inconstancy [5]. Although the process of EV manufacturing still creates emissions, these vehicles produce much less pollution on roads than conventional vehicles.

2 Hybrid electric vehicles

HEV is an intermediate stage between fully fuel-based vehicles and fully electric vehicles. Nonetheless, it can be classified under the electric vehicle category. HEV is mainly the integration of two or more sources of energy (a battery, a flywheel, etc.) and/or two or more sources of power onboard the vehicle (internal combustion engine, fuel cell, etc.). The HEV starts to operate using the energy supplies stored in the battery and then switches to the internal combustion engine as an energy source when the speed of the vehicle increases [4]. Honda and Toyota have already produced HEVs, which are now becoming very popular among consumers for their great performance and low emissions.

HEVs have many advantages, the most important of which is the availability of two sources of power, which means they can employ the appropriate power source when needed. However, HEVs have some disadvantages such as the environmental harm caused by the battery recycling process and GHG emissions. Although there are different types of HEVs, only gasoline HEVs are currently available. HEVs can be generally classified into three main categories [6,7]:

- Series HEVs

  In this type, the internal combustion engine drives the generator, which in turn supplies the electric motor. In HEVs, the motor alone drives the wheels of the vehicles instead of the engine. The generator (alternator) both charges the battery and supplies the electric motor with power. When large amounts of power are required, the motor draws electricity from both the batteries and the generator. A series HEV achieves lower fuel consumption by having the engine operating at the highest efficiency point during starting and stopping. The battery sizes can be reduced in this configuration, but at the same time the generator and the engine sizes need to be increased. The series HEV configuration is illustrated in Fig. 1. In general, a generator is integrated with an engine to

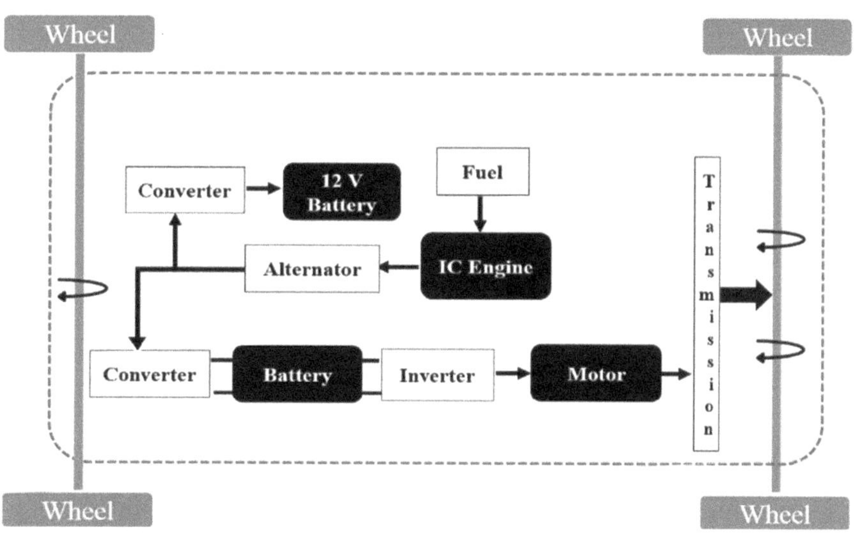

Fig. 1 Series HEV configuration.

generate power and charge the battery. The power electronics needed in the series HEV configuration are as follows:

- A converter used to convert the generator AC output into DC power to charge the battery.
- A DC to DC converter that charges the battery with 12 V power.
- An inverter to convert DC from the battery into AC power to supply the motor.

- Parallel HEVs

  In this configuration, an internal combustion engine and an electrical motor are connected in shunt and drive the vehicle's wheels together. A complex control system is needed in this type to integrate the engine with the motor and generator. These vehicles use the energy wasted in the braking process to recharge the battery, which reduces fuel consumption. Moreover, the battery can be recharged when the energy generated in the engine is greater than the energy required for propulsion. The internal combustion engine size can be reduced in parallel HEVs in comparison with series HEVs because the vehicle's wheels are driven by both the engine and the electric motor. Parallel HEV configuration is illustrated in Fig. 2.

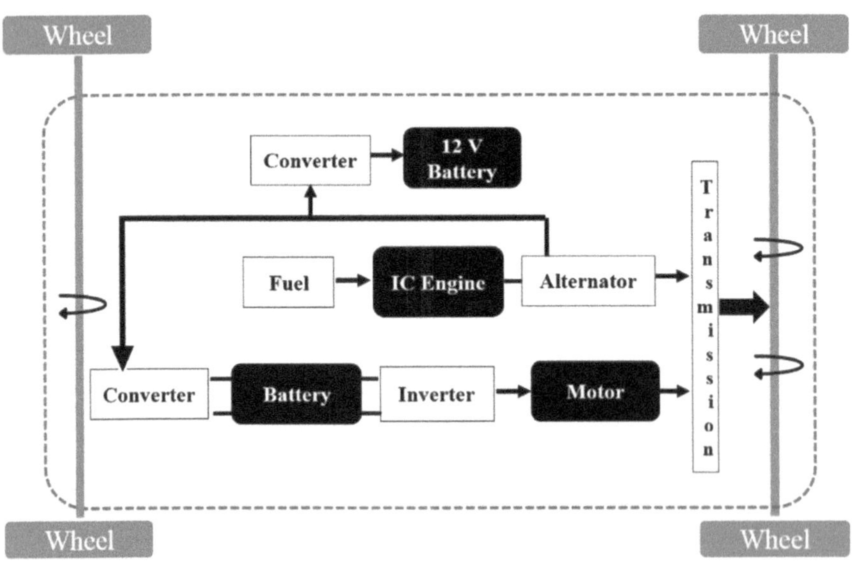

Fig. 2 Parallel HEV configuration.

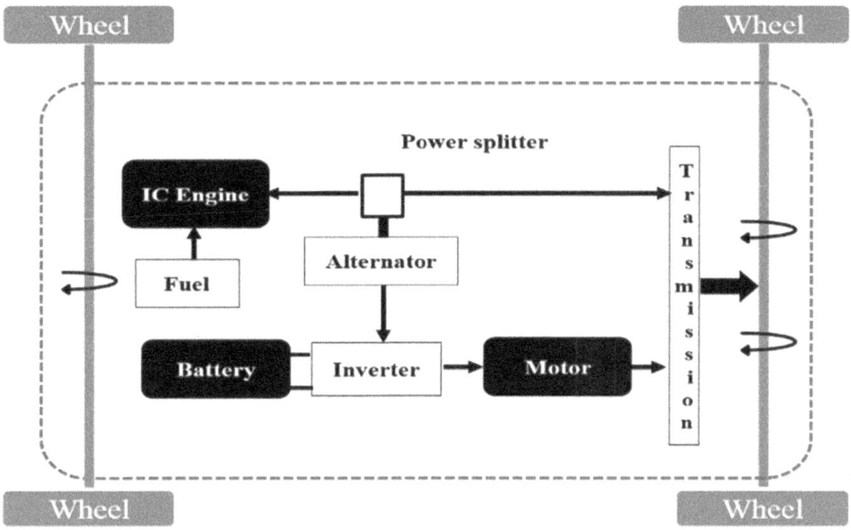

Fig. 3 Series-parallel HEV configuration.

- Series-parallel HEVs

  This type is a combination of series and parallel types, which means that the control system is more complex in this type. The engine output is divided into mechanical and electrical transmission paths using a power-split unit. This type of hybrid vehicle can provide high power when compared to series or parallel. Furthermore, the motor is smaller in this type. Fig. 3 shows the configuration of series-parallel HEVs.

3 Electric vehicles

Unlike an HEV, which integrates an internal combustion engine propulsion system with an electric motor propulsion system, an electric vehicle (EV) depends solely on an electrical motor for propulsion and uses only electrical energy to supply the vehicle, which means it has zero emissions. EVs are also called battery electric vehicles because they use the energy stored in batteries to supply the electric motor [7]. Electric vehicles have bigger batteries than HEVs because they depend on batteries as the sole source of energy. An EV battery is charged by plugging the vehicle into an external source of electricity. The capability to recharge the battery by regenerative braking is also available in EVs. The basic structure of an EV includes an electric motor, a battery, a DC-DC converter, a charger, and an electronic controller.

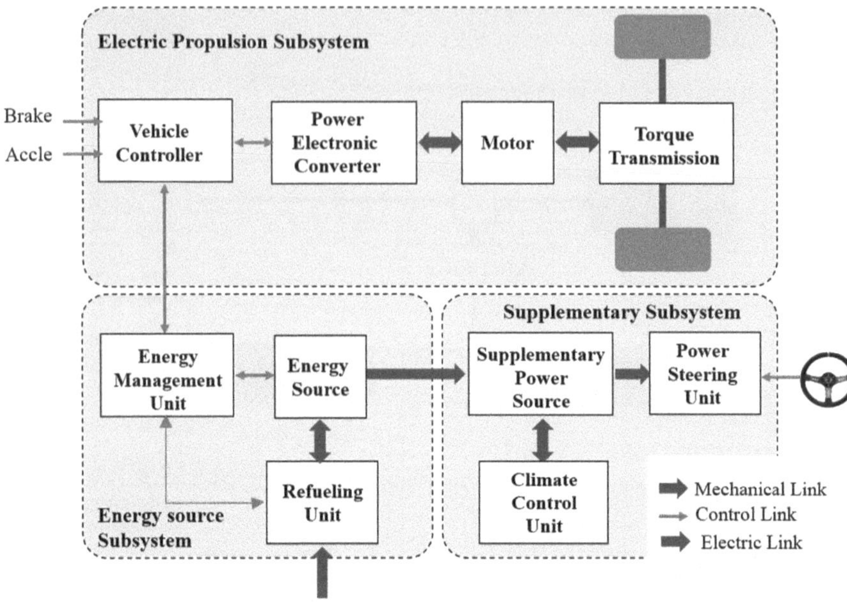

Fig. 4 The main structure of an EV.

The EV drive system includes three main subsystems: the energy source, the motor propulsion, and the supplementary subsystem. The energy source with its refueling unit and energy management unit form the energy source subsystem. The motor, converter, torque transmission, wheels, and controller together form the electric propulsion subsystem. The supplementary subsystem consists of a supplementary power source, a steering unit, and a climate control unit [8]. It's worth mentioning that EVs have a much smaller travel range due to the lower energy that can be stored in the battery. Fig. 4 shows the main structure of EVs.

There are several possible configurations for EVs based on the different types of clutch, gearbox, transmission, and differential as well as the number of motors, as illustrated in Fig. 5.

Fig. 5A shows an EV configuration that is a developed version of an internal combustion engine vehicle where the engine is replaced with an electric motor. This configuration also includes a clutch, a variable speed gearbox, and a differential on the front axle. In Fig. 5B, an EV configuration with a fixed gearbox and a differential on the front axle is illustrated. The clutch is absent in this type because the motor gives constant energy with a wide range of speeds, which means reduced size and weight. The EV configuration illustrated in Fig. 5C has the same components of that in Fig. 5B,

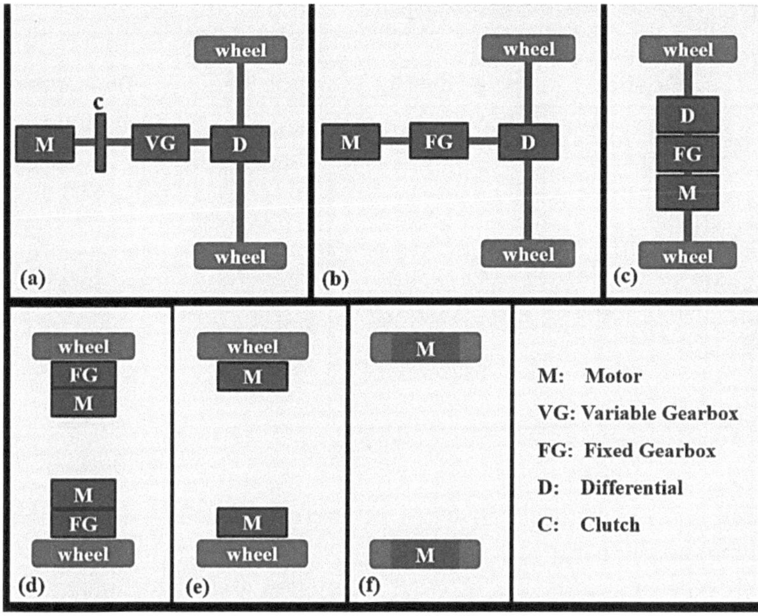

Fig. 5 The different configurations of EVs.

but these components are integrated together, leading to a simple driving system. In Fig. 5D, two traction motors are integrated with fixed gearboxes on the front axle for operation at different speeds on a curved road. The drive system in Fig. 5E is an in-wheel drive with two traction motors inside the EV's wheels with a thin gear. Similarly, the drive in Fig. 5F, includes two motors inside the wheels but no gear involved, which makes this structure simpler than that of Fig. 5E [9].

4 Energy storage devices used in EVs and HEVs

The energy storage devices in electric vehicle technology are the key component that distinguish these vehicles from conventional vehicles. Storage devices can take many forms such as a battery, a flywheel, an ultracapacitor, or hydrogen storage [10].

Batteries: Batteries are the most common energy storage device used in electric vehicles. The most popular battery types used in mobility applications are nickel batteries (nickel-metal hybrid, nickel-iron, nickel-cadmium), lithium batteries (Li-ion, Li-ion polymer, Li-iron sulfide, Li-iron phosphate), and lead-acid batteries. Table 1 shows a comparison between

Table 1 Comparison between battery types.

Battery type	Energy efficiency	Energy density	Cycle life	Discharge rate
Lead-acid	63% ÷ 90%	25 ÷ 50 Wh/kg	500–1500 cycles	5% per month (−20°C to 65°C)
Nickel	<80%	<50 Wh/kg	≈1000 cycles	20% per month (−20°C to 50°C)
Lithium	90 ÷ 100%	2 ÷ 4.5 kW/kg	<2000 cycles	3 ÷ 10% per month (−20°C to 60°C)

the three types of batteries in terms of energy efficiency, energy density, cycle life, and discharge rate [10,11].

Each type of battery has its own advantages and disadvantages, but in general, the lithium battery is the best type because it has the best energy efficiency, the highest energy density, and the highest cycle life when compared with nickel and lead-acid batteries. In addition, it has a moderate cost and discharge rate when compared with the other two types. One disadvantage of this type is that its lifetime can be reduced unexpectedly in high-temperature conditions and deep discharge, so it must be equipped with a protection circuit for safe operation [11].

To charge EV batteries, two methods can be used: conductive and inductive. Conductive charging is the common type in the current industry, and it is done by using a charger in the electric vehicle. The vehicle receives electrical energy by plugging into an AC supply (charging station) through the electrical charger [12].

Flywheels: As shown in Fig. 6, a flywheel consists of a rotating mass, an electric machine, bearings, a container, and a power electronic converter. It is also known as the mechanical battery because it stores kinetic energy in a high-speed rotating rotor attached to an electric machine's shaft and releases this energy when needed [13]. The charging process happens as the wheel's rotor speed increases by the electric machine (operating in motor mode) connected to it until it reaches a high speed. The discharging process happens when the speed of the wheel's rotor decreases and releases the stored kinetic energy in the form of electrical energy through the electric machine (operating in generator mode). Flywheels can be found in two forms:

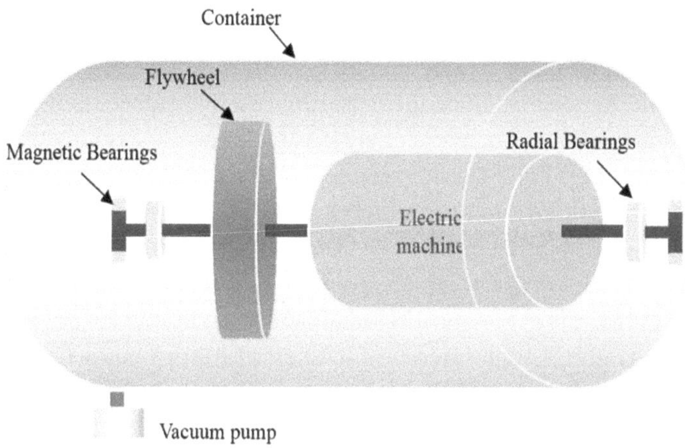

Fig. 6 Flywheel structure.

high-speed flywheels (speed between 10,000 and 100,000 rpm) and low-speed flywheels (speed <6000 rpm) [10].

Flywheels have many advantages, including a long lifetime, fast response, high energy efficiency (90 ÷ 95%), low maintenance costs, and an environmentally friendly nature. However, they also have many disadvantages including low energy density and high self-discharge losses. The flywheel friction losses can cause a self-discharge rate of up to 20% of the stored energy per hour. Therefore, a flywheel is not suitable for long-term storage applications [13].

Ultracapacitors: An ultracapacitor is similar to a conventional capacitor in structure, but it uses an electrolyte solution to store energy. The ultracapacitor energy storage capacity is 20 times that of a conventional capacitor because it has much greater capacity than that of a conventional capacitor [10]. The three main types of ultracapacitor are electric double-layer capacitors, pseudocapacitors, and hybrid capacitors. All these types share similar characteristics such as energy efficiency, lifetime, and energy density [14]. Ultracapacitors have a high life time, (1 ÷ 105 cycles for 40 years), in comparison to other storage devices. In addition, they have a high energy density that ranges from 1000 to 2000 kW/kg, and an energy efficiency of 85 ÷ 95%. Therefore, these capacitors have the ability to charge quickly and give energy without remarkable losses. However, ultracapacitors have a high self-discharge, which makes them unreliable for use as a single source of energy for vehicles. Moreover, their costs are high (>6000$/kWh), which makes them suitable to be used as supplementary energy sources in short-term energy storage applications [10,14].

Hydrogen storage system: A hydrogen storage system is a system that stores hydrogen and uses it when needed. This system depends on converting electrical energy into hydrogen, storing the hydrogen, then converting the hydrogen back into electrical energy using fuel cells with no emissions in the process [11,12]. Hydrogen is released through water electrolysis, which is a simple process that can give a high energy efficiency of more than 70%. This system is very environmentally friendly, but the conversion process consumes a huge amount of electricity. Nowadays, most hydrogen is obtained by reforming this gas using steam and a catalyst, which is less expensive and could give a high energy efficiency of more than 80%. However, this procedure is accompanied with CO_2 emissions [12,15]. A promising method of producing hydrogen is direct hydrogen production by photocatalytic water splitting, which relies on nanotechnology; however, this technology is still in the research phase [15].

Hydrogen is very difficult to handle in terms of carrying and storing because it is highly flammable and exists in a gaseous state at room temperature. There are three different solutions for hydrogen storage in EV tanks [10]:

- High-pressure tank: EVs on the roads today use a high-pressure tank of 35 MPa while high-pressure tanks of 70 MPa are still in the testing phase. These tanks have a simple structure and possess the characteristics of easy charging and discharging.
- Liquid hydrogen tank: gaseous hydrogen is transformed into liquid hydrogen when cooled to 20 K under atmospheric pressure. This process expands the driving range of the vehicle because the hydrogen stored in this state has more density than in the gaseous state. Moreover, liquid hydrogen is easier to store and transport.
- Hydrogen-absorbing alloy tank: The hydrogen stored in this type of tank has a higher density than liquid hydrogen, which means the tank is smaller and has the possibility of reversible hydrogen charge and discharge. The disadvantage of this type is the low gravimetric density.

5 Motor drives used in EVs

The motor drive system, including the electric motor, converter, and controller, is the core of an EV. An HEV is driven by both an electric motor and an internal combustion engine and an EV is driven by one or more electric motors. The EV motor drive must achieve high acceleration torque, fast torque response, high power density, high instantaneous power, high efficiency

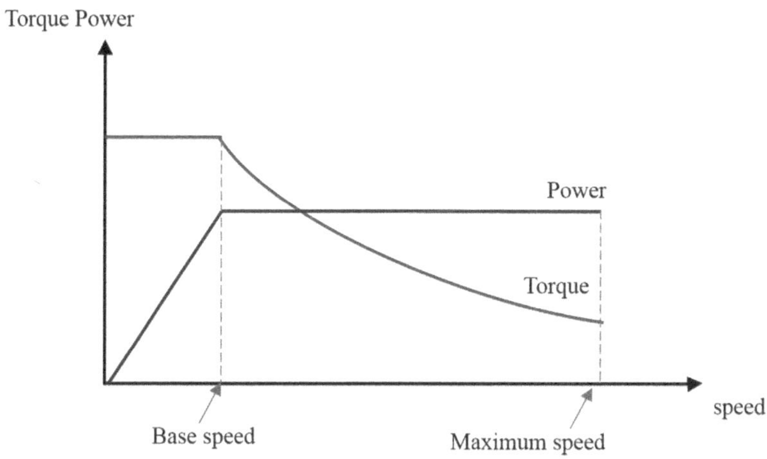

Fig. 7 Desired output characteristics for electric vehicle applications.

over a wide range of speeds, wide constant torque, constant power regions, and moderate costs [16]. Fig. 7 shows the desired output characteristics that meet the requirements for EV applications.

The five types of motors that can be used in HEVs and EVs are a DC motor, an induction motor, a brushless motor, a switched reluctance motor, and a synchronous motor; each has its advantages and disadvantages [17].

5.1 DC motors

DC motors have the ability to achieve high torques at low speeds, which is required in traction applications such as EV and HEV applications. In addition, the control system of these motors relies on changing the terminal voltage, which is considered to be a simple control system. There are two main types of DC motors: brushed and brushless. The brushed type can produce high torque at low speed with a torque-speed characteristic suitable for traction applications [16,18]. However, they have a bulky structure, low efficiency, high maintenance costs, and low reliability. Moreover, the maximum motor speeds are limited due to the continuous friction between the commutator and the brushes. For these reasons, the brushless DC motors are more favorable in EVs. A brushless DC motor has no brushes, which means it needs less maintenance. It also has a better efficiency and smaller size when compared with a brushed motor. Fig. 8 shows the torque-speed characteristics of the brushless DC motor [18].

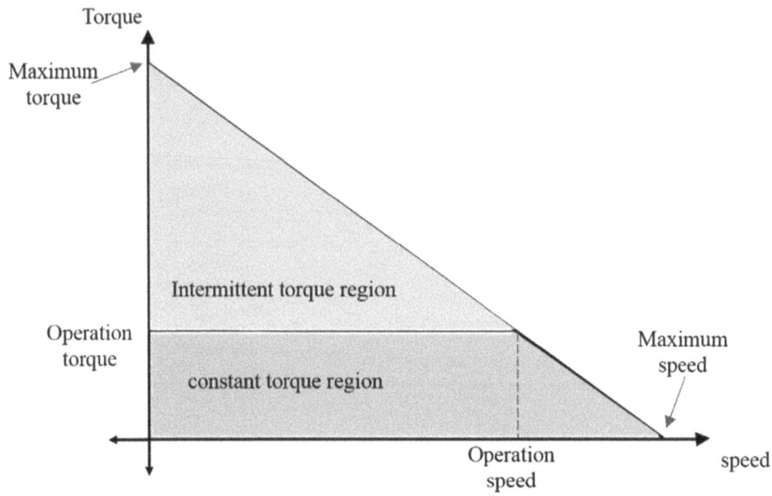

Fig. 8 The torque-speed characteristics of the DC brushless motor.

5.2 Induction motors

Having a very simple structure, low maintenance costs, reliability, and good maximum torque capability, induction motors are considered to be a very suitable choice to use in the drive systems of HEVs and EVs. Induction motors have an expanded constant power speed range when compared with DC motors because they have a commutator or brushes, which means there is no friction in these motors and increased maximum speed limits. The speed can be changed in an induction motor by changing the frequency of the voltage (voltage/frequency control). The field-oriented control is used in an induction motor to make it work as a separately excited DC motor [16]. In this case, the motor does not have speed limitation issues as in DC motors. The speed range can be expanded by weakening the flux in the constant power zone. Fig. 9 shows the speed-torque characteristics of the induction motor.

These characteristics show that in a high-speed region because of the big value of electromotive force, the stator current is reduced, which means the torque is inversely proportional to the speed square.

Induction motors have many disadvantages such as high copper losses, low power factor, low efficiency, and narrow constant power range. The control system in this motor is more complex than that in the DC motor [18,19].

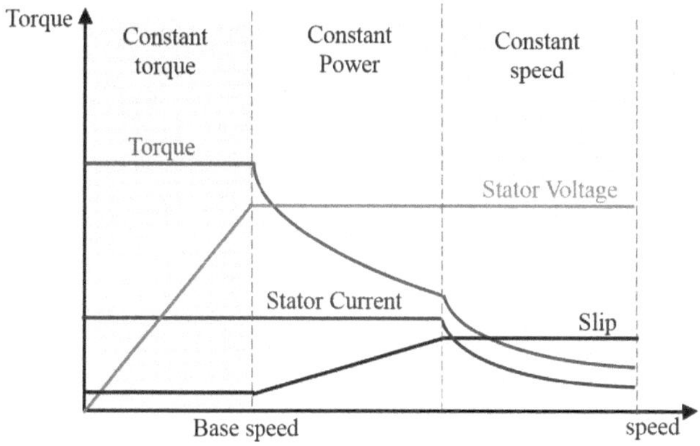

Fig. 9 Speed-torque characteristics in an induction motor.

5.3 Permanent magnet brushless DC motors (PMBLDC)

PMBLDC motors have high efficiency and high power density with a speed range that can be expanded to three or four times the base speed. These motors are better candidates than induction motors as propulsion motors. The main advantages of these motors are the highly reinforced magnetic field, which gives the ability to produce high power density with smaller size and reduced weight of the motor as well as the absence of copper losses due to the absence of rotor windings. These advantages make PMBLDC motors an attractive choice for EVs and HEVs [18,20]. However, there are many drawbacks related to this type of motor such as the relatively narrow range of constant power due to the limited capability of field weakening, which results from the strongly present field flux produced. To widen the maximum speed range and improve the motor efficiency, the conduction angle of the converter can be controlled at speeds higher than the base speed. However, at very high speeds, the efficiency may drop abruptly due to demagnetizing caused by the back electromotive force and possible force occurrence [16]. Furthermore, in high-temperature operation, the residual flux density decreases; therefore, the torque capacity of the machine also decreases. The speed-torque characteristic of this motor is shown in Fig. 10.

Based on the placement of the permanent magnet, PMBLDCs can be categorized into two main types: surface magnet (SPM) and interior magnet (IPM). A hybrid between the two types is the PM hybrid motor [19,21].

5.4 Switched reluctance motor drives (SRM)

The potentials of SRM motors make them more attractive for EV/HEV applications. These motors possess very good features such as a robust

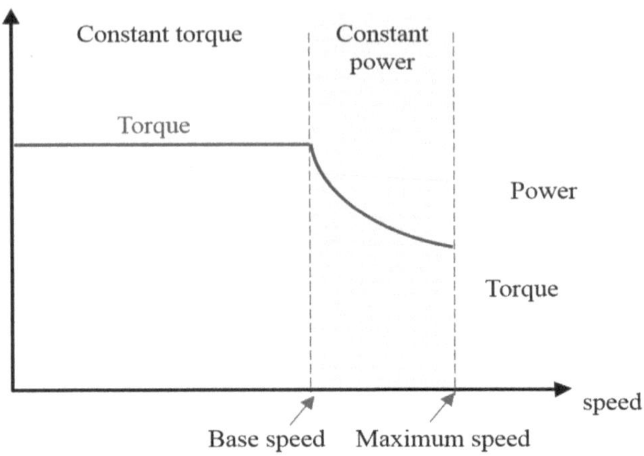

Fig. 10 Speed-torque characteristics of PMBLDC motors.

structure, low manufacturing cost, good speed-torque characteristics, and a broad constant power zone. The speed-torque characteristics almost match the vehicle load characteristics with a very high speed operation capability. In addition, SRMs have a high starting torque and a high torque-inertia ratio. The absence of rotor windings, commutators, magnets, and brushes makes the structure of this motor much simpler when compared with that of DC, induction, and permanent magnet motors [20]. Moreover, this motor has a high fault tolerance due to disconnection between the phases; therefore, the failure of one phase doesn't affect the motor. The main disadvantages of these motors are the acoustic noise, high torque ripples (due to salient-pole rotor and stator), big size and weight, and lower power density [21]. The speed-torque characteristics of SRMs are shown in Fig. 11.

Recent studies on this type of motor have suggested that EV and HEV applications need a high number of stator and rotor poles such as an 18-stator pole. Others suggested an SMR motor drive that has a combination of 18 and 12 poles. The major advantage of such a design is the higher produced torque, the higher reliability, and the higher fault tolerance [22].

5.5 Synchronous reluctance motor drives (SynRM)

These motors combine the benefits of both induction motors and permanent magnet motors. This type of motor possesses the advantages of both motors, which are the efficiency and robustness of induction motors as well as the synchronous speed operation of permanent magnet motors. The stator winding distribution in these motors is the same as induction and permanent

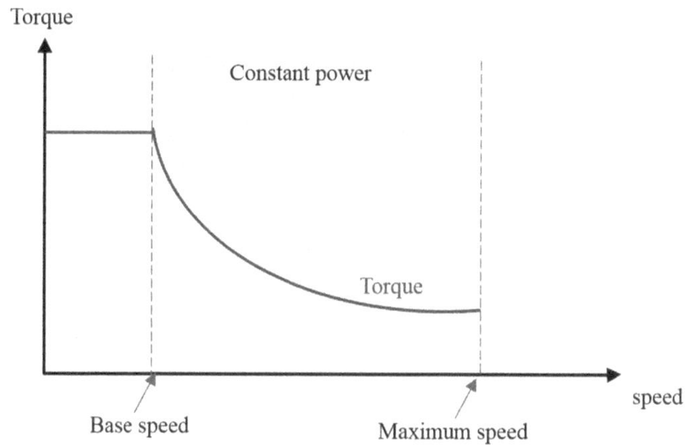

Fig. 11 Speed-torque characteristics of SRM motors.

magnet motors [16,18]. Moreover, the fault tolerance of this motor is as high as induction motors because the flux is absent when the stator windings are not supplied with power. The control system of this motor is similar to the permanent magnet control system due to the similarity in stator structure. The main drawback of this motor is the low power factor, which implies the need for an increased motor size and higher saliency ration. These terms are satisfied in machines with transversally and axially laminated rotor structures. The power factor is also improved by adding a small amount of permanent magnets to the machine [20].

6 Electronic architecture of EVs

Power electronics selection in electric and hybrid electric vehicles is substantial for the performance and operation efficiency of these vehicles. Power electronics include all semiconductor device-based circuits in the vehicle such as converters, inverters, and control and switching techniques [22]. The selection of power electronic devices for EV and HEV propulsion should meet the requirements of the voltage ratings, current ratings, switching frequencies, power losses, and dynamic characteristics. Voltage ratings depend on the nominal value of the battery voltage, the peak value of voltage during the charging period, and the peak value of voltage during the regenerative braking period. Current ratings depend on the motor maximum power ratings and how many devices are connected in parallel. Switching frequencies should be high in order to mitigate the acoustic noise, although when using very high switching frequencies, this will increase the switching

Table 2 Operation characteristics of GTO, MOSFET, IGBT, and BJT.

	Gating signal	Switching frequency (kHz)	Power capability
GTO	Current	Low	5000 V, 3000 A
MOSFET	Voltage	Very high	1000 V, 100 A
IGBT	Voltage	High	1200 V, 400 A
BJT	Current	Medium	1400 V, 800 A
MCT	Voltage	High	900 V, 150 A

losses [22,23]. Dynamic characteristics should enable high capability, easy driving, and easy paralleling. Among the available power electronics, Gate Turn-Off (GTO), Bipolar Junction Transistor (BJT), Metal-Oxide-Semiconductor Field-Effect Transistor (MOSFET), and Insulated-Gate Bipolar Transistor (IGBT) are particularly suitable for EV propulsion. The operating characteristics of these power electronics are provided in Table 2. The most popular type is IGBT because it has high input impedance, and high-speed characteristics. MCT is an attractive choice for EV propulsion because it possesses a high switching frequency, excellent dynamic characteristics, and high reliability [24].

Converters: Converters classified into DC-DC converters and DC-AC converters (inverters) are used for DC and AC motors, respectively. EV converters are employed as the inverter/generator, the main (HV) DC/DC converter, the auxiliary (HV/LV) DC-DC converter, and the auxiliary inverter [24]. Fig. 12 illustrates the types of converters in a typical EV.

Standard power converter topologies used for power conversion in the EV and HEV industry are two-level 3-ph power conversion topologies. An auxiliary HV/LV DC/DC converter is used to exchange energy between a high voltage (HV) DC power net and a 14 V DC (LV) power net. Different topologies, such as H-bridge and half-bridge topologies, with or without a resonant dc-link can be utilized for this converter [24,25].

Auxiliaries with high voltage must be considered. Air conditioning, oil pumping, cooling, HV power steering, charger, and HV/LV DC-DC converter are examples of auxiliaries. Most of them employ IGBT, except for the HV/LV DC-DC converter and the charger where MOSFET is used for high switching frequencies. The most-used pulse generation technique for converters is the PWM technique, which can be categorized intovoltage-based PWM and current-based PWM. Voltage-based PWM techniques are natural, regular, optimal, delta, and equal-area PWM. Current-based PWM techniques are hysteresis band PWM, instantaneous power technique, and space vector PWM (SVPWM) [24,26].

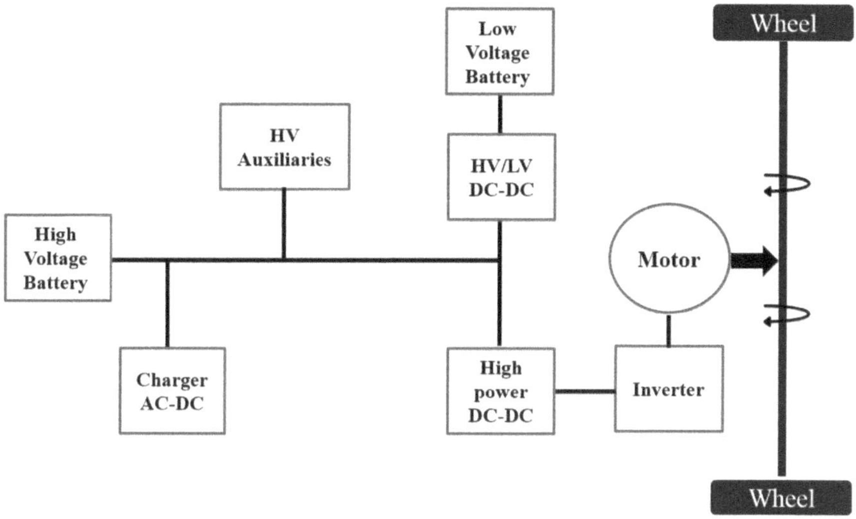

Fig. 12 Types of converters in EVs.

However, topologies with multilevel and multiphase could be future candidates in EV applications. Improvements in voltage and current waveforms as well as power capacities are among the benefits of such topologies. Multilevel topologies, on the other hand, offer a number of disadvantages for the electric car sector, including voltage fluctuations, complicated control systems, and higher prices. With this in mind, it's safe to assume that future HEV/EVs will continue to employ two-level power conversion topologies. When compared to 3-ph topologies, multiphase technologies depicted in Fig. 13 can provide various benefits such as better power capability, increased fault tolerance, higher power density, fewer torque ripples, and higher efficiency. Furthermore, an *n*-ph topology machine with the same features and performance may be simply generated from a 3-ph topology machine. Multiphase topologies are appealing future topologies in automotive applications for all the reasons stated above [25,26].

7 Challenges for the adoption of EV technology

Along with the many advantages of EV technology come many challenges that affect the adoption of this technology broadly. These challenges can be categorized into internal challenges that emerge from the technology itself and external challenges that affect the technology from the external factors [27].

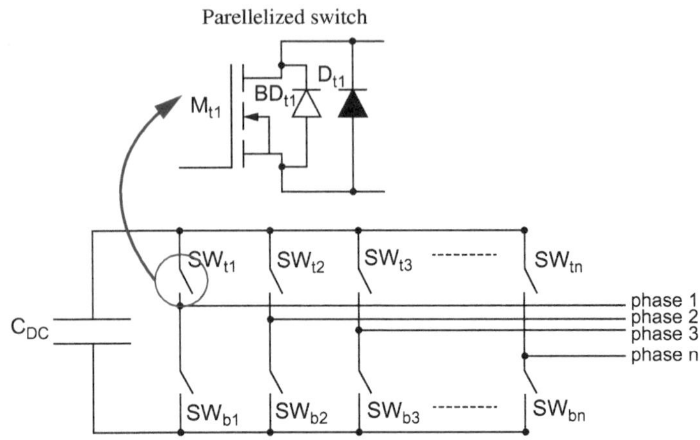

Fig. 13 Two-level multiphase topology with parallelized power switches for future HEV/EV electric drives.

7.1 Internal challenges

Three factors can be classified under this category: driving range, charging time, and high total cost of EV ownership.

The driving range of an electric vehicle in a single charge is a great challenge in the adoption of this technology. A survey conducted in large American cities found that 70% of the sample included in the survey thinks that driving range is a major disadvantage or at least a moderate disadvantage. In another survey, 33% of the respondents say that the battery range of EVs is their biggest concern [28]. Consumers seem to be willing to pay more for the additional distance that an EV could cover on a single charge. HEVs have a better driving range when compared with EVs, which makes them preferable.

From a consumer's point of view, the charge time of an electric vehicle is dead time. When the fast-charging feature is unavailable, people tend to pay more for expanded driving range. Hackbarth and Madlener did a survey in 2013 where the participants showed a tendency to pay between 5€ and 18€ to reduce the recharging time by 1 min, with a higher tendency to pay for EVs than plug-in hybrid electric vehicles. Reducing the charging time from 6 h to 10 min will cost 1750€–3500€ for plug-in hybrid electric vehicles and 3150€–6300€ for electric vehicles. Studies also showed that people are willing to pay nearly $8500 to reduce the charging time from 10 h to 10 min. It is obvious that the challenges in driving range and recharging time of EV technology are of great concern for consumers [27,28].

Regarding the costs of electric vehicles, studies have been conducted to understand the effect of vehicle prices and characteristics on the consumer's purchasing decision. The results of these studies showed that the purchase price of an electric vehicle in comparison with internal combustion engine vehicles is extremely important. In a study done within large cities in the United States in 2011, 55% of the study sample found that the purchase price of electric vehicles is the major disadvantage of this technology while 30% of the sample found that this is a moderate disadvantage. In another study that took place in the United Kingdom, 40 conventional car drivers were given an electric vehicle to drive for 1 week. The study showed that people were not comfortable with the upfront costs of an EV. Furthermore, electric vehicles have higher lifetime costs than conventional cars [29].

7.2 External challenges

The three main external factors that affect the adoption of electric vehicles are fuel price, availability of charging infrastructure, and consumer characteristics.

Studies showed that fuel prices in conventional and hybrid electric vehicles are higher than electric vehicles, which implies higher fuel cost savings in electric vehicles. Depending on the petroleum-based fuel prices, which are not constant between countries, the advantage of lower fuel costs of EVs will not be as important in countries where petroleum-based fuel prices are low [27]. Still, even in countries where petroleum-based fuel prices are high, the costs of electricity can still be a big challenge. In a 2012 study conducted by Prud'homme and Koning in France, they found that when comparing an electric vehicle with a conventional vehicle with similar features, the EV is considered to be economical only when petroleum fuel prices approach $470/barrel accompanied with decreased EV purchase prices [27].

The lack of an appropriate charging infrastructure can be a big challenge to the adoption of electric vehicles, especially when the EV driving range is limited. On the other hand, the number of EVs on the roads affects the investment in charging stations. A study done by Egbue and Long in 2012 that selected people with a positive attitude toward EV technology found that 17% of the sample considered the unavailability of charging infrastructure as their most important factor in EV adoption [30]. The adequacy of the charging infrastructure is determined by its location. For example, a study conducted in Ohio found that universities have a high service level (the number of EV chargers and amount of energy charged) when compared to workplace and shopping locations. So some users will prefer to charge at home (if they have quick charging capability) than to pay for a quick charging outside [27,29].

Consumer characteristics also can be a challenge to the adoption of EVs. The most important characteristics are education, number and type of cars owned, and level of environmentalism. Studies showed that the higher the level of education, the higher the likelihood of purchasing an EV. It was also noted that having multiple cars reduces the tendency toward purchasing an EV. In addition, it was found that some users will choose vehicle usefulness over environmental effects while most will choose to drive an EV because they experience less guilt. However, it doesn't mean that people don't value pollution reduction, but they value the performance more [29].

8 Future of EV technology

EV technology holds great promise for the future of transportation. The numerous benefits of this technology, ranging from reduced GHG emissions and improved driving experience to the integration of EVs into power systems, provide significant potential to improve mobility efficiency and economics. Several studies forecast the future of EV technology and its related aspects.

According to one study [31], by 2040, electric cars would account for 12%–28% of the worldwide fleet, up from 12% in 2016. Furthermore, even in a low-oil-price scenario, EVs will be preferred by 2040. These predictions were produced by prominent energy research organizations throughout the world: IRENA (2019), Bloomberg (2019), OECD/IEA (2018; 2019), OPEC (2018), and BP (2018). Furthermore, the petroleum product market share will fall from 93% in 2017 to 82%, or possibly 59% in the best case scenario. However, the widespread use of electric cars would result in a 7%–10% rise in worldwide power demand by 2040.

However, growth will be unevenly spread across the world's countries. Naturally, the most significant increase will come from nations that aggressively encourage EV technology, such as the United States, China, Japan, and the countries in the European Union [32]. Electricity demand from EVs alone is anticipated to account for more than a third of the country's total demand in some situations. However, the difficulty is adapting the power infrastructure to the rising demand peaks caused by EV charging patterns. To stay on track with "green" energy, world leaders in EV adoption must double down on the research and development of energy storage technologies; otherwise, EV adoption will result in increased fossil fuel use.

In terms of future fuel intensity development, it is expected that by 2050, all presently known efficiency improvements in hybrid electric automobiles would have been realized, resulting in a considerable reduction in energy input (kWh) per km [31]. According to studies, EVs may emit up to 97% fewer GHGs than conventional cars when using different main energy sources while HEVs can emit around 25% fewer GHGs than conventional vehicles [33].

Forecasting the future of EV technology is a difficult task with all the specifications that should be taken into account. Although it is obvious that the technology is not yet favorable when compared to conventional vehicles due to limited driving range, charging infrastructure, and battery performance, further technological improvements are needed while supportive measures for increasing the EV market share should be implemented to assure the spread of this technology.

9 Conclusion

Increasingly serious environmental issues such as global warming and environmental pollution have led to continuous searches for solutions to these issues. One solution that could solve the major problem of increased GHG emissions and depletion of energy resources is the electrification of the transportation system. EV technology is a promising solution that can mitigate emissions and reduce petroleum-based fuel consumption in internal combustion engine vehicles. Hybrid electric vehicles, which are an intermediate stage between fully petroleum-based conventional vehicles and fully electric vehicles, can be considered the first step toward full transformation into electrified roads. These vehicles are much better than conventional vehicles regarding the above-mentioned issues. Still, the ultimate goal is to rely fully on electric power in vehicles. Many countries set goals in terms of EV numbers on the road in upcoming years and this is indeed increasing. In the next two decades, studies have predicted that increased adoption of EV technology might reduce the consumption of petroleum products from 90% to 50% and reduce GHG emissions from vehicles by 95%. However, the full adoption of this technology still has many challenges regarding the driving range, batteries, and charging issues. These issues are still in the research phase and need more development to make this technology more suitable for consumer requirements and consequently more favorable when compared with conventional vehicles.

References

[1] G.A. Rhys-Tyler, W. Legassick, M.C. Bell, The significance of vehicle emissions standards for levels of exhaust pollution from light vehicles in an urban area, Atmos. Environ. 45 (2011) 3286–3293. Elsevier. 2011.

[2] F. Chen, N. Taylo, N. Kringos, Electrification of roads: opportunities and challenges, Appl. Energy 150 (2015) 109–119. Elsevier, 2015.

[3] M. Kebriaei, A.H. Niasar, B. Asaei, Hybrid electric vehicles: an overview, in: 2015 International Conference on Connected Vehicles and Expo (ICCVE), IEEE, 2015.

[4] M.A. Hannan, F.A. Azidin, A. Mohamed, Hybrid electric vehicles and their challenges: a review, Renew. Sustain. Energy Rev. 29 (2014) 135–150. Elsevier, 2013.

[5] C.C. Chan, Y.S. Wong, Electric Vehicles Charge Forward, IEEE Power & Energy Magazine, 2004.

[6] O.M. Govardhan, Fundamentals and classification of hybrid electric vehicles, Int. J. Eng. Tech. 3 (5) (2017) 194–198.

[7] M. Ehsani, Y. Gao, J.M. Miller, Hybrid electric vehicles: architecture and motor drives, Proc. IEEE 95 (4) (2007) 719–728.

[8] M. Nikowitz, Advanced Hybrid and Electric Vehicles, Lecture Notes in Mobility, Springer, 2016.

[9] K.V. Vidyanandan, Overview of electric and hybrid vehicles, Energy Scan 3 (2018) 7–14 (A House Journal of Corporate).

[10] Z. Li, A. Khajepour, J. Song, A comprehensive review of the key technologies for pure electric vehicles, Energy 182 (2019) 824–839. Elsevier.

[11] S. Sharma, A.K. Panwar, M.M. Tripathi, Storage technologies for electric vehicles, J. Traffic Transp. Eng. (Engl. Ed.) 7 (3) (2020) 340–361. Elsevier.

[12] J. Dixon, Energy storage for electric vehicles, in: 2010 IEEE International Conference on Industrial Technology, IEEE, Via del Mar, Chile, 2010.

[13] S.M. Mousavi, F. Faraji, A. Majazi, K. Al-Haddad, A comprehensive review of flywheel energy storage system technology, Renew. Sustain. Energy Rev. 67 (2017) 477–490. Elsevier, 2016.

[14] J. Libich, J. Máca, J. Vondrák, O. Čech, M. Sedlaříková, Supercapacitors: properties and applications, J. Energy Storage 17 (2018) 224–227. Elsevier.

[15] K. Kojima, H. Suzuki, K. Hirose, Recent challenges of hydrogen storage technologies for hydrogen- powered fuel cell vehicles, in: The 22nd International Battery, Hybrid and Fuel Cell Electric Vehicle Symposium & Exposition, EVS-22, Yokohama, Japan, Oct. 23–28, 2006.

[16] M. Yildirim, M. Polat, H. Kürüm, A survey on comparison of electric motor types and drives used for electric vehicles, in: 16th International Power Electronics and Motion Control Conference and Exposition, IEEE, 2014.

[17] A.M. Lulhe, T.N. Oate, A technology review paper for drives used in electrical vehicle (EV) & hybrid electrical vehicles (HEV), in: 2015 International Conference on Control, Instrumentation, Communication and Computational Technologies (lCCICCT), IEEE, 2015.

[18] D.U. Thakar, R.A. Patel, Comparison of advance and conventional motors for electric vehicle application, in: 2019 3rd International Conference on Recent Developments in Control, Automation & Power Engineering (RDCAPE), IEEE, Noida, India, 2020.

[19] S. Li, A Review of Electric Motor Drives for Applications in Electric and Hybrid Vehicles, Researchgate, March 2017.

[20] D.D.G. Dorrell, A.M. Knight, M. Popescu, L. Evans, D.A. Staton, Comparison of different motor design drives for hybrid electric vehicles, in: 2010 IEEE Energy Conversion Congress and Exposition, IEEE, Atlanta, GA, 2010.

[21] C. Chi-lan, W. Xiao-gang, B. Yue-wei, X. Yan-chun, L. Kai, Key technologies of EV motor drive system design, in: International Workshop on Automobile, Power and Energy Engineering, Elsevier, 2011.
[22] A. Emadi, Y.J. Lee, K. Rajashekara, Power electronics and motor drives in electric, hybrid electric, and plug-in hybrid electric vehicles, IEEE Trans. Ind. Electron. 55 (6) (2008) 2237–2245.
[23] C.C. Chan, K.T. Chau, An overview of power electronics in electric vehicles, IEEE Trans. Ind. Electron. 44 (1) (1997) 3–13.
[24] D. Graovac, A. Christmann, M. Münzer, Power semiconductors for hybrid and electric vehicles, in: 8th International Conference on Power Electronics—ECCE Asia, IEEE, 2011.
[25] A. Matallanaa, E. Ibarraa, I. Lopeza, J. Andreua, J.I. Garatea, X. Jordab, J. Rebollob, Power module electronics in HEV/EV applications: new trends in wide-bandgap semiconductor technologies and design aspects, Renew. Sustain. Energy Rev. 113 (2019) 109264. Elsevier, 2019.
[26] S. Chakraborty, H. Vu, M.M. Hasan, D. Tran, M. El Baghdadi, O. Hegazy, DC-DC converter topologies for electric vehicles, plug-in hybrid electric vehicles and fast charging stations: state of the art and future trends, Energies 12 (8) (2019) 43.
[27] M. Coffman, P. Bernstein, S. Wee, Electric vehicles revisited: a review of factors that affect adoption, Transp. Rev. (Taylor & Francis Online) 37 (1) (2017) 79–93.
[28] M. Wikström, L. Hansson, P. Alvfors, Investigating barriers for plug-in electric vehicle deployment in fleets, Transp. Res. D 49 (2016) 59–67. Elsevier, 2016.
[29] S. Hosseinpour, H. Chen, H. Tang, Barriers to the wide adoption of electric vehicles: a literature review based discussion, in: 2015 Portland International Conference on Management of Engineering and Technology (PICMET), IEEE, 2015.
[30] O. Egbue, S. Long, Barriers to widespread adoption of electric vehicles: an analysis of consumer attitudes and perceptions, Energy Policy 48 (2012) 717–729. Sep. 2012.
[31] A.A. Ajanovic, The future of electric vehicles: prospects and impediments, Wiley Interdiscip. Rev. Energy Environ. 4 (6) (2015) 521–536.
[32] D. Potoglou, P.S. Kanaroglou, Household demand and willingness to pay for clean vehicles, Transp. Res. Part D Transp. Environ. 12 (4) (2007) 264–274.
[33] A. Ajanovic, Renewable fuels—a comparative assessment from economic, energetic and ecological point-of-view up to 2050 in EU-countries, Renew. Energy 60 (2013) 733–738.

CHAPTER 7.2

Modeling of permanent magnet synchronous motor for zero-emission vehicles

R. Bharathwaaj[a], V. Mohanavel[b], Alagar Karthick[c], S. Vasanthaseelan[a], Manickam Ravichandran[d], T. Sakthi[e], and S. Rajkumar[f]

[a]Department of Mechanical Engineering, KPR Institute of Engineering and Technology, Coimbatore, Tamil Nadu, India
[b]Centre for Materials Engineering and Regenerative Medicine, Bharath Institute of Higher Education and Research, Chennai, Tamil Nadu, India
[c]Renewable Energy Lab, Department of Electrical and Electronics Engineering, KPR Institute of Engineering and Technology, Coimbatore, Tamil Nadu, India
[d]Department of Mechanical Engineering, K. Ramakrishnan College of Engineering, Trichy, Tamil Nadu, India
[e]Department of Mechanical Engineering, National Engineering College, Kovilpatti, Tamil Nadu, India
[f]Department of Mechanical Engineering, Faculty of Manufacturing, Institute of Technology, Hawassa University, Hawassa, Ethiopia

Contents

Active Electrical Distribution Network
https://doi.org/10.1016/B978-0-323-85169-5.00004-6

1 Introduction

The electric vehicle (EV) market is booming, thanks to more options, cheaper pricing, government subsidies, and ever-improving battery technology. Simply explained, instead of a gasoline or diesel engine, an EV is driven by an electric motor. This electric motor's rechargeable batteries may be charged using normal household electricity [1–5].

More than 1 million pure electric cars and vans were in operation globally as of September 2016. There are two types of EV technology: hybrid and pure electric. A pure electric vehicle's basic components are a rechargeable battery, a controller, and an electric motor. First, the battery must be charged. The controller then converts the DC to AC power so that the motor may use it. The motor converts electrical energy into mechanical energy [6–11]. Hybrid cars employ the same concept, but with the addition of a small gasoline engine that drives a generator. This drives the car at cruising speed, with batteries supplying additional power when the vehicle accelerates. The batteries can recharge themselves when the car is parked or decelerating [12–18].

Hybrid technology allows obtaining much greater mileage from your car, saving money and reducing the environmental impact. The most crucial reason for such technology is that environmental concerns are growing more severe at this time. Automotive exhaust carbon emissions have surpassed all other sources as the leading cause of air pollution, particularly in highly populated regions [19]. Air pollution is at dangerously high levels in many regions globally, posing a serious environmental risk to human health. Every year, poor air quality causes more than 6.5 million premature deaths worldwide, and around 3 million people die due to exposure to outdoor air pollution. One of the three major causes of outside air pollution is emissions produced by automobiles [20]. Transportation is among the most significant sources of greenhouse gas (GHG) emissions, accounting for around 90% of total emissions. Global CO_2 emissions in 2015 were 32.3 Gt, according to the International Energy Agency (IEA), with transportation responsible for 24% of total emissions and the road sector accounting for 75%. [21].

Global warming has the potential to raise human death rates due to heat stress, illness, and natural catastrophes as well as move the area for sustainable agriculture, disrupt habitats for animals and ecosystems, and alter the time and amount of water available [22]. The hazards of global warming have been communicated to the human race. According to some researchers, producing electricity for charging battery electric vehicle (BEV) or hydrogen fuel cell electric vehicles (FCEVs) could result in significant GHG emissions that are nowhere near those from internal combustion engine vehicles (ICEVs) based vehicles. However, this is an assessment predicated on the premise that electricity is generated utilizing fossil fuels such as coal and oil.

Indeed, a well-to-wheel (WTW) study revealed that gas emissions from electricity generation or hydrogen synthesis primarily rely on fundamental energy sources [23]. As power is created using nuclear power, hydropower, or renewable energy sources such as biomass, solar, and wind power, BEVs emit substantially less WTW gas than ICEVs. Solar electrolysis hydrogen generation can reduce gas emissions in FCEVs by 99.2% when compared to gasoline ICEVs.

It is worth noting that some of the world's most powerful nations, such as the United States, China, and the countries in the European Union, all appear to have strategies to expand the quantity of renewable energy in their energy production, enhancing plug in electric vehicle (PEV) environmental benefits considerably [24,25]. As a result, the argument's conclusion will be obvious: as technology and policy advance, PEVs will be able to exceed ICEVs in reducing GHG emissions. The growing urgency of the energy crisis caused by fossil fuel depletion is another significant motivator of PEV development.

Transportation, by far the most significant fuel-consuming business, uses 56% of global fuel consumption [20]. As a result, humans must adopt practical steps ahead of time to avert social and economic catastrophes when energy exhaustion occurs. Furthermore, the PEV energy efficiency is much greater than regular ICEVs. The BEV total energy efficiency ranges from 60% to 70%, including component efficiencies such as batteries, electric motors, recharge efficiency, and regenerative braking energy. Traditional ICEVs have a fuel efficiency of just 15%–18%, but FCEVs have an average conversion efficiency of 35%–55% [26].

Contrasted with ICEVs and traditional hybrid electric vehicles (HEVs) powered by a battery, PEVs have the potential to be used as disseminated energy storage structures that can connect to a smart grid. The power transfer over this connection may be bidirectional. Excess energy in PEV energy storage devices may be delivered back to the grid during moments of high peak demand or to compensate for renewable power supply unpredictability. Charging batteries or electrolyzing water to create hydrogen might be used to store excess grid energy. This vehicle-to-grid (V2G) option can enhance grid power quality and stability while reducing the effects of renewable generation intermittency [27–31].

2 The current state of PEVs

2.1 PEVs with a single source

Vehicles are propelled by a single energy source in single-source PEVs. The finest example of single-source PEVs is the first BEV, created in 1828. Fig. 1A illustrates the configuration of single-source PEVs. Electric cars have a limited operating duration while a significant charging time for batteries hampers their widespread use. BEVs have regained manufacturer and

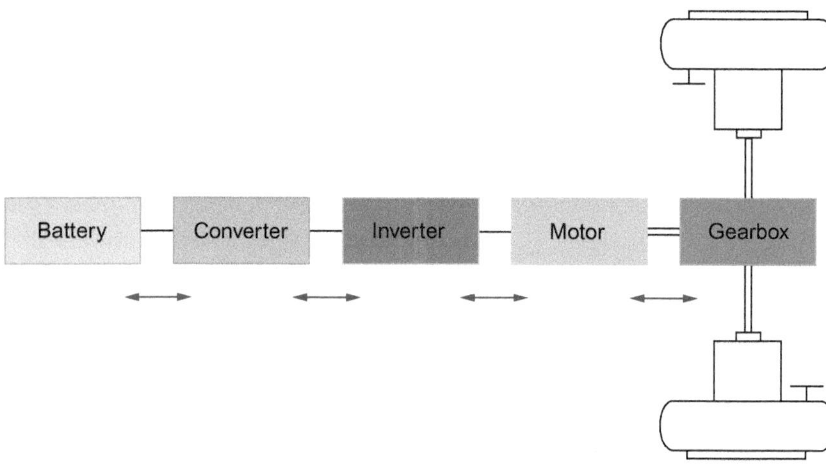

(a) Powertrain of BEVs

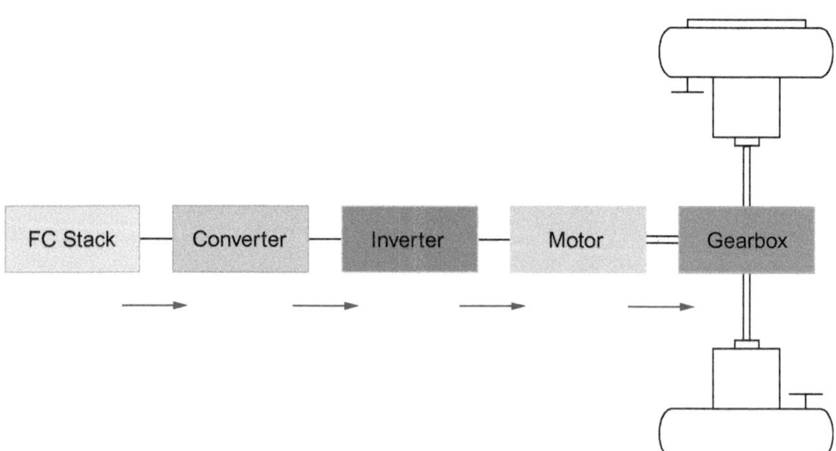

(b) Powertrain of FCEVs

Fig. 1 Single-source PEVs.

customer attention in recent years due to substantial environmental concerns and energy problems created by ICEVs. BEVs emit no emissions, are very efficient, and are quieter. During vehicle deceleration, the electric motor may also act as a generator, allowing BEVs to collect regenerative braking energy. Repeated charging and discharging, on the other hand, may cause the batteries to overheat, reducing their lifespan.

FCEVs are other PEVs with single-source types, as seen in Fig. 1. An fuel cell (FC) stack (with a hydrogen tank), a power converter, an inverter, and an electric motor are all part of powertrain of FCEV as presented in Fig. 1B. The proton-exchange membrane fuel cells (PEMFCs) are the most promising automotive applications due to their low operating temperature, high power density, and ability to run on ordinary air. FCEVs use oxygen and hydrogen to create electricity, with clean water as a chemical byproduct. FCEVs are known for their high performance, recovery, and long-term energy production as well as their quiet operation. Liquid hydrogen is also transportable, similar to the fuel tank of an ICEV. Due to the lack of bidirectional power flow in the engine, FCEVs cannot recover braking energy compared to BEVs.

2.2 PEVs with two sources

The vehicle propulsion system of a double-source PEV incorporates two energy sources, reducing the drawbacks of using only one. Only a few instances of dual-source pairings were discovered in the literature, including battery and super capacitors (SC), battery and flywheel, battery and hydroxypropyl trimethyl ammonium chloride chitosan reaction (HACC), battery and FC, and FC and SC. In battery-SC PEVs, as shown in Fig. 2A, the battery serves as the major energy source for supplying traction energy while the SC serves as an auxiliary energy source for satisfying the vehicle's system features and recovering regeneration energy [1,2]. Similarly, the battery serves as the principal energy source for fulfilling power needs in the battery and flywheel hybrid energy storage system illustrated in Fig. 2B. In contrast, the flywheel acts as an auxiliary energy storage system for storing regenerative energy in transitory situations. During deceleration, the flywheel accumulates brake energy and releases it during vehicle acceleration. As a result, battery capacity might be increased, and vehicle performance and efficiency may improve [3].

The HACC, unlike the SC and the flywheel, can transform kinetic energy into fluid power fast during vehicle deceleration and then release

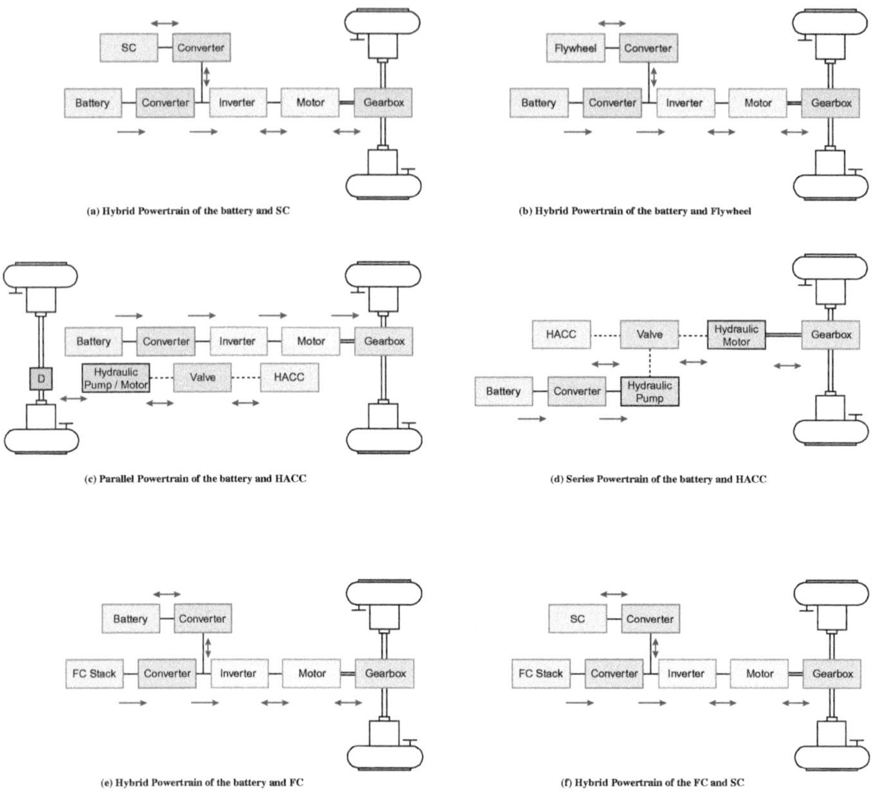

Fig. 2 Dual-source PEVs.

it when the vehicle need it. The battery and the HACC powertrain can be built in parallel or series, as shown in Fig. 2A, B, and C. The battery may charge the HACC in series, allowing the HACC to achieve its maximum capacity and assuring the vehicle's dynamic performance. In a series design, however, improved energy conversion may result in increased energy losses and worse overall system efficiency [4]. The battery is utilized to replace the deficit or absorb the surplus of FC power in the drivetrain, as shown in Fig. 2D.

Furthermore, the battery can store braking energy during deceleration [5,6]. Fig. 2E and F depicts a PEV configuration that incorporates both the FC and SC. The FC provides the primary energy and traction force source while the SC assists with acceleration and gathers regenerative energy during deceleration [7,8].

2.3 PEVs with many sources

The vehicle engine in PEVs with multiple energy sources has at least three energy sources. By combining several sources of energy, they may attain their maximum potential and increase dynamic performance. The multisource set-up increases energy storage capacity while allowing each advantage to be fully optimized to enhance system efficiency, robustness, and equipment lifespan. Furthermore, the best design across all energy sources saves money and prevents the use of oversized or overweight components. As a consequence, combining several sources of energy is a cost-effective way to put PEVs into practice. The different multisource powertrain concepts that have been documented in the literature [9–14] are depicted in Fig. 3. The FC stack is the primary energy source for delivering traction energy in these configurations, whereas the SC acts as a backup source of energy for conserving regenerative energy during deceleration and stopping. As illustrated in Fig. 3C, the battery may be used as both a primary and auxiliary energy source for storing surplus regenerative energy. It's worth noting that the PV cell is utilized as a backup energy source to create electricity while also conserving energy.

2.4 Impacts of PEVs on the environment

PEV environmental impacts may be investigated using life cycle analysis (LCA), which can explicitly quantify resource use and environmental discharges over the course of a product's full life cycle. A WTW analysis, for example, is a common LCA tool that summarizes energy consumption and pollutants from the point of primary energy source extraction (well) to the point of use (wheels). Broad assessment criteria must be provided to make a more accurate comparison between various powertrains. Despite the fact that kilometers are not the best functional unit for LCA, they are the most widely used in the literature.

In addition, CO_2 emissions are the most commonly measured result to analyze environmental consequences, and other emissions such as water vapor are not taken into account in this study. Emissions are thus measured in g CO_2 equivalent per kilometer [15,16]. In commercial applications, single-source PEVs are the most common vehicle type; however, alternative PEVs are also produced.

Natural gas is commonly utilized to generate hydrogen due to its inexpensive cost. BEVs use power converted from natural gas. Renewable energy sources such as solar and wind energy may also be used to create

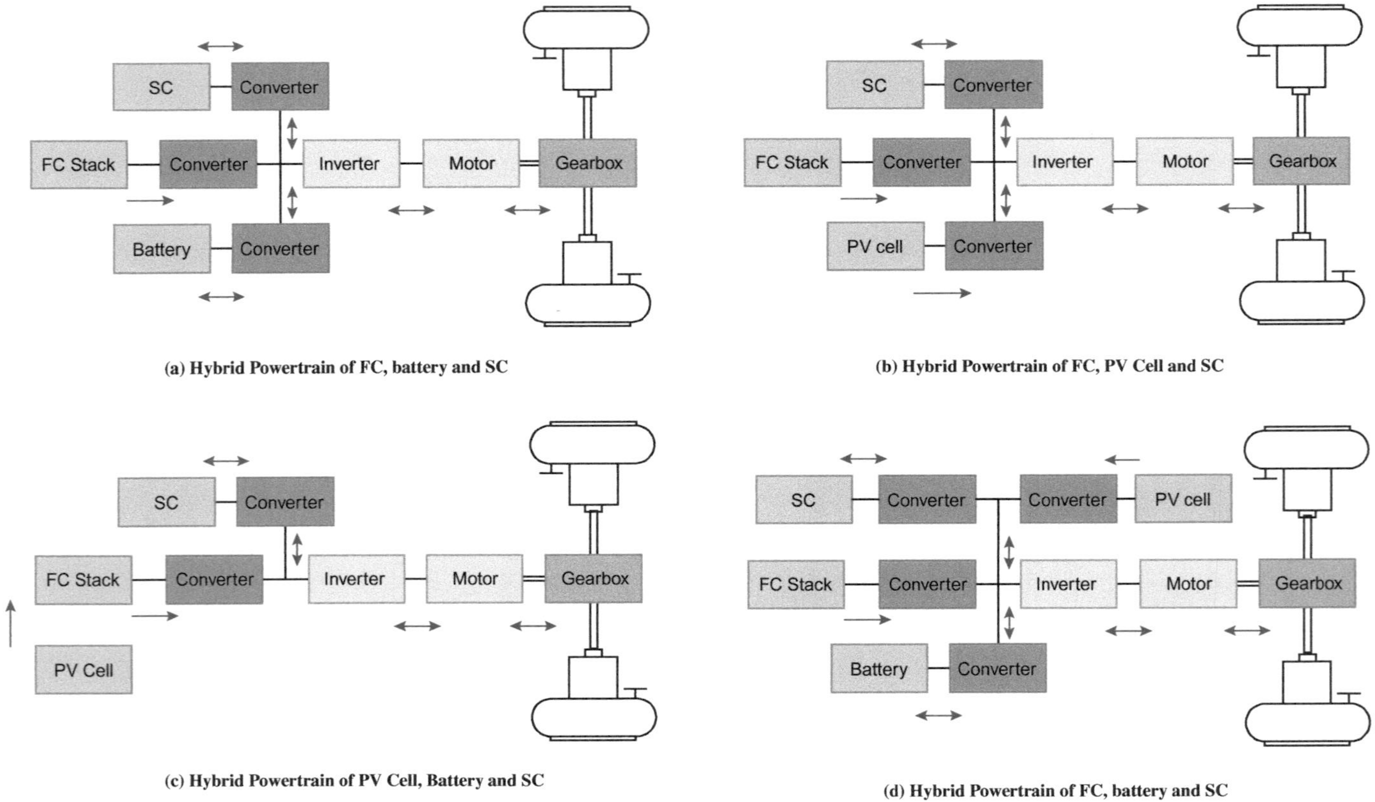

Fig. 3 Multisource PEVs.

Table 1 Energy consumption.

Vehicle types with energy sources	Fuel consumption (l_gas_eq/100 km)	Emissions (g_CO_2_eq/km)
Battery electric vehicle with Li-ion battery based on natural gas	3.02	58.83
Fuel cell electric vehicles with hydrogen based on natural gas	4.00	74.21
Li-ion battery-BEVs based on solar/wind energy	1.54	0.00
Fuel cells with hydrogen based on solar/wind energy	4.44	2.99
Petrol IC engine vehicles	6.00	144.00
Diesel IC engine vehicles	4.70	133.00

hydrogen and electricity. Table 1 compares traditional ICEVs to lithium-ion battery BEVs and hydrogen FCEVs, which run on natural gas and solar/wind energy, respectively. According to the data, FCEVs driven by pure hydrogen have greater WTW fuel consumption and emissions than BEVs, regardless of whether they are powered by natural gas or solar/wind energy. BEVs use converted electricity directly, but hydrogen generation requires a lot of energy conversion and has a large carbon impact. BEVs and FCEVs, on the other hand, have significantly lower WTW fuel consumption and emissions than ICEVs.

Using renewable energy sources allows BEVs and FCEVs to emit extremely low WTW while simultaneously offering the highest environmental benefits. BEVs that run on electricity generated by solar or wind power, in particular, emit no pollutants. It's also worth mentioning that FCEVs fuelled by solar/wind energy have roughly three times the WTW fuel consumption of BEVs. This is because of the FC's low efficiency and the electrolysis process, which produces hydrogen using solar/wind energy. On the other hand, FCEVs have technological advantages over BEVs, including a more extended driving range per charge, a faster refueling/recharging time, and the ability to recover waste heat for heating fuel.

2.5 Model description

Several inbuilt models in the SIMULINK program are used to create a real-time model of an electric car. The four significant models are motor drive, high-voltage battery, longitudinal vehicle model, and driver and environment model. Finally, a measurement and scope model is created to visualize the simulation findings. The overall model block diagram is shown in Fig. 4.

2.5.1 Driver and environment model

A driving cycle is a collection of data points that represents the speed of a vehicle over time. Driving cycles are developed by many nations and organizations to assess automobile performance in various characteristics such as fuel economy, electric vehicle autonomy, and hazardous pollutants. The model includes a driver and environment module to mimic driving conditions, as shown in Fig. 5.

A typical drive cycle of FTP75, which is utilized for light-duty cars, is chosen as the standard drive model from the Simscape model. The model in question has a length of 2474 s with a predetermined speed variation. A controller is used to keep the motor speed consistent, as seen in Fig. 6.

2.5.2 High-voltage battery model

A high-voltage battery provides the power source for the EV system. The battery for the EV system is taken from the standard models available in the Simulink library. A realistic DC-link current profile, derived from a dynamic driving cycle, is used in the model. DC voltage source and internal resistance are included in the battery model. The settings that were considered in the simulation are listed in Table 2. The voltage and current sensors are positioned in the proper locations to monitor the battery's vital parameters (Fig. 7).

2.5.3 PMSM motor model

An electric motor is the heart of any electric vehicle. A variety of electric motors are utilized as prime movers in the contemporary environment, with BLDC motors being the most popular due to their high torque generating capacity and compact size. As befitting their numerous benefits over ordinary DC motors, AC motors in electric vehicles have received the most attention in recent years. Permanent magnet synchronous motors (PMSM) are currently employed in such pursuits. Fig. 8 shows the Simulink model for the prime mover, which includes a PMSM controller, a three-phase inverter, and a motor. A PMSM field-oriented controller is created and built to synchronize the rotors and maintain the drive's speed, as illustrated in Fig. 9.

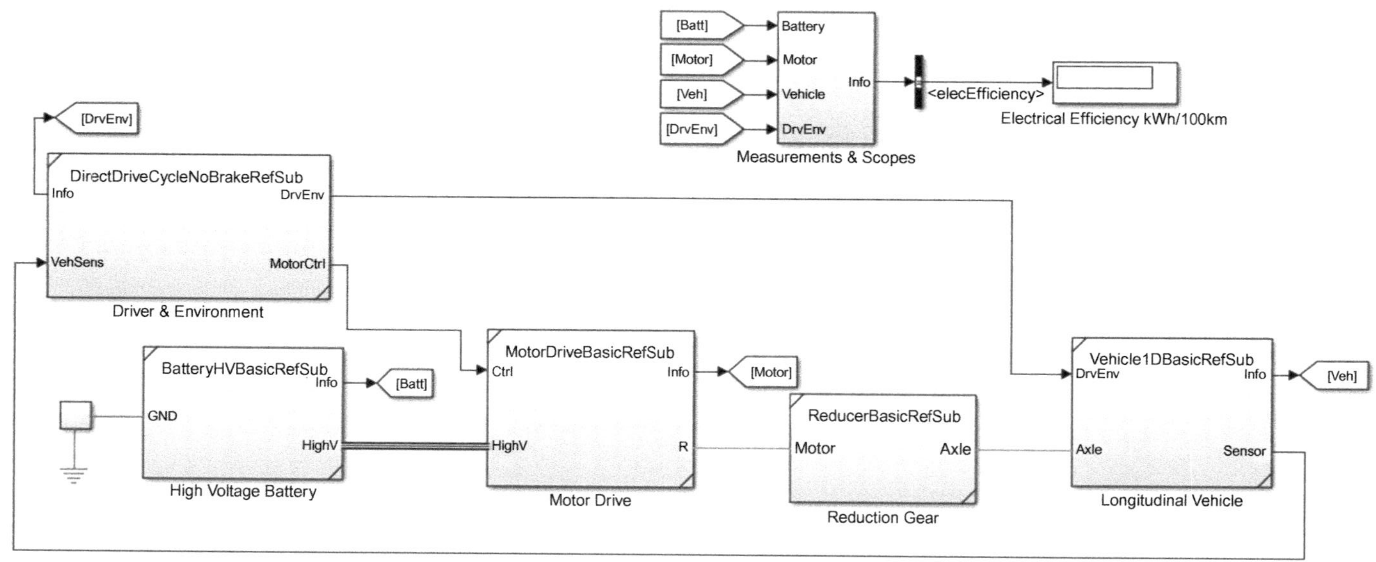

Fig. 4 MATLAB-SIMULINK model of an electric vehicle.

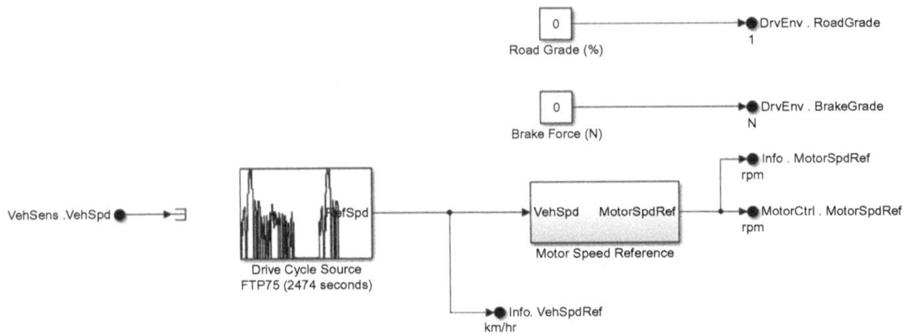

Fig. 5 Driver and environment model.

The controller's inputs come from a real-time model. The controller operates by sending a gate signal to a three-phase inverter based on inputs from the speed sensor, which gives the motor's current operational speed, and the bus voltage sensor, which provides information on the inverter's input voltage. The actual speed value is compared to the reference speed that must be maintained in the final drive, and the controller calculates the necessary adjustment to keep the motor running at the appropriate speed. The controller is backed up by a line voltage sensor and a line current sensor to ensure appropriate motor functioning.

3 Results and discussions

In this part, the findings of the simulation studies are explained with appropriate examples. Fig. 10 depicts the speed plotted as a function of time for various torques. A set of 25 values has been computed based on the variation of 5 speed and 5 torque ranges. Fig. 11 depicts the electrical losses in the system for each of the 25 trials performed.

From Fig. 10, it can be seen that the uniform speed is obtained in the high-speed and high-torque operation of the motor for the given duty cycle of vehicle operation. Furthermore, at the lower speeds of motor operation, it is found to be pretty challenging to maintain constant speed for all the ranges of torque. It can be seen from the plots presented in the first column of Fig. 10. As the target speed increases, the speed curve tends to be smoother, showing the flawless operation of the motor in high-speed range for all the operational loads of the motor. This is also on par with the fact that AC motors are efficient and smooth at high operation speeds. This clearly shows that the electric vehicle operating with an AC motor as a prime mover can

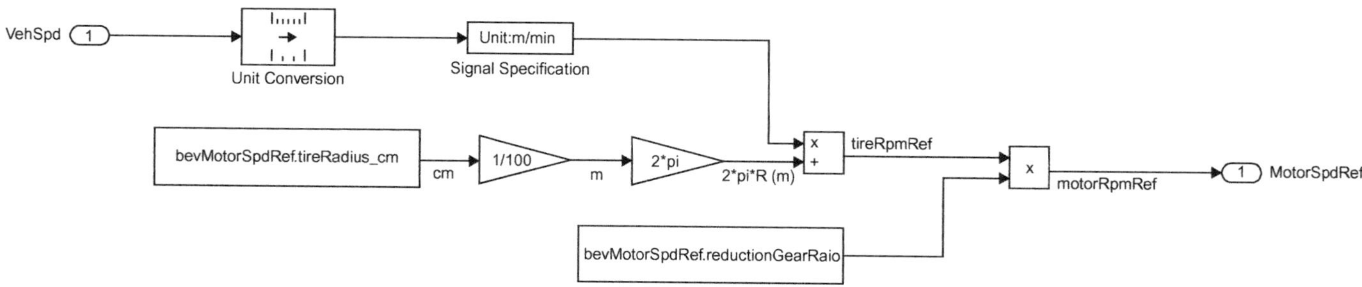

Fig. 6 Motor speed reference model.

Table 2 Simulation parameters.

Sl no	Parameter	Value	Unit
Vehicle parameters			
1	Vehicle mass	1600	kg
2	Tire rolling radius	30	cm
Battery parameters			
3	Battery voltage	350	V
4	Battery internal resistance	0.002	Ohms
Bevel gear parameters			
5	Bevel gear ratio	7.0	–
6	Bevel gear efficiency	0.98	–

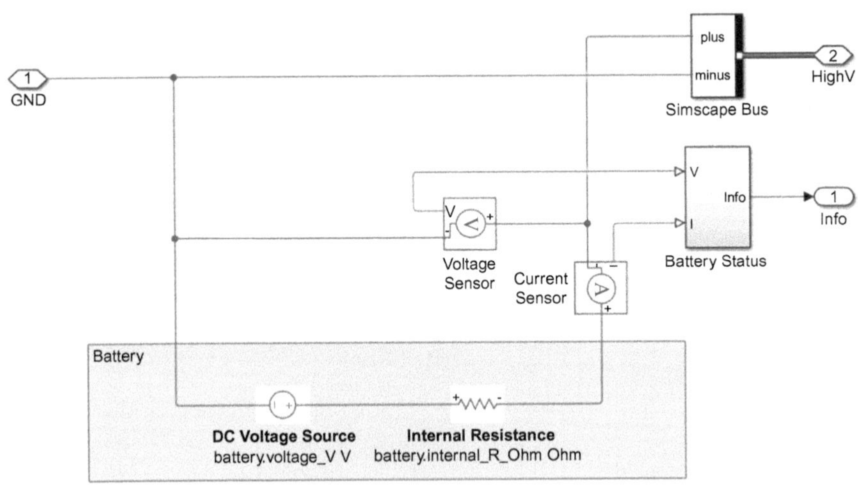

Fig. 7 High-voltage battery model.

easily attain higher speeds with lower efforts. On the other hand, the initial torque that has to be given to propel the vehicle at lower speeds will be difficult because the controller has to handle high fluctuations in the voltage given to the motor. Due to this, the controller may suffer a high rate of failure. To prevent this, the controller must be optimized to handle both lower and higher speeds of vehicle operation. Because the vehicle is a system in which more variations in operation can be found, it is better to design a controller that can best suit the purpose.

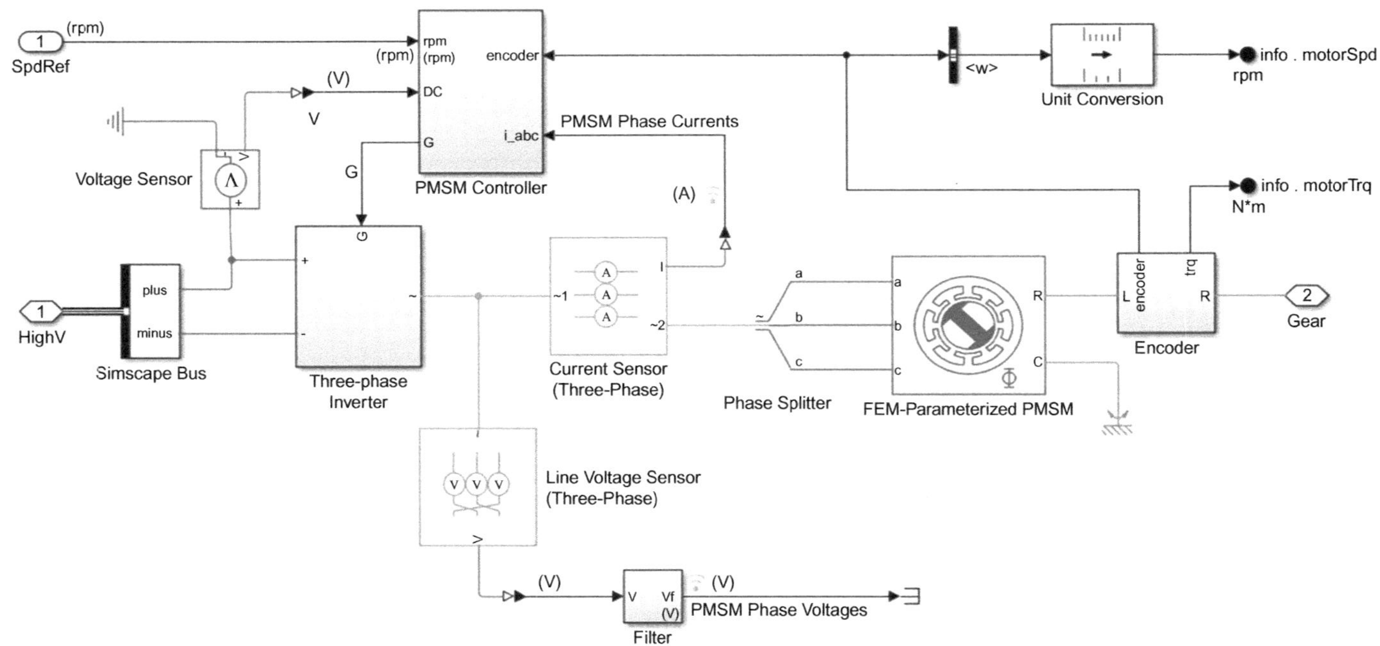

Fig. 8 Motor drive model using PMSM.

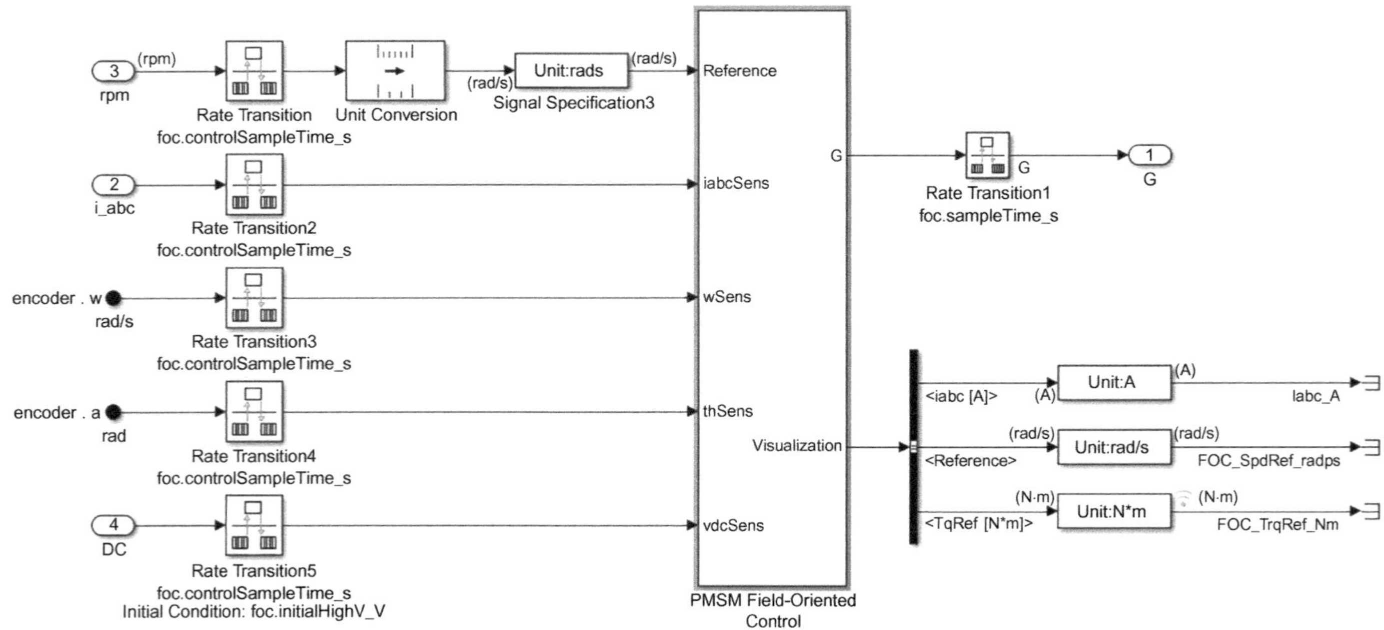

Fig. 9 PMSM controller.

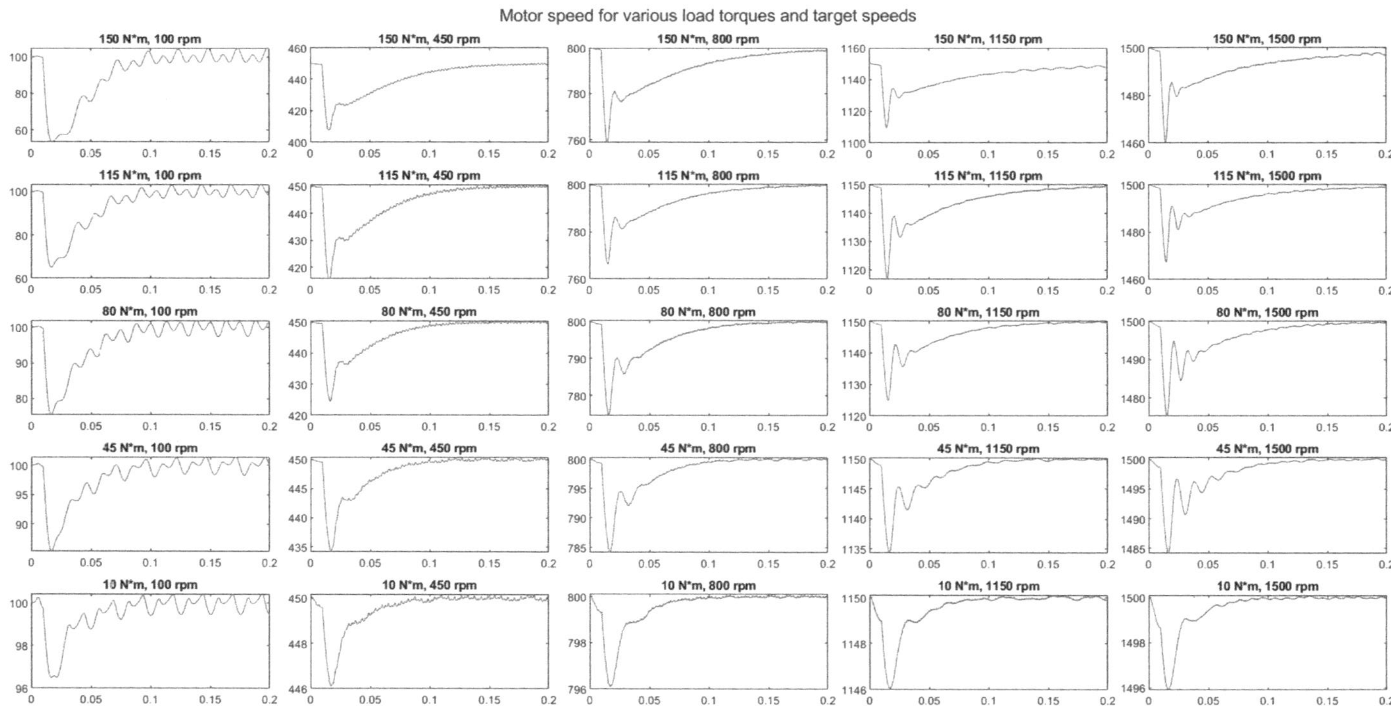

Fig. 10 Motor speeds for various loads and torque.

To have a better controller, an electronic inverter-based field operation controller can be used. In all vehicle loading scenarios, the inverter-based controllers enter a steady state with constant rotation speed after a transitory time of less than 0.4 s. This demonstrates that the inverter-based controllers can be utilized in an electric vehicle [17]. In the *d–q* axis, mathematical models for both steady-state and transient analysis, the most common approach for investigating the electric motor control of the PMSM type, are used. From the stator variables, the Park transform is applied to create the two-axis variables [18].

The electrical losses for varied speed and torque characteristics are shown in Fig. 11. In PM machines, there are three types of losses: (a) stator loss, (b) rotor eddy current loss, and (c) windage loss. Stator loss is made up of two elements: copper and iron. Traditional I2R loss as well as stray load loss due to skin effect and proximity influence are included in the copper loss. Because eddy current losses are more significant at high speeds and low loads, the electrical losses are more significant for the motor's high speed and low torque operation, as shown in the diagram. Because eddy current losses in the rotor are minimized, the lowest electrical losses are observed during low speed and low torque motor operation.

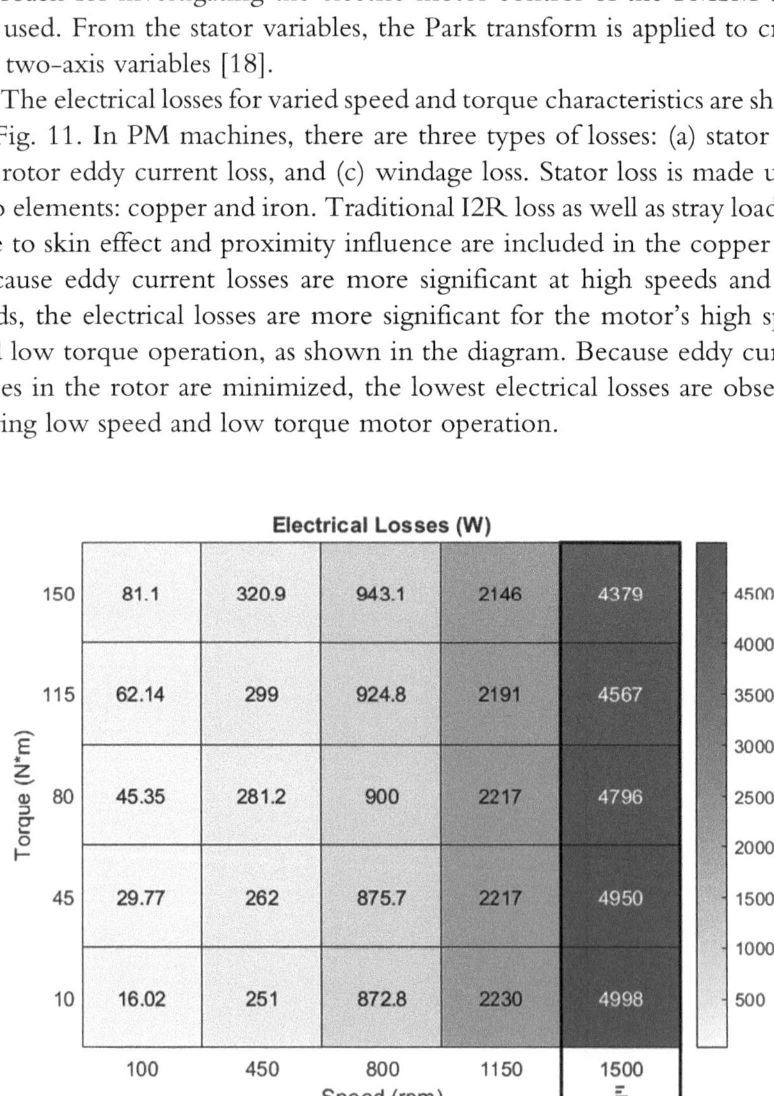

Fig. 11 Electrical losses.

4 Conclusion

A mathematical model is developed in SIMULINK to simulate the actual driving conditions by varying the target speed and torque. The simulation results confirm the promising electric vehicle operation with a PMSM motor with a field-operated controller. The future of transportation solely relies on the electric vehicle in all possible aspects. Affordability is one big hurdle that hinders the development of a commercial electric vehicle. In such cases, an AC-based electric drive can be a better power source.

A Appendix

A.1 Set up

```
% Open model and set up simulation
mdl = "MotorDrive_testHarness";
if not(bdIsLoaded(mdl)), open_system(mdl); end
set_param(mdl, ...
 'SimscapeLogOpenViewer', 'off',...
 'EnablePacing', 'off',...
 'SolverType', 'Variable-step',...
 'Solver', 'VariableStepAuto',...
 'SaveTime', 'on',...
 'TimeSaveName', 'tout',...
 'SignalLogging', 'on',...
 'SignalLoggingName', 'logsout',...
 'SimscapeLogType', 'local',...
 'SimscapeLogSimulationStatistic
s', 'off',...
 'SimscapeLogToSDI', 'off',...
 'SimscapeLogName', 'logsoutssc',...
 'SimscapeLogDecimation', 1,...
 'SimscapeLogLimitData', 'off',...
 'ReturnWorkspaceOutputs', 'on',...
 'ReturnWorkspaceOutputsName', 'out');
set_param(mdl+"/Motor Drive/FEM-Parameterized PMSM", "LogSimula-
tionData", "on")
 % necessary for electrical loss calculation after simulation
tMax = 0.2; % Simulation time > 60/rpm0/N
set_param(mdl, 'StopTime',num2str(tMax))
```

```
set_param(mdl, 'FastRestart', 'on')
save_system(mdl)
```

A.2 Run simulations

A.2.1 Define torque and speed conditions

```
% The ranges specified below are rather narrow for electric vehicle motors.
% This is due to the lack of sufficient motor data for FEM-Parameterized PMSM block
% as well as to the limitation of the FOC controller used in the model.
SpeedVec = [100, 450, 800, 1150, 1500]; % rpm
TorqueVec = [10, 45, 80, 115, 150]; % N*m
```

Build simulation input data set and run simulations

```
nS = length(SpeedVec);
nT = length(TorqueVec);
clear in
in(1:nS, 1:nT) = Simulink.SimulationInput(mdl);
for i_spd = 1:nS
 for i_trq = 1:nT
  rpm0 = SpeedVec(i_spd);
  torque0 = TorqueVec(i_trq);
  % First two setBlockParameter calls are used by Simulation Manager's
  % default Figure window to show the progress of simulations.
  % First one for horizontal, left to right.
  % Second one for vertical, bottom to top.
  in(i_spd,i_trq) = in(i_spd,i_trq).setBlockParameter(mdl+"/Constant Target Speed", 'Value',num2str(rpm0));
  in(i_spd,i_trq) = in(i_spd,i_trq).setBlockParameter(mdl+"/Step Load Torque", 'After',num2str(torque0));
  in(i_spd,i_trq) = in(i_spd,i_trq).setBlockParameter(mdl+"/Load inertia", 'w',num2str(rpm0));
  in(i_spd,i_trq) = in(i_spd,i_trq).setBlockParameter(mdl+"/Motor Drive/FEM-Parameterized PMSM", ...
    'angular_velocity',num2str(rpm0));
  in(i_spd,i_trq) = in(i_spd,i_trq).setBlockParameter(mdl+"/Motor Drive/FEM-Parameterized PMSM", ...
    'angular_velocity_priority','High');
 end
```

```
end
%out = sim(in, "ShowSimulationManager","on");
out = parsim(in, "ShowSimulationManager","on"); % need Parallel
Computing Toolbox
```

A.3 Visually inspect motor speed behavior

```
clear result
result(1:nS,1:nT) = struct('spd',0, 'trq',0, 'log',timeseries);
for i_spd = 1:nS % speed
 for i_trq = 1:nT % torque
  spd = SpeedVec(i_spd);
  trq = TorqueVec(i_trq);
  result(i_spd,i_trq).spd = spd;
  result(i_spd,i_trq).trq = trq;
  result(i_spd,i_trq).log = out(i_spd,i_trq).logsout.get
("<motorSpd>").Values;
 end
end
f = figure;
f.Position(3:4) = [800 600];
tl = tiledlayout(nT, nS);
tl.TileSpacing = "compact";
tl.Padding = "compact";
title(tl, "Motor speed for various load torques and target speeds")
for i_trq = 1:nT % torque
 for i_spd = 1:nS % speed
  nexttile
  irev_trq = nT - i_trq + 1;
  plot(result(i_spd,irev_trq).log.Time, result(i_spd,irev_trq).
log.Data)
  title(num2str(result(i_spd,irev_trq).trq) + " N*m, " + num2str
(result(i_spd,irev_trq).spd) + " rpm")
 end
end
```

A.4 Obtain electrical losses

```
Losses = zeros(nS,nT);
for i_spd = 1:nS
 for i_trq = 1:nT
```

```
  rpm0_abs = abs(SpeedVec(i_spd));
  % Time range for last full AC cycle
  tMin = tMax - 60/rpm0_abs/PmsmFem.NumPolePairs;
  assert(tMin < tMax)
  % Extract losses
  tabulatedLosses = ee_getPowerLossSummary(out(i_spd,i_trq).log-
soutssc, tMin, tMax);
  Losses(i_spd,i_trq) = sum(tabulatedLosses.Power);
 end
end
```

A.4.1 Visualize

```
SpeedVecLabels = cell(1,nS);
for idx = 1:nS
 SpeedVecLabels{idx} = num2str(SpeedVec(idx));
end
TorqueVecLabels = cell(1,nT);
for idx = 1:nT
 TorqueVecLabels{idx} = num2str(TorqueVec(idx));
end
figure
hmap = heatmap(SpeedVecLabels, TorqueVecLabels, Losses);
hmap.YDisplayData = flip(TorqueVecLabels);
hmap.Title = "Electrical Losses (W)";
hmap.XLabel = "Speed (rpm)";
hmap.YLabel = "Torque (N*m)";
```

References

[1] M. Ehsani, Y. Gao, J.M. Miller, Hybrid electric vehicles: architecture and motor drives, Proc. IEEE 95 (4) (2007) 719–728.
[2] J.P. Trovão, P.G. Pereirinha, H.M. Jorge, C.H. Antunes, A multi-level energy management system for multi-source electric vehicles—an integrated rule-based meta-heuristic approach, Appl. Energy 105 (2) (2013) 304–318.
[3] X. Fu, A novel design for flywheel battery of electric vehicles, in: International Conference on Intelligent System Design and Engineering Application, IEEE Computer Society, 2010, pp. 107–111.
[4] G. Niu, F. Shang, M. Krishnamurthy, C.M. Garcia, Evaluation and selection of accumulator size in electric-hydraulic hybrid (EH2) powertrain, in: IEEE Transportation Electrification Conference and Expo, Dearborn, MI, USA, Jun. 27–29, 2016.
[5] W. Wu, R.W.G. Bucknall, Conceptual evaluation of a fuel-cell-hybrid powered bus, in: IEEE Power Engineering Conference, Dublin, Ireland, Sept. 2–5, 2013.

[6] H. Fathabadi, Combining a proton exchange membrane fuel cell (PEMFC) stack with a Li-ion battery to supply the power needs of a hybrid electric vehicle, Renew. Energy 130 (2019) 714–724.
[7] P. Thounthong, S. Raël, B. Davat, Control strategy of fuel cell/supercapacitors hybrid power sources for electric vehicle, J. Power Sources 158 (1) (2006) 806–814.
[8] H. Fathabadi, Fuel cell hybrid electric vehicle (FCHEV): novel fuel cell/SC hybrid power generation system, Energ. Conver. Manage. 156 (2018) 192–201.
[9] M.C. Kisacikoglu, M. Uzunoglu, M.S. Alam, Fuzzy logic control of a fuel cell/ultracapacitor hybrid vehicular power system, in: Vehicle Power and Propulsion Conference, 2006 (VPPC '06), 2007, pp. 1–5.
[10] J. Bauman, M. Kazerani, A comparative study of fuel-cell-battery, fuel-cell--ultracapacitor, and fuel-cell-battery-ultracapacitor vehicles, IEEE Trans. Veh. Technol. 57 (2) (2008) 760–769.
[11] P. Thounthong, V. Chunkag, P. Sethakul, et al., Energy management of fuel cell/solar cell/SC hybrid power source, J. Power Sources 196 (1) (2011) 313–324.
[12] B. Wu, F. Zhuo, F. Long, et al., A management strategy for solar panel-battery-SC hybrid energy system in solar car, in: IEEE International Conference on Power Electronics and Ecce Asia, 2011, pp. 1682–1687.
[13] M.B. Burnett, L.J. Borle, A power system combining batteries and SCs in a solar/hydrogen hybrid electric vehicle, in: Vehicle Power and Propulsion, 2005 IEEE Conference, 2005, pp. 709–715.
[14] H. Fathabadi, Novel fuel cell/battery/supercapacitor hybrid power source for fuel cell hybrid electric vehicles, Energy 143 (2018) 467–477.
[15] T.R. Hawkins, O.M. Gausen, A.H. Strømman, Environmental impacts of hybrid and electric vehicles-a review, Int. J. Life Cycle Assess. 17 (8) (2012) 997–1014.
[16] S. Ramachandran, U. Stimming, Well to wheel analysis of low carbon alternatives for road traffic, Energ. Environ. Sci. 8 (11) (2015) 3313–3324.
[17] Z. Bitar, S. Al Jabi, Studying the performances of induction motor used in electric car, Energy Procedia 50 (2014) 342–351.
[18] A. Assegaf, A. Purwadi, Y. Haroen, Dynamic response analysis of permanent magnet synchronous motor drives for city electric car, in: 2013 International Conference on Information Technology and Electrical Engineering (ICITEE), IEEE, 2013.
[19] Ministry of Ecology and Environment of the People's Republic of China, China Vehicle Environmental Management Annual Report, 2018.
[20] International Energy Agency, World Energy Outlook, 2016.
[21] International Energy Agency, CO_2 Emissions from Fuel Combustion: Overview, 2017.
[22] M.Z. Jacobson, Review of solutions to global warming, air pollution and energy security, Energ. Environ. Sci. 2 (2) (2009) 148–173.
[23] M. Yazdanie, F. Noembrini, S. Heinen, et al., Well-to-wheel costs, primary energy demand, and greenhouse gas emissions for the production and operation of conventional and alternative vehicles, Transp. Res. Part D: Transp. Environ. 48 (2016) 63–84.
[24] J.R. Woo, H. Choi, J. Ahn, Well-to-wheel analysis of greenhouse gas emissions for electric vehicles based on electricity generation mix: a global perspective, Transp. Res. Part D: Transp. Environ. 51 (2017) 340–350.
[25] J.J. Hwang, J.K. Kuo, W. Wu, et al., Lifecycle performance assessment of fuel cell/battery electric vehicles, Int. J. Hydrogen Energy 38 (8) (2013) 3433–3446.
[26] K. Jorgensen, Technologies for electric, hybrid and hydrogen vehicles: electricity from renewable energy sources in transport, Util. Policy 16 (2) (2008) 72–79.
[27] V. Nian, M.P. Hari, J. Yuan, A new business model for encouraging the adoption of electric vehicles in the absence of policy support, Appl. Energy 235 (2019) 1106–1117.
[28] N. Shaukat, B. Khan, S.M. Ali, et al., A survey on electric vehicle transportation within smart grid system, Renew. Sustain. Energy Rev. 81 (2018) 1329–1349.

[29] D.B. Richardson, Electric vehicles and the electric grid: a review of modeling approaches, impacts, and renewable energy integration, Renew. Sustain. Energy Rev. 19 (19) (2013) 247–254.
[30] A. Zakariazadeh, S. Jadid, P. Siano, Integrated operation of electric vehicles and renewable generation in a smart distribution system, Energ. Conver. Manage. 89 (2015) 99–110.
[31] K.M. Tan, V.K. Ramachandaramurthy, J.Y. Yong, Integration of electric vehicles in smart grid: a review on vehicle to grid technologies and optimization techniques, Renew. Sustain. Energy Rev. 53 (2016) 720–732.

CHAPTER 7.3

Mitigating the effects of PEV fast charging on grids with renewable and storage systems

Mulusew Ayalew and Damot Gebregziabher
Department of Electrical and Computer Engineering, Hawassa University, Hawassa, Ethiopia

Contents

1 Introduction

Plug-in electric vehicles (PEVs) have a great role in emissions reduction and reducing the dependence of electric vehicles on fuel to increase the economy.

Active Electrical Distribution Network
https://doi.org/10.1016/B978-0-323-85169-5.00006-X

The author in [1] proposed smart charging of PEVs. The study considered 7.5 million people in the United States. In PEVs, using smart charging has a great role in enhancing the efficiency of electric power utilities.

The author in [2] presented PHEV load modeling and load flow analysis for the analysis of power loss and voltage deviation of a distribution network. In this paper, by considering the load characteristics at 2030, the effects of the PEV on transformer overloading, peak demand, and voltage drops were studied.

The author in [3] proposed parking for PEV charging. PEVs draw electric power during the day, and night, and this is the main cause of increases in electric power peak demand.

In [4], the authors proposed the Monte Carlo simulation-based PEV charging method under uncertain conditions. In this paper, the arrival time, departure time, and travel time of the vehicles are calculated by using a probability density function. Also the impacts on the distribution transformer load were analyzed.

The author in [5] proposed a smart charging technique of PEVs to minimize the distribution network power loss, total drawn power, and charging cost. The state of charge (SOC) is supposed to be with in the ranges from 20% to 90%. In this paper, the influence of PEV on the electric distribution network voltage and current profile were illustrated by using deterministic analysis.

The author in [6] presented the uncertainties of the state of charge and starting time of PEVs. The effects of PEV charging on electric distribution network power loss, voltage deviation, and economy were analyzed by Monte Carlo simulation methods.

The author in [7] proposed the stochastic ideal for appropriate PEV charging plans. The simulation studies numerous penetration stages for the upcoming years. In this paper, the comparisons of uncontrolled and controlled charging scenarios were illustrated. During the study period at 2030 of PEV uncontrolled charging, there was a 30% probability of transformers being loaded, whereas with smart charging, 13% of the transformers were loaded. Numerous other studies are described for PEV smart charging strategies [8–11].

The author in [12] proposed a probabilistic problem design of distributed energy resource (DER) based on uncertainties of PEV driving time. In this paper, the DER influence on total cost and the effect of PEVs were presented.

The author in [13] proposed mathematical modeling and the optimal sizing and placement of PEVs to minimize the overall cost. In this paper, the optimal sizing and placement of PEVs were studied.

The author in [14] proposed the optimal sizing and placement of PEVs to minimize the overall cost.

2 Electric vehicles

There are different forms of PEVs accessible today. They are classified based upon whether the electric vehicles (EVs) run solely on electricity, if they use a combination of ICE and energy, if they have plug-in capabilities, and so forth. The many forms of EVs are mentioned below [15,16].

2.1 Battery-based electric vehicle

A battery-based EV only gets power from a rechargeable electric battery; this type of vehicle doesn't need an internal combustion engine. The electric batteries of this vehicle are charged from an external source of the electric grid. Currently, fully charged battery-based EVs can go a distance of 140–180 km. These types of vehicles have a great role in reducing emissions and hence are greatly encouraged by governments.

2.2 Hybrid electric vehicle (HEV)

Most of the electric vehicles are grouped under this category. Hybrid electric vehicles are driven from a combination of a chargeable electric battery and an internal combustion engine (ICE). In some situations, these EVs can be driven independently. In an HEV, the chargeable electric battery and ICE can be set in either series or parallel operating conditions. In series connection mode, the task of the ICE is only to recharge the electric battery. In parallel connection mode, the electric battery and ICE supply power. HEV batteries can't be plugged directly into the electric grid; rather, regenerative braking techniques are used to charge batteries from the electric power grid.

2.3 PHEV

PHEVs, like HEVs, are powered by either a battery or an ICE. To take power from the grid, these vehicles may be connected to outlets. Furthermore, these cars, like HEVs, may be recharged by braking.

A. PEV charging characteristics

The types of batteries, capacity, and types of vehicles are the basic criteria for PEV charging requirements. Based on the standards of the US Society of Automotive Engineers (SAE), vehicles having a fully charged

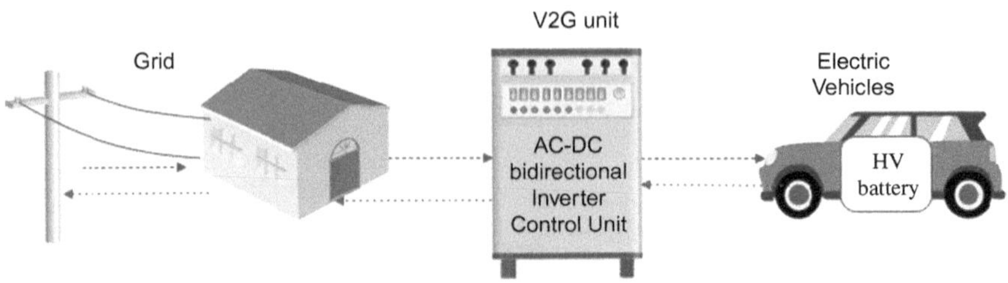

Fig. 1 Electric vehicles with V2G technology.

battery can run using only electric batteries until the state of charge of the battery drops to 20%–30%. Beyond this minimum threshold level, vehicles have to add ICE to operate as PHEVs to maintain the state of charge of the battery.

The V2G is environmentally friendly and can help the PEV reduce energy costs and stabilize demands on the power grid. As shown in Fig. 1, in vehicle-to-grid technology, the unused powers of the car are supplied to the smart grid. V2G, which is also known as vehicle-grid integration (VGI), can help the energy of the electric grid during peak load period.

The connection of electric vehicles with the utility poses several problems in terms of power quality. V2G technology enables peak shaving for the load profile distribution system by allowing electricity generation during peak hours. When a fault occurs, this disrupts the frequency stability. V2G at a high charging rate is safe for power systems in terms of stability and power quality (Fig. 2).

For grid-connected fast PEV charging, a series of EV power electronic convertors and control systems are the basic parts (Fig. 3).

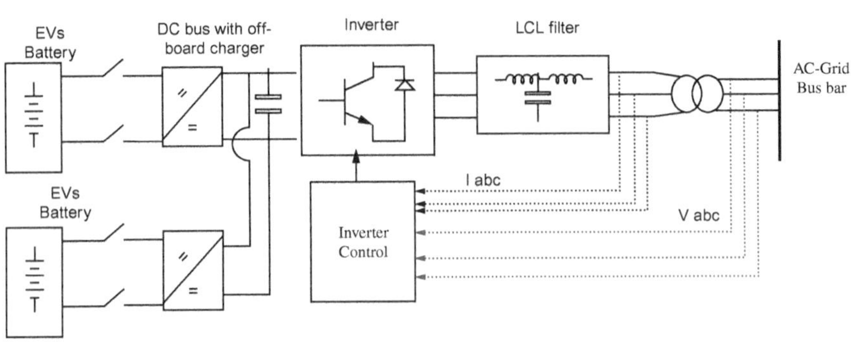

Fig. 2 Grid-connected electric vehicle converters and controllers.

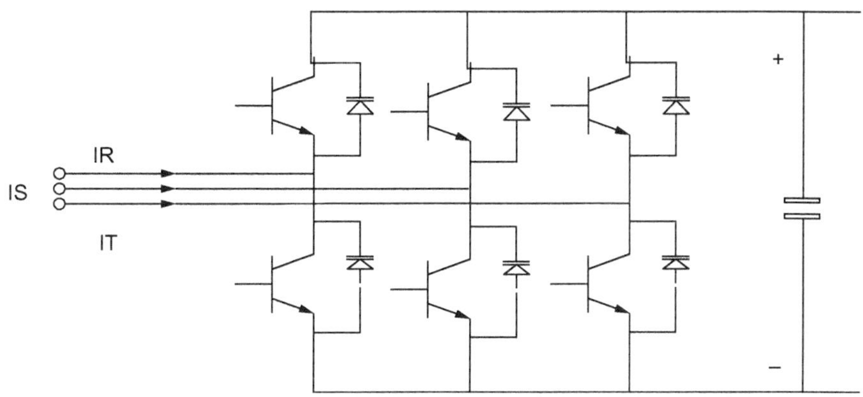

Fig. 3 Three-phase converter unit block diagram.

The three-phase converter unit of Fig. 4 is a structure of DC charging station parts of electric vehicles. An LCL filter and transformer are used to connect the converter to the distribution grid. The slots in the DC bus are used to link the electric vehicles.

The power converting system in grid-connected EVs is made up of two types of pulse width modulation (PWM) power electronic converters: current-regulated and voltage-source. The charging methods for electric vehicles are categorized into three kinds, according to the SAE: AC charging level 1, AC charging level 2, and AC charging level 3 [6,17] (Table 1).

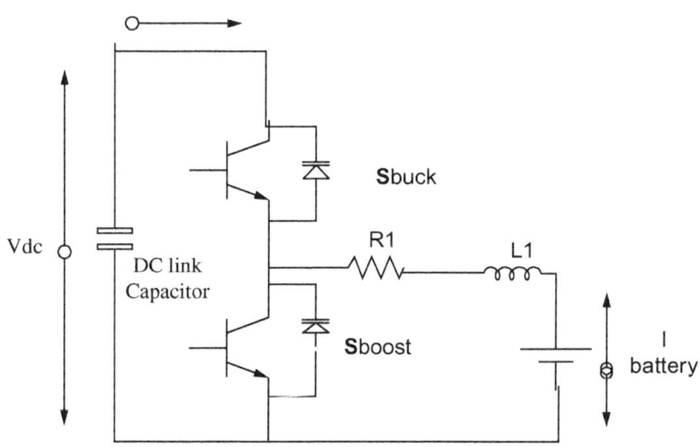

Fig. 4 Power converting circuits.

Table 1 AC charger levels of electric vehicles.

Item	AC level 1 charger	AC level 2 charger	AC level 3 charger
Vehicle charging capacity	120 V/16 A, 2 kW	208–240 V/80 A, 15 kW	208–600 V/400 A, 96 kW
Charging time to fully charge the vehicle battery	5–8 h	2.5–4 h	10–15 min
Application areas	Not feasible for commercial and public charging. It is ideal for residential charging	Applicable for residential and commercial charging areas	Applicable for residential and commercial charging areas

B. Typical battery sizes of vehicles

Different types of PEVs have different structures with different battery sizes. Table 2 [18] shows the energy needs of several types of PEVs based on data from the Electric Power Research Institute (EPRI) and the Pacific Northwest National Laboratory (PNNL).

Battery capacity rating is sized based on the energy requirements per kilometer driven. The rating of the battery capacity is designed according to the PEV all electric range (AER). AERs of PEVs can be defined as the vehicle capacity if only running in electric power mode. After the EVs cross the AERs, then ICE charges the vehicle until it can be recharged at the nearest charging station. The battery capacity of PEVs is determined by using Eq. (1) by considering different levels of AER, and is illustrated in Table 2.

Table 2 Battery sizes of different types of vehicles [19].

Type of vehicle	.PHEV-30	.PHEV-40	.PHEV-60
.Compact sedan	7.8	10.4	15.6
.Mid-size sedan	9	12	18
.Mid-size SUV	11.4	15.2	22.8
.Full-size SUV	13.8	18.4	27.6

$$C_B = \text{AER (km)} \times \text{energy consumption per km (kWh/km)} \tag{1}$$

Where C_B = battery capacity of PEVs.

To recharge the battery, its energy requirements are determined by the battery capacity (C_B). The state of charge (SOC) of the battery is the percentage of energy stored in a battery that is measured at a particular time. Thus, the SOC of the battery can be calculated, as given in Eq. (2), from the required length to be driven by the PEVs, the AER, and the capacity of the battery.

The SOC of the PEVs when arriving at home after the end of a trip is given in Eq. (2):

$$\text{SOC} = \begin{Bmatrix} (1 - \{D/AER\}), D \leq AER \\ 0, D \geq AER \end{Bmatrix} \tag{2}$$

$$D = \gamma d \tag{3}$$

Where γ is the percentage of the distance driven in PEV electric mode. d is the total distance driven by the PEVs.

The electric cars run in electric mode until the battery level drops to zero, at which point they switch to ICE power. When the cars exceed AER limitations, the battery's level of charge decreases to zero.

The PEV battery energy requirements to make it fully charged are calculated as in Eq. (4):

$$hE = (1 - \text{SOC})C_B \tag{4}$$

Where, E is the actual required energy to fully charge the PEV battery.

The PEV battery charges until the SOC becomes 100%. The required energy (E) of the PEV battery is supplied from the substation and the amount of energy depends on the size of the battery.

3 Problem formulation of PEV charging/discharging

3.1 Objective function

The charge/discharge scheduling of a PEV needs to smooth the electric power grid load profile. The main aim of the optimization problem is to smooth the peak load period and fill the valley periods of the load curve characteristics. Thus, optimal PEV charging/discharging scheduling can diminish the gap of the base load and peak load characteristics curve. To solve this optimization problem, the squared deviation between the desired load curve and the base load curve is formulated in Eqs. (5), (6).

$$\mathrm{Min} ft\, s \cdot t \tag{5}$$

$$ft = \left(P_{\mathrm{b,t}} - \mathrm{P}_{\mathrm{target}}, t\right)^2 \quad \text{for} \quad t = 1:T \tag{6}$$

Where, θ is the set of constraints equations, T is the total iteration number, $P_{\mathrm{b,t}}$ is the base load demand at the specific time instant t.

3.2 Constraints

The vehicle-related constraints, power flow constraints, and power grid technical constraints are the main PEV charge/discharge scheduling problems. The electric power flow equation at each bus, the minimum and maximum voltage limits, and the thermal limits of the distribution network are considered to be the electric power flow and distribution network technical constraints. The vehicle-related constraint PEV scheduling algorithm is illustrated as follows.

(1) Eq. (7) is the PEV maximum and minimum value of charging and discharging rate constraints, so as to avoid the avoid battery degradation.

$$P_{\mathrm{min}} \le P_{\mathrm{charger}} \le P_{\mathrm{max}} \tag{7}$$

(2) Eq. (8) is the PEV maximum and minimum SOC inequality constrained equation.

$$\mathrm{SOC}_{\mathrm{PEV,\,min}} \le \mathrm{SOC}_{\mathrm{PEV}} \le \mathrm{SOC}_{\mathrm{PEV,\,max}} \tag{8}$$

If the battery DOD is 60%, this indicates that the SOC will not decrease below 40%.

3.3 PEV modeling

The battery initial SOC time and the battery charging start times are the basic requirements of PEV modeling. The battery initial SOC time depends on the vehicle's travel distance. The SOC of the PEV battery is updated at each time instant by using Eq. (9).

$$SOC_n^t = SOC_n^{t-1} + \mathbf{h}_{ch,n}\, P_{ch,n}^t\, \mathrm{Dt}\, C_n^t - \mathbf{h}_{disch,n}\, P_{disch,n}^t\, \mathrm{Dt}\, D_n^t \tag{9}$$

Where, SOC_n^t is the state of charge of the nth PEVs at time instant t, $P_{ch,n}^t$ is the charging power of the nth PEVs at time instant t, $P_{disch,n}^t$ is the discharging power of the nth PEVs at time instant t, $\mathbf{h}_{ch,n}$ is the charging efficiency, $\mathbf{h}_{disch,n}$ is the discharging efficiency, C_n^t is the PEV charging plans at time instant t (C_n^t, $D_n^t \in \{0,1\}$), D_n^t is the PEV discharging plans at time instant t (C_n^t, $D_n^t \in \{0,1\}$), $C_n^t \times D_n^t = 0$ because the charging/discharging of the PEV battery cannot be performed at the same time.

The optimization problems of PEV energy requirements, charging/discharging power, and the whole amount of energy in the charging station are given by Eqs. (10), (11):

$$E_{req,n}^t = C \times \mathrm{Dt}\left(SOC_{\text{full}} - SOC_n^t\right) \tag{10}$$

$$E_{avail,n}^t = C \times \mathrm{Dt}\left(SOC_n^t - DOD\right) \tag{11}$$

$$P_{ch,n}^t = \frac{E_{req,n}^t}{\sum_{j=1}^{m} \sum_{req,j}^{t} .} \times P_{cs}^t \tag{12}$$

$$P_{disch,n}^t = \frac{E_{avail,n}^t}{\sum_{j=1}^{m} \sum_{avail,j}^{t} .} \times P_{cs}^t \tag{13}$$

Where, $E_{req,\ n}^t$ is the PEV required energy at time instant, $E_{avail,n}^t$ is the PEV available energy at time instant, $P_{ch,n}^t$ is the charging power of the nth PEV at time instant, $P_{disch,n}^t$ is the PEV discharging power at time instant, M is the PEV number in the charging station (CS), P_{cs}^t is the CS total injected or absorbed power.

4 Effects of PEV on electricity capacity and demand

4.1 Effects

An interconnected electric power grid contains generation, transmission, and distribution systems. The electric power generated from the generation station consists of different electric power sources such as renewable and nonrenewable energy sources. The electric power transmission system contains transmission networks that transfer electricity from the generation station to distribution networks [20]. The distribution network consists of substations, feeders, and step-down transformers to deliver electric power to the end user. Charging PEVs from the electric grid has two basic impacts on the electric power system:

(1) PEV battery penetration levels.

(2) PEV charging time, location, and duration.

4.2 Influence on power generation

A high penetration of PEVs causes extra energy requirements that have to be supplied by generation stations. Due to the scarcity of available storage devices, an electric grid will face challenges of load and generation balance [21]. Because the PEV charging time duration is not constant, this can lead to increases in the peak load demand of a system. Thus, generation plants have to generate power at their rated capacity, which leads to increased cost and reduced system reliability [22].

4.3 Impact on transmission

Due to the additional energy requirements of PEV charging, transmission networks have to be upgraded to carry these energy requirements [23].

Also, transformers may be overloaded for extended periods due to the additional energy requirement of PEV charging. Thus, the transformer lifetime will be reduced [17,24].

4.4 Impact on distribution

Increases in the penetration level of PEVs have a huge influence on the distribution network more than the generation system and transmission networks. A distribution network can be affected by the size, location, time, and duration of PEV charging. The distribution network capacity such as the feeder, protective devices, and transformers has to be related with the penetration level of PEVs to reduce distribution system overloading [25]. A low-voltage distribution network is unable to hold simultaneously varying charging situations. When the PEV charger draws energy simultaneously from a local distribution substation, it will cause a considerable influence on local distribution networks, equipment, and peak demand [26,27].

4.5 Voltage unbalance indication

Voltage disruption is a power quality problem are produced by the irregular distribution of single-phase and three-phase loads on the distribution network. Mostly, it happens on weak distribution networks such as on industrial and distribution networks. The PEV charger can also causes a voltage unbalance at the distribution network. Voltage unbalance can be defined as the ratio of negative sequence voltage to the positive sequence voltage [28]. The maximum voltage deviation (%PVU) of three-phase voltage is expressed by the National Electrical Manufacturer Association (NEMA), as shown in Eq. (14) below.

$$\%PVU = \frac{\text{Maximum voltage deviation from the average voltage}}{\text{Average voltage}} \times 100 \tag{14}$$

Based on NEMA standards, a maximum unbalanced voltage at the motor terminal can be tolerated up to 1% deviation. This rating varies for different motor classes, sizes, and load types [29].

5 Integration of EVs with renewable energy sources

Advancements in using renewable energy sources (RES) have a quite attractive effect on electric power systems. The power from renewable energy sources, especially power from wind and photovoltaics, suffers from the unpredictable and intermittent nature of wind speed and sun radiation [30]. Renewable energy power sources are weather-dependent and vary with time, which makes it difficult to control and plan the power system [31]. An energy storage system (ESS) is the favorable solution to balance the grid-connected electricity generation from these RES [32].

The PEV batteries also have a promising effect to balance the grid-connected renewable energy sources. Thus, PEV chargers can draw a surplus of power from renewable energy. and vehicles can deliver back to the electric grid through different charging stations by a V2G system [33]. Thus, PEVs have secondary services and can act as energy buffers for grid power regulations.

In [34], the author proposed the integration of a power grid, distributed RES (power from wind and PV solar), and PEVs having a V2G service. The main aim of this paper was reducing greenhouse gases and maintaining energy security by achieving the integration of PEVs and RES (Fig. 5).

5.1 PV solar energy with EV

Solar power generation is very encouraging and feasible. It has a great advantage for PEV battery charging and supporting the electric grid.

The authors in [35,36] proposed the placement of solar photovoltaic arrays on the roofs of parking, which is quite attractive to charge the PEV battery. In this paper, an energy economics emission analysis was done at the working location. The V2G applications are also viable in grid-connected PV solar power [37]. They also enhance grid operation and reduce operating costs; generation scheduling is needed [38].

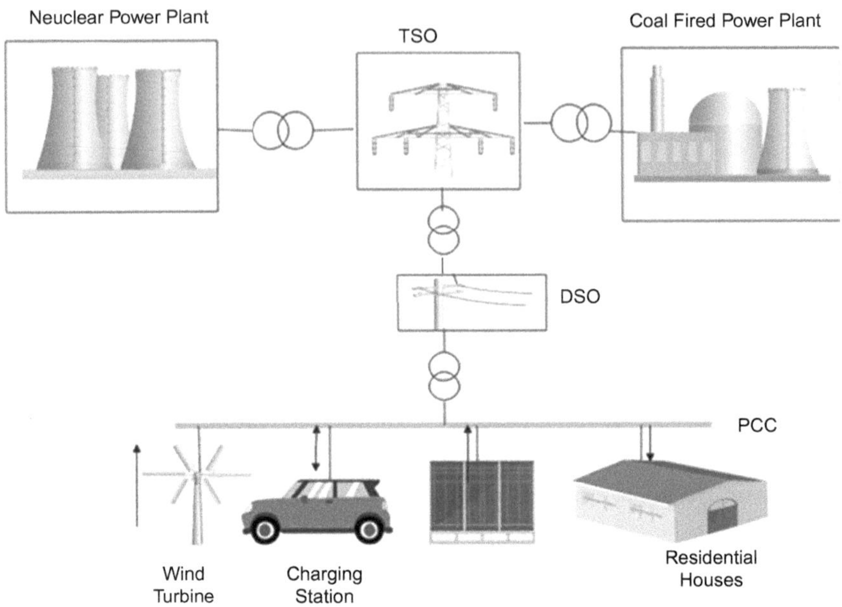

Fig. 5 Integration of wind, photovoltaics, and PEVs with the electric grid.

The author in [38] proposed the coordination and scheduling of PEVs and industrial microgrids (IMG). In this paper, the scheduling of distributed RES (with PV solar) and combined heat and power generation with PEVs was done to reduce the overall operating and charging costs of the system. However, the intermittent nature of the PV output power needs a controller and compensation. Fig. 6 illustrates a solar charging plant integrated to an electric power grid through a bidirectional DC/AC power converter.

5.2 Wind energy with EVs

Electric power generation from wind energy is a prevalent and feasible way for power production.

The authors in [39,40] proposed an investigation on a wind energy conversion system (WECS) and the PEV impact and feasibility on the electric grid. Due to the intermittent nature of wind speed, power generation from wind energy mostly suffers from frequency fluctuation. The V2G service of PEVs can be considered frequency regulation to balance or mitigate the imbalance of load and power generation.

The author in [41] proposed an integration of wind energy with a microgrid (MG) and plug-in hybrid electric vehicles (PHEVs). To meet the

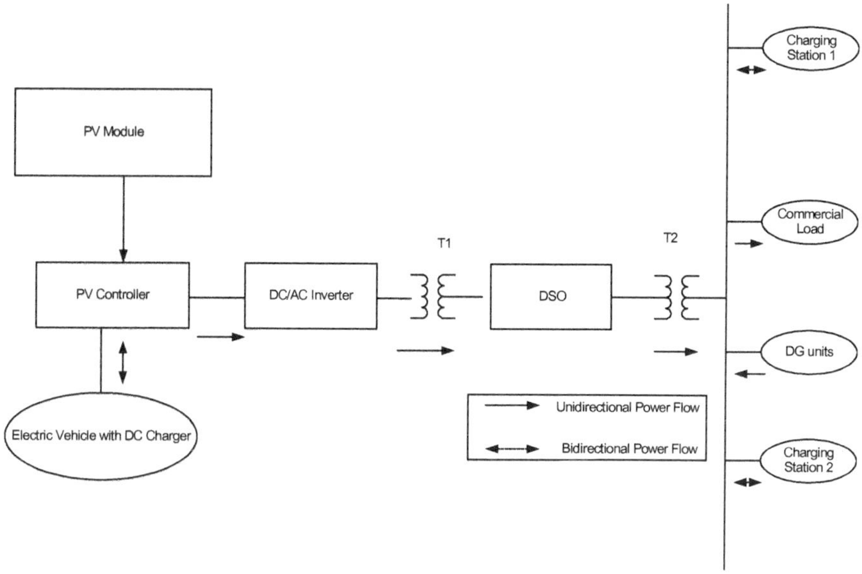

Fig. 6 Grid-connected PV solar PEV charging station.

generation load balance, the coordination of the wind-PEV arrangement can improve the consumption of these power sources. In this paper, the V2G concepts are also proposed to balance the PEV charging and discharging (V2G) power profile.

In [42], the author suggested a stochastic unit commitment model for combinations of thermal generating units, PHEVs, and wind power. By employing PHEVs, the major goal of this schedule was to reduce power system running costs while also balancing the intermittent nature of wind energy. The battery life duration, PHEV capacity, and driving patterns were used as constraint equations in this study.

6 Cost analysis on interacting EVs and RESs

In this section, the constraint problem solutions used to optimize the cost analysis of interfaces of EVs and RESs are considered in detail (Table 3).

6.1 Minimizing operational cost

The charging provider considers the operational cost from two perspectives. The interaction of PEVs with RESs is taken into account for the problem formulation on minimizing the operational cost of the system. The main

Table 3 Cost analysis interface according to different perspectives.

	Cost-aware interaction on different viewpoints
Charging service providers (battery manufacturers) Consumers (EV owners) Electric grid owners	✓ to decrease operational cost ✓ to increase the amount of profit ✓ to lower battery charging cost ✓ electricity generation cost ✓ system lifecycle cost ✓ energy transmission cost

tasks of charging providers are only on problem formulation on operation cost minimization.

The authors in [38,43–46] proposed nonlinear programing to solve the problems of minimizing operational cost and formulating mathematical optimizing problems.

The authors in [34,47,48] proposed mixed integer programming to solve the problems of minimizing operational cost and formulating mathematical optimizing problems.

The authors in [38,43,44] studied operation cost minimization of a MG with PEVs interacting in RES. A grid-connected microgrid needs to have a microgrid controller that is liable for scheduling all its subsystems such as distributed generators, battery energy storage systems, (BESS), PEV charging appliances, and PEV charging stations.

The author in [43] proposed scheduling the power generation of RESs, PEVs, and battery swapping stations (BSS). The PEVs and BSS will draw electricity when an RES generates a huge amount of power. When the grid has a shortage of power, the PEV battery and BSS can discharge as electricity sources.

The author in [44] presented the total cost minimization of a microgrid by considering PV solar power. The author proposed a radial basis function network (RBFN) method to predict the PV solar power output. The mean absolute percentage error (MAPE) was used to check the performance of the radial basis function network.

$$\mathrm{MAPE} = \frac{1}{N}\sum_{i=1}^{N}\frac{\left|P_i^f - P_i^\alpha\right|}{P_i^\alpha} \times 100\% \tag{15}$$

Where, N is the total number of forecasting data, P_i^f is the forecasted value, P_i^α is the actual value, i is the index of the dataset.

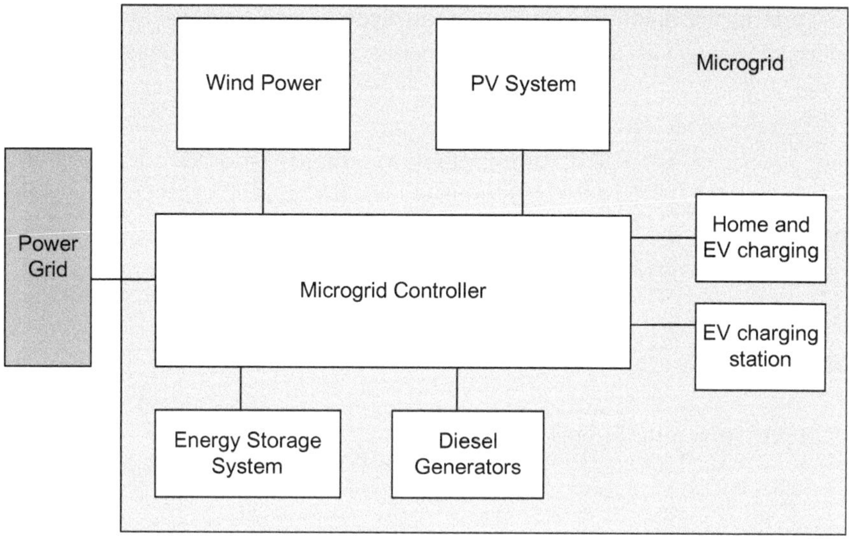

Fig. 7 PEVs in a grid-connected microgrid.

A genetic algorithm is used to solve the optimization problem of Eq. (15) and to minimize the total cost of the microgrid (Fig. 7).

The author in [44] proposed a unit commitment (UC) problem model to minimize the operational cost of a power system. The basic definition of a unit commitment problem is given by [47].

$$OC = \sum_{i=1}^{N} \sum_{t=1}^{T} \left[FC_{it(Pit)} + MC_{it(Pit)} \right] + ST_{it} + SD_{it} \tag{16}$$

Where, OC is the system operational cost, $FC_{it(Pit)}$ is the fuel cost, $MC_{it(Pit)}$ is the maintenance cost, ST_{it} is the startup cost, SD_{it} is the shutdown cost.

The formulation of operational cost minimization problem, based on unit commitment problem is also proposed in [34,44,48].

7 Conclusion

Because PEVs are environmentally beneficial energy solutions, they are expected to become a way of electrification of road transportation technology. The introduction of these cars will clash with the current power grid configuration, which includes generation, transmission, and one distribution network. Increased use of PEVs and their recharging in an uncoordinated way may result in additional peaks in the electric power system, which might overburden power distribution elements and generate energy pollution issues.

The use of electric vehicles to integrate renewable energy sources has also been proposed. Electric cars are thought to be able to provide additional grid services such as voltage and frequency control, peak power leverage, and reactive power assistance, which can enhance operational efficiency, protect the grid, and lower power system running costs.

References

[1] M.M. Collins, G.H. Mader, The timing of EV recharging and its effect on utilities, IEEE Trans. Veh. Technol. 32 (1) (1983) 90–97.

[2] K. Clement-Nyns, E. Haesen, J. Driesen, The impact of charging plug-in hybrid electric vehicles on a residential distribution grid, IEEE Trans. Power Syst. 25 (1) (2010) 371–380.

[3] D. Wu, D.C. Aliprantis, K. Gkritza, Electric energy and power consumption by light-duty plug-in electric vehicles, IEEE Trans. Power Syst. 26 (2) (2011) 738–746.

[4] E. Pashajavid, M.A. Golkar, Charging of plug-in electric vehicles: stochastic modelling of load demand within domestic grids, in: Proc. 20th Iranian Conference on Electrical Engineering (ICEE), IEEE, Tehran, Iran, 2012.

[5] I. Sharma, C.A. Cañizares, K. Bhattacharya, Modeling and impacts of smart charging PEVs in residential distribution systems, in: Proc. Power and Energy Society General Meeting, IEEE, 2012, pp. 1–8.

[6] I. Sharma, C. Canizares, K. Bhattacharya, Smart charging of PEVs penetrating into residential distribution systems, IEEE Trans. Smart Grid 5 (3) (2014) 1196–1209.

[7] R. Scharrenberg, B. Vonk, P. Nguyen, EV stochastic modelling and its impacts on the Dutch distribution network, in: Proc. International Conference on Probabilistic Methods Applied to Power Systems (PMAPS), IEEE, Durham, 2014.

[8] A. Lojowska, D. Kurowicka, G. Papaefthymiou, L. Van der Sluis, Stochastic modeling of power demand due to EVs using copula, IEEE Trans. Power Syst. 27 (4) (2012) 1960–1968.

[9] R.A. Verzijlbergh, Z. Lukszo, J.G. Slootweg, M.D. Ilic, The impact of controlled electric vehicle charging on residential low voltage networks, in: Proc. International Conference on Networking, Sensing and Control (ICNSC), IEEE, Delft, Netherlands, 2011.

[10] J. Zheng, X. Wang, K. Men, C. Zhu, S. Zhu, Aggregation model-based optimization for electric vehicle charging strategy, IEEE Trans. Smart Grid 4 (2) (2013) 1058–1066.

[11] M.F. Shaaban, Y.M. Atwa, E.F. El-Saadany, PEVs modeling and impacts mitigation in distribution networks, IEEE Trans. Power Syst. 28 (2) (2013) 1122–1131.

[12] G. Cardoso, M. Stadler, M. Chehreghani Bozchalui, R. Sharma, C. Marnay, A. Barbosa-Povoa, P. Ferrao, Stochastic programming of vehicle to building interactions with uncertainty in PEVs driving for a medium office building, in: Proc. Industrial Electronics Society, IECON 2013-39th Annual Conference of the IEEE, 2013. Vienna, Austria.

[13] Z. Liu, F. Wen, G. Ledwich, Optimal planning of electric-vehicle charging stations in distribution systems, IEEE Trans. Power Delivery 28 (1) (2013) 102–110.

[14] L. Jia, Z. Hu, Y. Song, Z. Luo, Optimal siting and sizing of electric vehicle charging stations, in: Proc. Electric Vehicle Conference (IEVC), IEEE International, Greenville, SC, 2012.

[15] US Department of Energy, Alternative Fuels Data Center, 2021. Available from: http://www.afdc.energy.gov/vehicles/electric.html (Accessed 16 July 2021).
[16] CAA Electric Vehicles, [Online]. Available http://electricvehicles.caa.ca/types-ofelectric-vehicles/.
[17] S. Blumsack, C. Samaras, P. Hines, Long-term electric system investments to support plug-in hybrid electric vehicles, in: Power and Energy Society General Meeting—Conversion and Delivery of Electrical Energy in the 21st Century, 2008 IEEE, 2008, pp. 1–6.
[18] M. Kintner-Meyer, K. Schneider, R. Pratt, Impacts Assessment of Plug-in Hybrid Vehicles on Electric Utilities and Regional US Power Grids, Part 1: Technical Analysis, Pacific Northwest National Laboratory (a), 2007, pp. 1–20.
[19] Greenhouse, National Inventory Report 1990–2013: Greenhouse Gas Sources and Sinks in Canada Executive Summary, 2022.
[20] L.L. Grigsby, Electric Power Generation, Transmission, and Distribution, CRC Press, May 2007.
[21] W. Kempton, J. Tomic, Vehicle-to-grid power fundamentals: calculating capacity and net revenue, J. Power Sources 144 (2005) 268–279.
[22] S. Hadley, in: Oak Ridge National Laboratory, U.S. Department of Energy (Eds.), Potential Impacts of Plug-in Hybrid Electric Vehicles on Regional Power Generation, 2008. Tech. Rep.
[23] P. Denholm, W. Short, An Evaluation of Utility System Impacts and Benefits of Optimally Dispatched Plug-in Hybrid Electric Vehicles, National Renewable Energy Laboratory, October 2006. Tech. Rep.
[24] C. Camus, C.M. Silva, T.L. Farias, J. Esteves, Impact of plug-in hybrid electric vehicles in the Portuguese electric utility system, in: Power Engineering, Energy and Electrical Drives, 2009. POWERENG '09. International Conference on, 2009, p. 285290.
[25] C. Farmer, P. Hines, J. Dowds, S. Blumsack, Modeling the impact of increasing PHEV loads on the distribution infrastructure, in: System Sciences (HICSS), 2010 43rd Hawaii International Conference on, 2010, pp. 1–10.
[26] J. Taylor, A. Maitra, M. Alexander, D. Brooks, M. Duvall, Evaluation of the impact of plug-in electric vehicle loading on distribution system operations, in: Power & Energy Society General Meeting, 2009. PES '09, IEEE, 2009, pp. 1–6.
[27] A. Maitra, J. Taylor, D. Brooks, M. Alexander, M. Duvall, Integrating plug-in- electric vehicles with the distribution system, in: Electricity Distribution—Part 1, 2009. CIRED 2009. 20th International Conference and Exhibition on, 2009, pp. 1–5.
[28] A. Eltom, M.A. Aziz, The economics of energy efficient motors during unbalanced voltage conditions, in: Power Engineering Society Inaugural Conference and Exposition in Africa, 2005 IEEE, 2005, pp. 378–384.
[29] N.S. Moharari, Microprocessor Based Protection of Induction Motors, Using Thermal, Mechanical, and Skin Effect Electrical Models to Predict Motor Temperature Rise, Master of Science in Engineering, Electrical Engineering, The University of Tennessee, Chattanooga, 1989.
[30] International Electrotechnical Commission (IEC), Grid Integration of Largecapacity Renewable Energy Sources and Use of Large-Capacity Electrical Energy Storage, White paper 3, 2012.
[31] D. Dallinger, S. Gerda, M. Wietschel, Integration of intermittent renewable power supply using grid-connected vehicles—a 2030 case study for California and Germany, Appl. Energy 104 (2013) 666–682.
[32] B. Battke, T.S. Schmidt, D. Grosspietsch, V.H. Hoffmann, A review and probalistic model of lifecycle costs of stationary batteries in multiple applications, Renew. Sustain. Energy Rev. 25 (2013) 240–250.

[33] D.B. Richardson, Electric vehicles and electric grid: a review of modeling approaches, impacts and renewable energy integration, Renew. Sustain. Energy Rev. 19 (2013) 247–254.
[34] A.Y. Saber, G.K. Venayagamoorthy, Plug-in vehicles and renewable energy sources for cost and emission reductions, IEEE Trans. Ind. Electron. 58 (4) (2011) 1229–1239.
[35] P. Tulpule, V. Marano, S. Yurkovich, G. Rizzoni, Economic and environmental impacts of a PV powered workplace parking garage charging station, Appl. Energy 108 (2013) 323–332.
[36] D.P. Birnnie III, Solar-to-vehicle (S2V) systems for powering commuters of the future, J. Power Sources 186 (2009) 539–542.
[37] J. Traube, F. Lu, D. Maksimovic, Photovoltaic power system with integrated electric vehicle DC charger and enhanced grid support, in: Proceedings of the 15th International Power Electronics and Motion Control Conference (EPE/PEMC), 2012.
[38] S.Y. Derakhshandeh, A.S. Masoum, S. Deilami, M.A. Masoum, G.M. Hamedani, Coordination of generation scheduling with PEVs charging in industrial microgrids, IEEE Trans. Power Syst. 28 (3) (2013) 3451–3461.
[39] H. Lund, W. Kempton, Integration of renewable energy into the transport and electricity sectors through V2G, Energy Policy 36 (2008) 3578–3587.
[40] J.A.P. Lopes, P.M.R. Almeida, F.J. Soares, Using vehicle-to-grid to maximize the integration of intermittent renewable energy resources in islanded electric grids, in: Proceedings of the International Conference on Clean Electrical Power, 2009, pp. 290–295.
[41] T. Wu, Q. Yang, Z. Bao, W. Yan, Coordinated energy dispatching in microgrid with wind power generation and plug-in electric vehicles, IEEE Trans. Smart Grid 4 (3) (2013) 1453–1463.
[42] C. Liu, J. Wang, A. Botterud, Y. Zhou, A. Vyas, Assessment of impacts of PHEV charging patterns on wind-thermal scheduling by stochastic unit commitment, IEEE Trans. Smart Grid 3 (2) (2012) 675–683.
[43] M. Zhang, J. Chen, The energy management and optimized operation of electric vehicles based on microgrid, IEEE Trans. Power Delivery 29 (2014) 1427–1435.
[44] C. Chen, S. Duan, Optimal integration of plug-in hybrid electric vehicles in microgrids, IEEE Trans. Ind. Inf. 10 (2014) 1917–1926.
[45] Z. Tan, P. Yang, A. Nehorai, An optimal and distributed demand response strategy with electric vehicles in the smart grid, IEEE Trans. Smart Grid 5 (2014) 861–869.
[46] A.Y. Saber, G.K. Venayagamoorthy, Resource scheduling under uncertainty in a smart grid with renewables and plug-in vehicles, IEEE Syst. J. 6 (2012) 103–109.
[47] N.P. Padhy, Unit commitment-a bibliographical survey, IEEE Trans. Power Syst. 19 (2004) 1196–1205.
[48] J. Wang, C. Liu, D. Ton, Y. Zhou, J. Kim, A. Vyas, Impact of plug-in hybrid electric vehicles on power systems with demand response and wind power, Energy Policy 39 (2011) 4016–4021.

CHAPTER 7.4

State of charge prediction of battery management system for electric vehicles

Alagar Karthick[a], V. Mohanavel[b], V. Kumar Chinnaiyan[a], J. Karpagam[a], I. Baranilingesan[a], and S. Rajkumar[c]
[a]Renewable Energy Lab, Department of Electrical and Electronics Engineering, KPR Institute of Engineering and Technology, Coimbatore, Tamil Nadu, India
[b]Centre for Materials Engineering and Regenerative Medicine, Bharath Institute of Higher Education and Research, Chennai, Tamil Nadu, India
[c]Department of Mechanical Engineering, Faculty of Manufacturing, Institute of Technology, Hawassa University, Hawassa, Ethiopia

Contents

1 Introduction

Future energy storage technology is critical to electric vehicles (EVs) and the growth of smart grids [1,2]. Due to many EV accidents, several discussions have taken place on electric vehicles in recent years. This also increases the demands and challenges for the battery management system (BMS) [3,4]. As part of electric cars, BMS is the path to the market for EV industrialization. As a result, high-performance BMS development for lithium-ion batteries has become a hot topic of research [5,6]. Battery models can be categorized as physics-based models, electrical-equivalent circuit models, and artificial intelligence models [7]. Evidence indicates that data-driven approaches have been commonly used in the design of a battery model. This was combined with a data-driven modeling algorithm to create a broad data-based battery maintenance model [8].The Monte Carlo phenomenological kinetic model

Active Electrical Distribution Network
https://doi.org/10.1016/B978-0-323-85169-5.00012-5

will accurately represent the structural evolution of the batteries. The three-dimensional kinetic model of Monte Carlo is used to estimate the kinetic parameters during the lithium-sulfur battery discharge reaction [9]. The results obtained are perfect, but the methods need to be optimized. The state of charge (SOC) is analyzed using two well-known methods: model-based and data-driven [10]. SOCs with a shorter development time, limited knowledge of the battery material's characteristics, and complex chemical reactions can also be calculated using a data-driven method [11,12]. The SOC models are made up of several chemical, physical, and dynamic equations on the theoretical front [13]. Significant design challenges include limited domain awareness and longer model development times. Data-driven SOC estimation methods do not require mathematical and semantic analysis of source and target variables for fuzzy systems [14]. The EV battery performance is analyzed by developing the internal resistance model to predict the performance [15]. The sensitivity-based SOC estimation of the battery is investigated based on the extended Kalman filter. It is reported that the numerical and the extended Kalman filter have slight deviations in the aging [16]. The fractional-based model for battery cell aging has been developed [17]. The MATLAB Simulink model is designed to predict the lithium-ion battery's characteristics based on temperature and capacity fading [18]. The particular state algorithm is developed to investigate the parameters of the lithium-ion battery in the online mode. The predicted results have good agreement with the experimental results [19]. The model is developed to investigate SOC's mismatch in the lithium-ion battery based on the joint moving horizon approach [20–22].

The linear NN battery model is used to detect the OCV-SOC ratio with the second back-propagation neural network (BPNN) to determine the first- or second-order electrochemical model parameters. The radial basis function neural network (RBFNN) quantification algorithm is a developed model to estimate the battery pack's SOC. The SOC load-classification NN model has been developed [23]. Model structure and postprocessing improves overfitting [24,25]. A BPNN-based SOC estimation technique is implemented in [26]. The battery component is analyzed, and particle swarm optimization (PSO) is also used to improve the prediction accuracy and stability. A new deep-feed NN approach for estimating SOC batteries was used in [27], and battery measurements are mapped directly to SOC. A nonlinear RBFNN observer for estimating SOC with an inclusive equivalent circuit model [28]. An improved nonlinear autoregressive exogenous neural network (NARXNN) algorithm for SOC battery estimation has been developed [29]. The lighting search algorithm is used to determine the best value for input delay, retroactivity delay, and hidden layer neurons. The Levenberg-Marquardt (L-M) algorithm for the NNN multilayer wavelet model is suggested [30]. The PSO algorithm is used to optimize wavelet neural network (WNN) for SOC estimation. The SOC recurrent neural network (RNN) battery is built with gated recurring modules [31]. A stacked LSTM network for modeling and estimating lithium-ion phosphate battery dynamics have been proposed in [32]. Although the initial SOCs are inaccurate, the developed model showed rapid convergence toward true SOCs. A fuzzy NN (FNN) battery model is built to capture nonlinear battery dynamics [33]. For additional SOC estimation, the genetic algorithm (GA) is typically used to evaluate the battery model's parameters [34]. A new SOC approach is built based on the gray model and gas [35]. The use of a genetic algorithm has resulted in greater precision and recurrence. The support vector machine (SVM) cluster of similar supervised learning methods can perform any multivariate function to any degree of accuracy. An improved SVM is proposed for the SOC estimation system based on regression. This method is more straightforward and more effective than artificial neural networks. An adaptive unscented Kalman filter (AUKF) and most miniature square support vector machines (LSSVM) were combined to estimate the SOC battery [36]. The classification of the SOC prediction algorithms is shown in Fig. 1. The battery model can be designed and revised consistently, even with limited training samples. The minimum SVM battery power forecast algorithm has been developed [37].

According to the literature studies, the battery's performance prediction is made in the data-driven method concerning the available dataset. The battery's operation in the EV needs to be analyzed based on emission cycle transport mode. The emission cycle is classified as the Urban Dynamometer Driving Schedule (UDDS), the LA92 period, and the US06 Supplementary Federal Test Method (SFTP). The battery's performance prediction for the emission cycle needs to be investigated and compared with the data-driven model. In this work, the UDDS emission cycle, the LA92 period, and the SFTP test method were investigated with the obtained dataset for predicting battery SOC. The battery is a complex nonlinear system with several state variables, so accurate battery status measurement is the key to the battery management system. The SOC estimation of the various emission cycles is predicted based on a machine learning algorithm such as the feedforward neural network, ensemble bagging, and ensemble boosting.

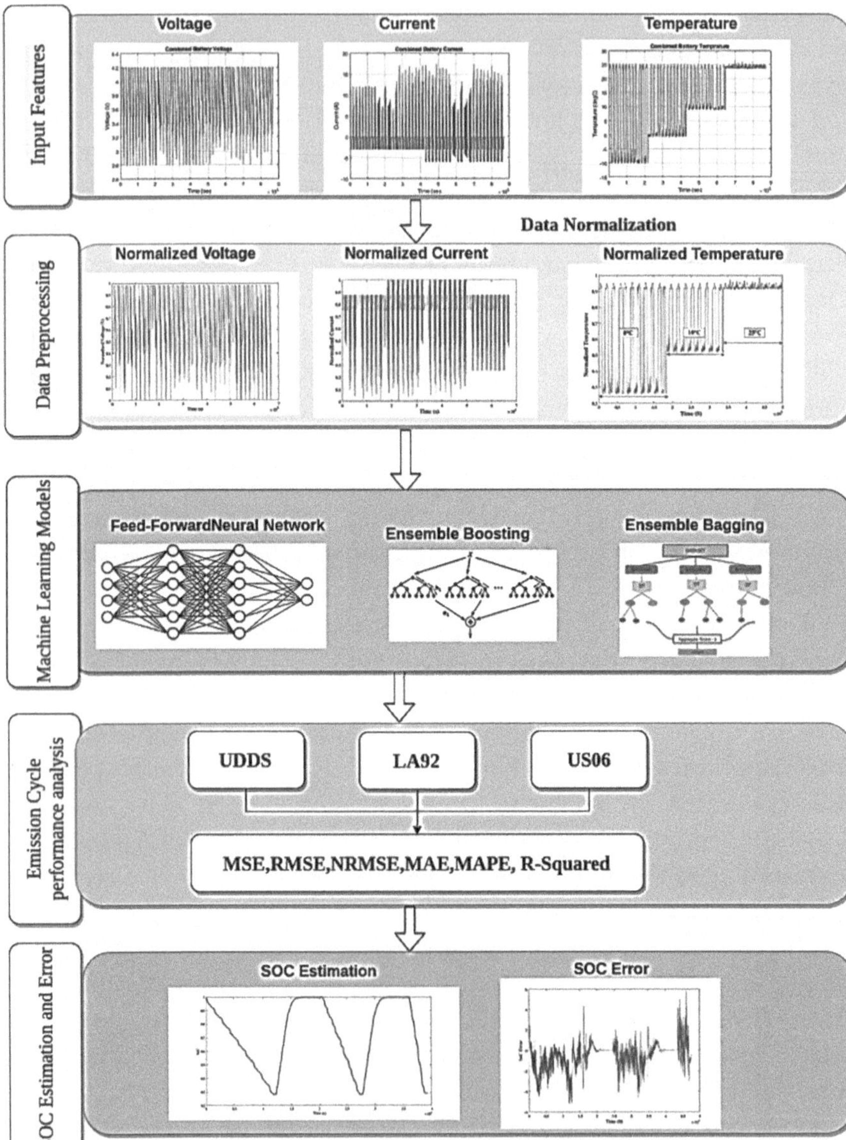

Fig. 1 The methodology adopted in the study for SOC prediction.

2 Materials and method

2.1 Dataset description

The dataset of the study was obtained from McMaster University in Hamilton, Ontario [38]. In the SOC estimator design process, these data are used.

The specification of the battery is listed in Table 1. The dataset is split into training, validation, and testing, as shown in Table 2. The SOC data are calculated by dividing the Ah data by the nominal cell capacity. The obtained battery datasets are trained, validated, and tested using the matrix laboratory (MATLAB 2020b version) with a campus-wide license. The toolboxes utilized to implement the same are the neural network toolbox, the regression toolbox, statistics, and the fitting toolbox.

3 Results and discussion

The SOC estimation of the battery for the EV is the critical parameter of the operation. The step involved in the analysis of the battery is presented in Fig. 1. The dataset was obtained from McMaster University in Hamilton, Ontario, Canada. The features selected as the battery parameters are voltage, current, and temperature concerning time. The obtained battery voltage, battery current, and battery temperature from the datasets are normalized due to reduced fluctuation during training and speeding up the training process. The EV speed characteristics are considered in this study for the analysis. The emission cycles use a UDDS, the LA92 cycle, and SFTP for research to check the accuracy of the proposed SOC estimation process. The battery SOC is predicted using various algorithms such as the feedforward neural network, ensemble bagging, and ensemble boosting. To analyze the performance of the proposed algorithms, different regression metrics are adopted such as root mean square error (RMSE), mean square error (MSE),

Table 1 Battery dataset specification.

Parameter	Specification
Battery	LG HG2
Capacity	3 Ah
Nominal voltage	3.6 V
Energy density	240 (Wh/kg)
Format	Cylindrical 18,650
Test profiles	UDDS, LA92, SFTP, CC

Table 2 Distribution of dataset.

Dataset	Training	Validation	Testing
LG HG2	669,956	39,293	44,284

normalized root mean square error (NRMSE), mean absolute error (MAE), mean fundamental percentage error (MAPE), and R squared.

The operating conditions of UDDS, LA92, and SFTP can illustrate the actual usage of power batteries and are also used to check the precision of the estimate of the SOC battery. Under UDDS, LA92, and SFTP, the car also has to speed up or split, and the current battery shifts relatively. The input feature of the normalized battery voltage is shown in Fig. 2. The variation in the battery current is shown in Fig. 3. If the current rate is out of control, the limit value should be substituted. The battery cell's current and voltage are determined during the test in real time by the current and voltage sensors. The battery surface is connected to a thermocouple whose other end is associated with the battery module test system to transmit temperature information. The test assumes that the EV's lithium battery is charged in a fixed environment and operates at varying ambient temperatures. The battery is then loaded at a 10°C (15.5 A) rate at 25°C to the maximum limit voltage (4.2 V). As the temperature considered ranges from 0°C to 25°C, the typical lithium battery cell testing temperatures are 0°C, 10°C, and 25°C, as shown in Fig. 4. The battery module test device computes SOC reference data. The surface temperatures of the initial battery cell are almost equal to ambient temperatures of 25°C periods. The battery temperature increases differently

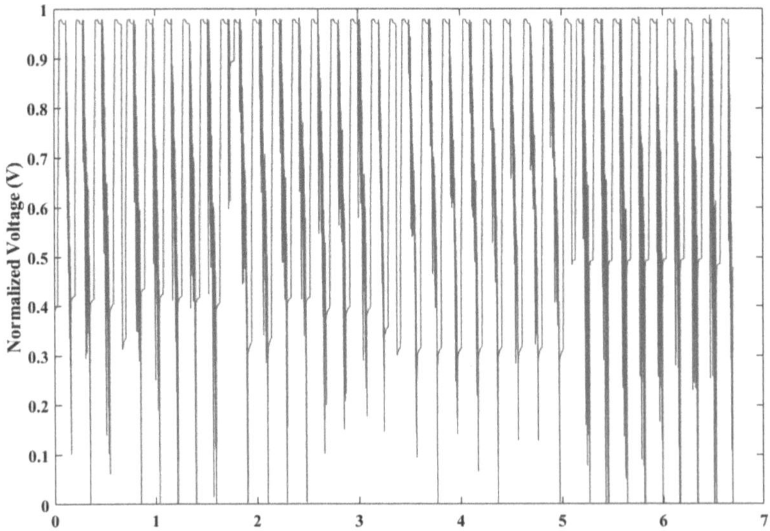

Fig. 2 Battery voltage curve during operation of UDDS, LA92, and SFTP emission cycle mode.

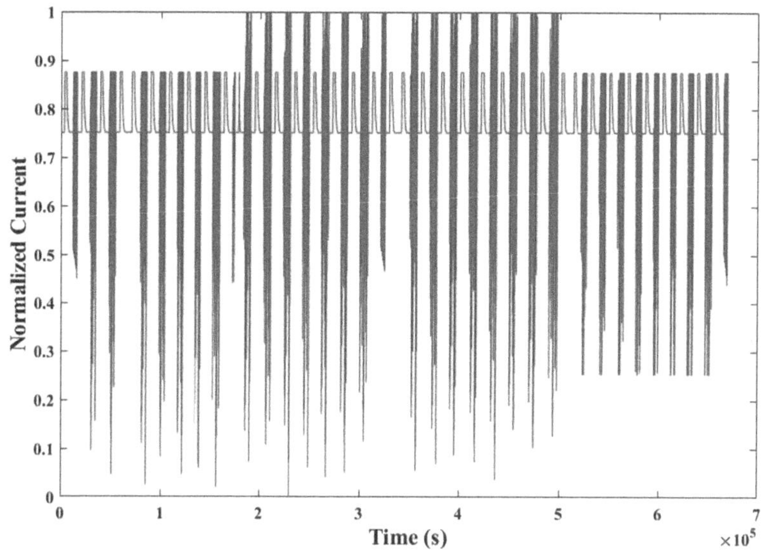

Fig. 3 Battery current curve during operation of UDDS, LA92, and SFTP emission cycle mode.

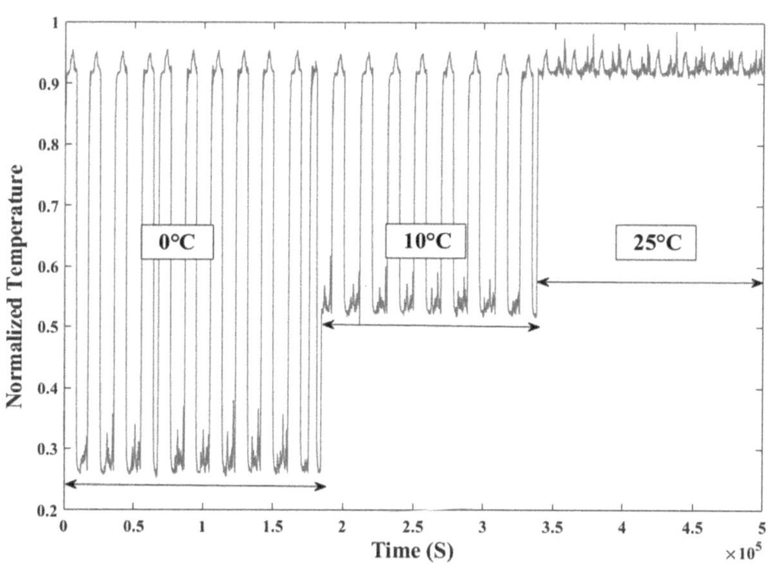

Fig. 4 Battery temperature at various operating conditions.

if the current rate of discharge is high. The temperature and temperature changes are not similar. The most noticeable thing is a rise of 10°C, but the battery temperature range is reasonably fixed within a single operating state of the UDDS, LA92, and SFTP. The battery temperature is elevated to a higher level if the SOC level is reduced.

The SOC estimates of the effects of UDDs, LA92, and SFTP during operating temperatures of 0°C for algorithms FNN, EBOG, and EBAG are shown in Fig. 5. MAE, RMSE, and SD are also selected and specified in Table 3 to analyze SOC estimation errors. As shown in Fig. 5, with different algorithms, the calculated SOC results perform very well at 0°C. Still, without considering the temperature factor during UDDS mode, the SOC estimation error increases over time. The results of the SOC calculation show a more apparent difference of 0°C.

In contrast, the FNN evaluation method results in a minor error, as the battery discharge characteristics have caused substantial changes at low temperatures. During the whole test, the SOC estimate during the LA 92 error obtained with the proposed combination method can be easily maintained and remain at 0°C at 1.5%. The SOC error of the EBAG estimator is 3%, similar to the previous one. The mistake of the EBOG method increases over time for the SFTP drive loop. The suggested estimation method, therefore, allows the estimator to better track the real SOC value. Within the acceptable spectrum of errors, the FNN estimator also has many advantages. Measurement processes that do not understand the temperature's impact cause unavoidable errors and result in mistakes accumulating during the discharge phase.

The SOC estimates of the effects of UDDS, LA92, and SFTP at an operating temperature of 10°C for the FNN, EBOG, and EBAG algorithms are shown in the test in Fig. 6. For analysis of SOC estimation errors, MAE, RMSE, and SD are also chosen and specified in Table 4. As seen in Fig. 7, the calculated SOC results essentially perform very well at 10°C with different algorithms. Still, the SOC estimation error increases over time without considering the temperature factor during UDDS mode. The results of the SOC calculation show a more apparent difference of 10°C.

In contrast, the FNN evaluation method results in a minor error, as the battery discharge characteristics have created substantial changes at low temperatures. During the entire test, the SOC estimate obtained during the LA92 error with the proposed combination method can be easily preserved and remain within 0.7% at 10°C. The EBAG estimator's SOC error stands at 2% compared to the previous one. For the SFTP drive loop, the EBOG

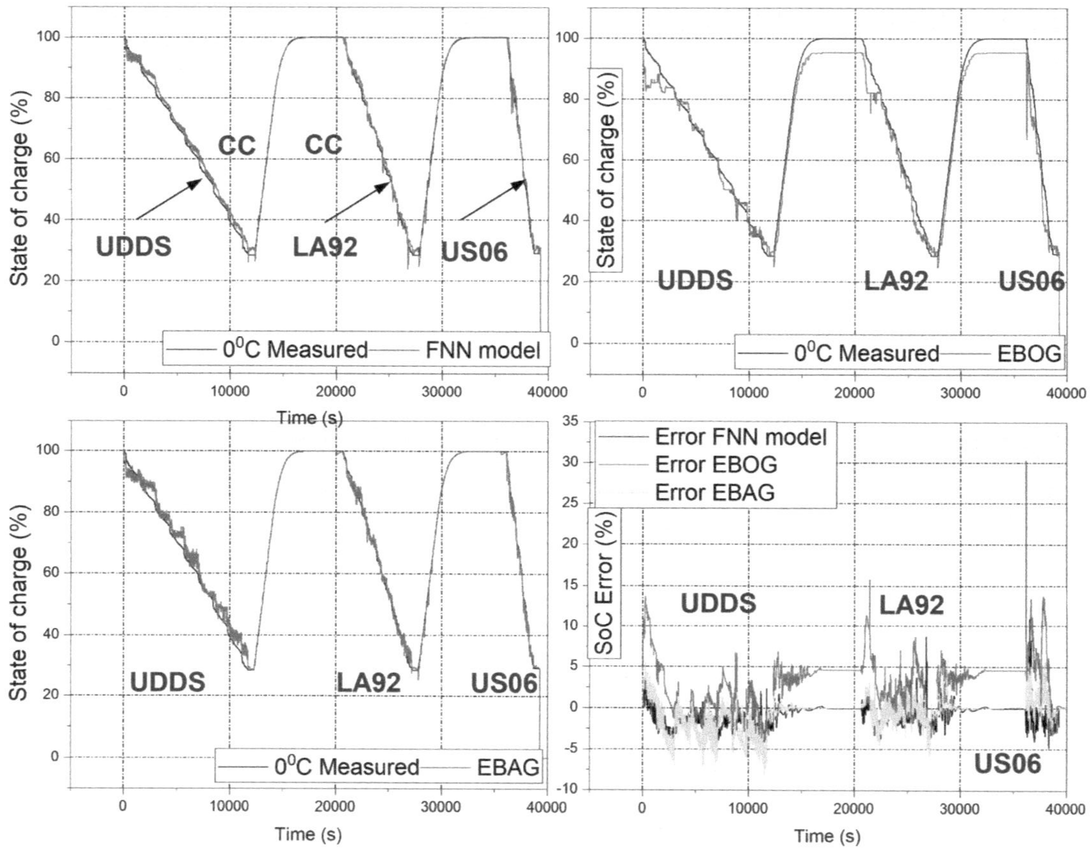

Fig. 5 Prediction SOC of battery estimation at 0°C.

Table 3 Performance metrics of battery operating at 0°C.

Drive cycles	Model	MSE	RMSE	NRMSE	MAE	MAPE	R^2
UDDS	FNN	0.03190	0.17861	0.24127	0.01595	0.00020	0.99680
	EBOG	0.02658	0.16304	0.22439	0.01329	0.00017	0.97785
	EBAG	0.02565	0.16304	0.22767	0.01329	0.00017	0.99018
LA92	FNN	0.01677	0.12951	0.18108	0.00414	0.00021	0.99600
	EBOG	0.62118	0.24920	0.43580	0.01533	0.00080	0.97580
	EBAG	0.01015	0.10070	0.15610	0.00251	0.00010	0.99470
US06	FNN	0.02174	0.14744	0.20923	0.01087	0.00060	0.99510
	EBOG	0.05144	0.22680	0.34578	0.05262	0.00327	0.96327
	EBAG	0.00403	0.06347	0.09019	0.00412	0.00023	0.99537

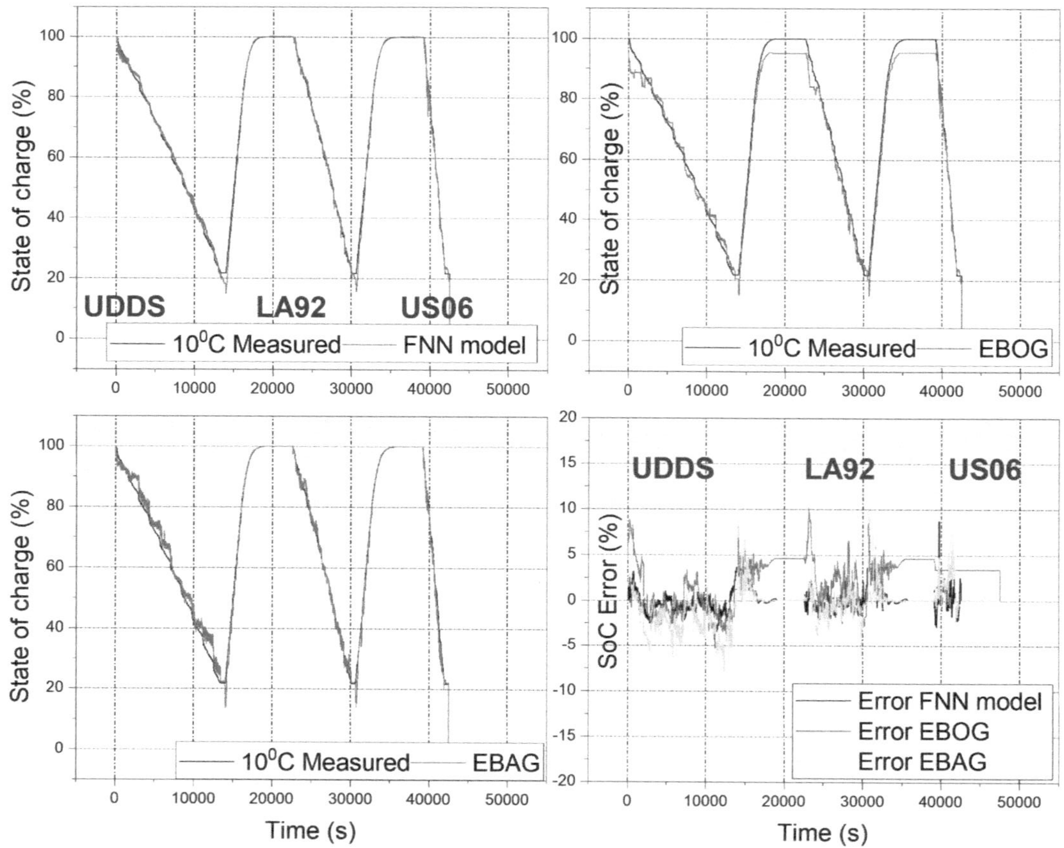

Fig. 6 Prediction SOC of battery estimation at 10°C.

Table 4 Performance metrics of battery operating at 10°C.

Drive cycles	Model	MSE	RMSE	NRMSE	MAE	MAPE	R^2
UDDS	FNN	0.02700	0.16432	0.20032	0.01350	0.00016	0.99585
	EBOG	0.02051	0.14321	0.19104	0.01025	0.00013	0.98362
	EBAG	0.01707	0.13066	0.16541	0.00854	0.00010	0.99199
LA92	FNN	0.00696	0.08344	0.10369	0.00348	0.00007	0.99890
	EBOG	0.04193	0.20477	0.28023	0.02097	0.00042	0.99266
	EBAG	0.01004	0.10022	0.12789	0.00502	0.00010	0.99723
US06	FNN	0.00159	0.03992	0.02473	0.00080	0.00004	0.99664
	EBOG	0.03792	0.19472	0.00623	0.01896	0.00107	0.99324
	EBAG	0.03333	0.18257	0.00598	0.01667	0.00094	0.99388

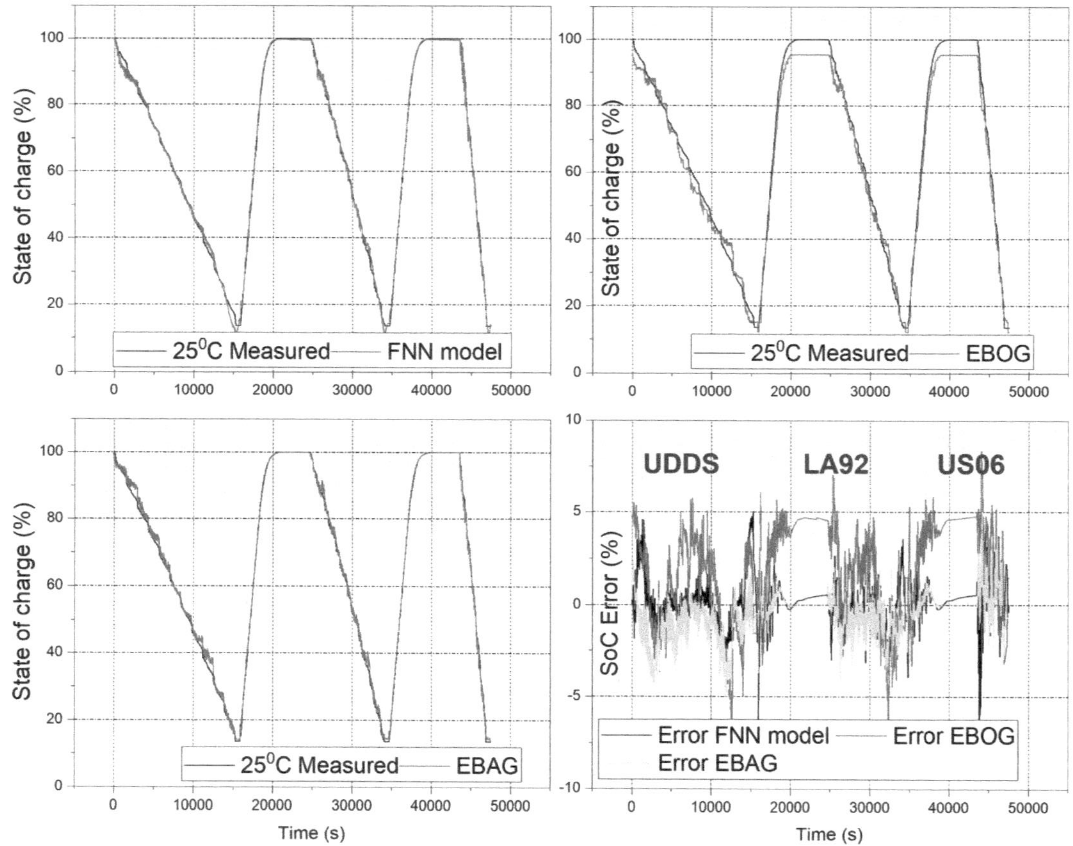

Fig. 7 Prediction SOC of battery estimation at 25°C.

method error gets more significant over time. The suggested method of estimation, therefore, allows the estimator to better track the real SOC value. FNN is indicated due to its efficient training.

FNN methods can handle nonlinear systems, but the drawbacks are overfitting, local minima, random initial data, long training data requirements, and increased complexity to multilayered structures. Boosting is suitable for handling the missing data in the dataset. Owing to the increased complexity of the algorithm, boosting is challenging to implement in real time. The versatility of these techniques results in multiple parameter statistics that directly affect the model's behavior. Bagging has the advantages of offering reliability and improves the algorithm's accuracy for machine learning used in predictive regression. But in this case, if it has not been accurately modeled, it may result in high bias. It, therefore, can lead to underfitting because measurement processes that do not understand the temperature impact cause unavoidable errors and result in mistakes accumulating during the period of discharge.

The SOC estimates of the effects of UDDS, LA92, and SFTP during operating temperatures of 25°C for algorithms FNN, EBOG, and EBAG are shown in Fig. 6. MAE, RMSE, and SD are also selected and specified in Table 5 to analyze SOC estimation errors. As shown in Fig. 7, with different algorithms, the calculated SOC results perform very well at 25°C. Still, without considering the temperature factor during UDDS mode, the SOC estimation error increases over time. The results of the SOC calculation show a more apparent difference of 25°C. In contrast, the FNN evaluation method results in a minor error, as the battery discharge characteristics have caused substantial changes at low temperatures. During the whole test, the SOC estimate during the LA92 error obtained with the proposed combination method can be easily maintained and remain at 25°C at 0.2%. The SOC error of the EBAG estimator is 0.2%, similar to the previous one. The error of the EBOG method increases over time for the SFTP drive cycle. The suggested estimation method, therefore, allows the estimator to better track the real SOC value. Within the acceptable spectrum of errors, the FNN estimator also has many advantages because measurement processes that do not understand the temperature's impact cause unavoidable errors and accumulate during the discharge phase.

The EBOG algorithm appears to have a higher SOC error while at average temperatures, the FNN will decrease the estimated error. Under the suggested protocol, the mistake shows almost the best results at moderate temperatures.

Table 5 Performance metrics of battery operating at 25°C.

Drive cycles	Model	MSE	RMSE	NRMSE	MAE	MAPE	R^2
UDDS	FNN	0.00323	0.05680	0.06942	0.00161	0.00002	0.99799
	EBOG	0.01478	0.12159	0.16220	0.00739	0.00009	0.98802
	EBAG	0.02355	0.15346	0.19429	0.01177	0.00013	0.99696
LA92	FNN	0.00980	0.09900	0.03113	0.00490	0.00009	0.99808
	EBOG	0.01850	0.13602	0.01316	0.00925	0.00018	0.99236
	EBAG	0.01668	0.12914	0.01279	0.00834	0.00016	0.99877
US06	FNN	0.00030	0.00545	0.00465	0.01676	0.00000	0.99529
	EBOG	0.05403	0.23245	0.03137	0.03049	0.00141	0.99506
	EBAG	0.01228	0.11080	0.01688	0.00693	0.00031	0.99693

4 Conclusions

The change in the battery system's ambient temperature would ultimately influence the SOC value in the current operating conditions of electric vehicles. This study analyzes the relationship between the battery's power and the temperature concerning the emission cycle. It then proposes a feedforward neural network, ensemble boosting, and ensemble bagging based on the temperature model.

References

[1] H. Rahimi-Eichi, U. Ojha, F. Baronti, M.-Y. Chow, Battery management system: an overview of its application in the smart grid and electric vehicles, IEEE Ind. Electron. Mag. 7 (2) (2013) 4–16, https://doi.org/10.1109/mie.2013.2250351 (Online).
[2] P. Ramanan, K. Kalidasa Murugavel, A. Karthick, K. Sudhakar, Performance evaluation of building-integrated photovoltaic systems for residential buildings in southern India, Build. Serv. Eng. Res. Technol. (2019), 014362441988174.
[3] L.J. Aaldering, J. Leker, C.H. Song, Analysis of technological knowledge stock and prediction of its future development potential: the case of lithium-ion batteries, J. Clean. Prod. 223 (2019) 301–311, https://doi.org/10.1016/j.jclepro.2019.03.174 (Online).
[4] X. Feng, M. Ouyang, X. Liu, et al., Thermal runaway mechanism of lithium ion battery for electric vehicles: a review, Energy Storage Mater. 10 (2018) 246–267.
[5] Q. Zhang, D. Wang, B. Yang, X. Cui, X. Li, Electrochemical model of lithium-ion battery for wide frequency range applications, Electrochim. Acta 343 (2020) 136094.
[6] A. Bhattacharjee, R.K. Mohanty, A. Ghosh, Design of an optimized thermal management system for Li-ion batteries under different discharging conditions, Energies (2020), https://doi.org/10.3390/en13215695 (Online).
[7] Y. Wang, D. Yang, X. Zhang, Z. Chen, Probability based remaining capacity estimation using data-driven and neural network model, J. Power Sources 315 (2016) 199–208.
[8] S. Li, H. He, J. Li, Big data driven lithium-ion battery modeling method based on SDAE-ELM algorithm and data pre-processing technology, Appl. Energy 242 (2019) 1259–1273.
[9] V. Thangavel, O.X. Guerrero, M. Quiroga, et al., A three dimensional kinetic Monte Carlo model for simulating the carbon/sulfur mesostructural evolutions of discharging lithium sulfur bat- teries, Energy Storage Mater. 24 (2020) 472–485, https://doi.org/10.1016/j.ensm.2019.07.011 (Online).
[10] R. Xiong, L. Li, Q. Yu, Q. Jin, R. Yang, A set membership theory based parameter and state of charge co-estimation method for all-climate batteries, J. Clean. Prod. 249 (2020) 119380.
[11] Q. Lin, J. Wang, R. Xiong, W. Shen, H. He, Towards a smarter battery management system: a critical review on optimal charging methods of lithium ion batteries, Energy (2019), https://doi.org/10.1016/j.energy.2019.06.128 (Online).
[12] Q. Liu, J. Liu, W. Le, Z. Guo, Z. He, Data-driven intelligent location of public charging stations for electric vehicles, J. Clean. Prod. 232 (2019) 531–541.
[13] X. Hu, L. Xu, X. Lin, M. Pecht, Battery lifetime prognostics, Joule 4 (2020) 310–346.
[14] F. Yang, X. Song, F. Xu, K.-L. Tsui, State-of-charge estimation of lithium-ion batteries via long short-term memory network, IEEE Access 7 (2019) 53792–53799, https://doi.org/10.1109/access.2019.2912803 (Online).

[15] J. Remmlinger, M. Buchholz, M. Meiler, P. Bernreuter, K. Dietmayer, State-of- health monitoring of lithium-ion batteries in electric vehicles by on-board internal resistance estimation, J. Power Sources 196 (12) (2011) 5357–5363, https://doi.org/10.1016/j.jpowsour.2010.08.035 (Online).
[16] A. Rauh, L. Senkel, H. Aschemann, Sensitivity-based state and parameter estimation for fuel cell systems, IFAC Proc. Vol. 45 (13) (2012) 57–62, https://doi.org/10.3182/20120620-3-dk-2025.00071 (Online).
[17] O. Erdinc, B. Vural, M. Uzunoglu, A dynamic lithium-ion battery model considering the effects of temperature and capacity fading, in: International Conference on Clean Electrical Power, 2009, pp. 383–386.
[18] J. Reuter, E. Mank, H. Aschemann, A. Rauh, Battery state observation and condition monitoring using online minimization, in: 21st International Conference on Methods and Models in Automation and Robotics (MMAR), 2016, pp. 1223–1228.
[19] J.N. Shen, J.J. Shen, Y.J. He, Z.F. Ma, Accurate state of charge estimation with model mismatch for Li-ion batteries: a joint moving horizon estimation approach, IEEE Trans. Power Electron. 34 (2019) 4329–4342.
[20] K. Liu, Y. Shang, Q. Ouyang, W.D. Widanage, A data-driven approach with uncertainty quantification for predicting future capacities and remaining useful life of lithium-ion battery, IEEE Trans. Ind. Electron. 68 (4) (2021) 3170–3180, https://doi.org/10.1109/tie.2020.2973876 (Online).
[21] X. Dang, L. Yan, K. Xu, et al., Open-circuit voltage-based state of charge estimation of lithium-ion battery using dual neural network fusion battery model, Electrochim. Acta 188 (2016) 356–366.
[22] S. Tong, J.H. Lacap, J.W. Park, Battery state of charge estimation using a load-classifying neural network, J. Energy Storage 7 (2016) 236–243, https://doi.org/10.1016/j.est.2016.07.002 (Online).
[23] F. Sun, R. Xiong, H. He, A systematic state-of-charge estimation framework for multi-cell battery pack in electric vehicles using bias correction technique, Appl. Energy 162 (2016) 1399–1409, https://doi.org/10.1016/j.apenergy.2014.12.021 (Online).
[24] E. Chemali, P.J. Kollmeyer, M. Preindl, R. Ahmed, A. Emadi, Long short-term memory networks for accurate state-of-charge estimation of Li-ion batteries, IEEE Trans. Ind. Electron. 65 (8) (2018) 6730–6739, https://doi.org/10.1109/tie.2017.2787586 (Online).
[25] E. Chemali, P.J. Kollmeyer, M. Preindl, A. Emadi, State-of-charge estimation of Li-ion batteries using deep neural networks: a machine learning approach, J. Power Sources 400 (2018) 242–255, https://doi.org/10.1016/j.jpowsour. 2018.06.104 (Online).
[26] J. Chen, Q. Ouyang, C. Xu, H. Su, Neural network-based state of charge observer design for lithium-ion batteries, IEEE Trans. Control Syst. Technol. 26 (1) (2018) 313–320, https://doi.org/10.1109/tcst.2017.2664726 (Online).
[27] M.S.H. Lipu, M.A. Hannan, A. Hussain, et al., State of charge estimation for lithium-ion battery using recurrent NARX neural network model based lighting search algorithm, IEEE Access 6 (2018) 28150–28161, https://doi.org/10.1109/access.2018.2837156 (Online).
[28] B. Xia, D. Cui, Z. Sun, et al., State of charge estimation of lithium-ion batteries using optimized Levenberg-Marquardt wavelet neural network, Energy 153 (2018), https://doi.org/10.1016/j.energy.2018.04.085. Elsevier BV. [Online].
[29] F. Cadini, C. Sbarufatti, F. Cancelliere, M. Giglio, State-of-life prognosis and diagnosis of lithium-ion batteries by data-driven particle filters, Appl. Energy 235 (2019) 661–672, https://doi.org/10.1016/j.apenergy.2018.10.095 (Online).
[30] C. Burgos, D. Sáez, M.E. Orchard, R. Cárdenas, Fuzzy modelling for the state-of-charge estimation of lead-acid batteries, J. Power Sources 274 (2015) 355–366, https://doi.org/10.1016/j.jpowsour.2014.10.036 (Online).

[31] T.O. Ting, K.L. Man, E.G. Lim, M. Leach, Tuning of kalman filter parameters via genetic algorithm for state-of-charge estimation in battery management system, Scientific World Journal 2014 (2014) 1–11, https://doi.org/10.1155/2014/176052 (Online).
[32] L. Chen, Z. Wang, Z. Lu, et al., A novel state-of-charge estimation method of lithium-ion batteries combining the grey model and genetic algorithms, IEEE Trans. Power Electron. 33 (2018) 8797–8807.
[33] J. Meng, G. Luo, F. Gao, Lithium polymer battery state-of-charge estimation based on adaptive unscented kalman filter and support vector machine, IEEE Trans. Power Electron. 31 (3) (2016) 2226–2238, https://doi.org/10.1109/tpel.2015.2439578 (Online).
[34] Z. Deng, Y. Lin, Y. Cai, H. Deng, L. Sun, Online available capacity prediction and state of charge estimation based on advanced data-driven algorithms for lithium iron phosphate battery, Energy 112 (2016), https://doi.org/10.1016/j.energy.2016.06.130. Elsevier BV. [Online].
[35] C. Vidal, P. Kollmeyer, E. Chemali, A. Emadi, Li-ion battery state of charge estimation using long short-term memory recurrent neural network with transfer learning, in: 2019 IEEE Transportation Electrification Conference and Expo (ITEC), 2019.
[36] J. Kalawoun, K. Biletska, F. Suard, M. Montaru, From a novel classification of the battery state of charge estimators toward a conception of an ideal one, J. Power Sources 279 (2015) 694–706, https://doi.org/10.1016/j.jpowsour. 2015.01.038 (Online).
[37] D. Zhou, W. Zheng, F. Ping, X. Pan, Research on online estimation of available capacity of lithium batteries based on daily charging data, J. Power Sources 451 (2020) 227713, https://doi.org/10.1016/j.jpowsour.2020.227713 (Online).
[38] D. Yang, X. Zhang, R. Pan, Y. Wang, Z. Chen, A novel Gaussian process regression model for state-of-health estimation of lithium-ion battery using charging curve, J. Power Sources 384 (2018) 387–395.

CHAPTER 7.5

Modified SVPWM technique for speed control of PMSM in electric vehicles

Ashish Kumar[a], Ram Niwash Mahia[a], and Om Prakash Mahela[b]
[a]Department of Electrical Engineering, NIT Hamirpur, Hamirpur, India
[b]Power System Planning Division, Rajasthan Rajya Vidyut Prasaran Nigam Ltd., Jaipur, Rajasthan, India

Contents

1 Introduction

Traffic is increasingly expanding in metropolitan areas, destroying the environment at an alarming rate. With this in mind, there has been an increase in demand for environmentally friendly automobiles that are also fuel efficient, as fuel is rapidly depleting, leading to an increase in gasoline prices. The automobile manufacturers have reduced greenhouse gas emissions such as carbon dioxide, carbon monoxide, hydrocarbons, and others to some extent, they have yet to build a zero-emission vehicle until they make electric vehicles.

Different types of electric vehicles (EVs) such as hybrid electric vehicles (HEVs), plug-in hybrid electric vehicles (PHEVs), and fuel cell vehicles are

Active Electrical Distribution Network
https://doi.org/10.1016/B978-0-323-85169-5.00002-2

all being developed by automakers. Electric components and electronic features are used in these vehicles, which improve vehicle performance, reduce emissions, enhance fuel economy, and increase passenger comfort and safety. They consist of many power electronics components such as converters (DC-DC converters, for example) so that they can match the fuel cell voltage to the battery pack voltage. In addition, an inverter with desirable control system techniques is required to deal with even the most unfavorable environmental conditions. Furthermore, power electronics improves the overall system reliability, reduces size and costs, etc. Furthermore, electric motors are the major part of the vehicle that decides the dynamic performance of cars, and converters are used to manage the vehicle's operating parameters.

Induction motors are used in most industrial applications, and they are mainly used in areas where field weakening is a major technique to avoid drive oversizing. However, permanent magnet synchronous motors (PMSM) are becoming the favorite for electric vehicles [1–3]. They have several properties such as a high torque-to-inertia ratio, greater power factor, higher efficiency, and robustness. Different modulation techniques can be implemented to control PMSM but space vector PWM (SVPWM) is considered to be the best method, as it has higher DC bus utilization. It is also easy to implement with much less commutation time. SVPWM output is fed to voltage source inverters (VSI), which provide power to the PMSM. SVPWM is used to vary the voltage and frequency to a desired value with less total harmonic distortion (THD). The speed of PMSM can be varied by changing the voltage and frequency. For this work, SVPWM is best suited [4–6].

2 Hybrid electric vehicle

Onboard HEVs, energy is supplied from two or more sources. Other energy sources, such as a flywheel or a battery, can be used. A hybrid vehicle can be powered by an engine, an ultracapacitor, a battery, a fuel cell, or other things. The vehicle is built in such a way that two or more of these energy sources can be used at the same time. An internal combustion engine (ICE) and a motor are integrated in this approach to provide the best performance possible.

2.1 Series hybrid vehicles

A series HEV is basically an electric car that can use a power source to charge onboard. In most cases, an engine is connected to a generator and the generator provides the necessary electricity to charge the cells. It is also feasible to build such a system as the generator serves even in load-leveling device as well as a source of thrust power. In this situation, the battery size can be small, but the generator and engine sizes would need to be increased.

2.2 Parallel hybrid vehicles

Parallel HEVs can provide the most cost-effective solution while also allowing for the use of existing production capacity for batteries, motors, and engines. The parallel structure of the HEV system is designed in such a way that it allows the ICE and the electric motor to produce desirable torque to the driving wheels, either in a separate way or concurrently. The ICE is aided by increased torque in working areas where the engine isn't as efficient as it could be.

3 Space vector pulse width modulation

The three-phase axes of a balanced system are R, Y, and B. The V_{RN}, V_{YN}, and V_{BN} are three-phase line-to-neutral voltages applied along the respective axes, which can be transformed into a "voltage source vector," Vs.

$$Vs(t) = V_{RN}(t) + V_{YN}(t)\, e^{j(2\Pi/3)} + V_{BN}(t)\, e^{j(4\Pi/3)} \tag{1}$$

(a and b) are the voltage space vector that is resolved along two mutually perpendicular axes. The (a-axis) is along the R-phase axis and another axis (b-axis) that leads the axis by 90 degrees. V_a is the component of the Vs vector along the a-axis and V_b is the component of the Vs vector along the b-axis [7–10].

$$V_a(t) = \frac{3}{2} V_{RN}(t) \tag{2}$$

$$V_b(t) = \frac{\sqrt{3}}{2}\left(V_{YN}(t) - V_{BN}(t)\right) \tag{3}$$

3.1 Voltage vectors of a voltage source inverter

Voltage source inverters (VSI) have a single pole-double throw (SPDT) switch in each phase (leg). Voltage measured from a pole to the midpoint of the DC bus is termed "pole voltage" [11–13]. V_{RO}, V_{YO}, and V_{BO} are the pole voltages of the R, Y, and B phases, respectively. At a given instant, when the top devices of the phase are ON, the pole voltage is equal to $+(V_{DC}/2)$. When the bottom devices are ON, the pole voltage is equal to $-(V_{DC}/2)$.

The instantaneous line-to-line voltages ($V_{RY}(t)$, $V_{YB}(t)$, $V_{BR}(t)$) are obtained using pole voltages, as shown in Eqs. (4a)–(4c).

$$V_{RY}(t) = V_{RO}(t) - V_{YO}(t) \tag{4a}$$

$$V_{YB}(t) = V_{YO}(t) - V_{BO}(t) \tag{4b}$$

$$V_{BR}(t) = V_{BO}(t) - V_{RO}(t) \tag{4c}$$

The inverter feeds a three-phase PMSM. Assuming a balanced and star-connected motor, the relations between instantaneous line-to-neutral voltages and instantaneous line-to-line (L-L) voltages at motor terminals are expressed below.

$$V_{RN}(t) = \frac{1}{3}(V_{RY}(t) - V_{BR}(t)) \tag{5a}$$

$$V_{YN}(t) = \frac{1}{3}(V_{YB}(t) - V_{RY}(t)) \tag{5b}$$

$$V_{BN}(t) = \frac{1}{3}(V_{BR}(t) - V_{YB}(t)) \tag{5c}$$

The inverter has 2^3 or eight switch states as either top or bottom devices can be ON in every phase. There are two states that produces zero output voltages (+ + + (7), − − − (0)); they are known as zero states. The other six states that produce nonzero output (+ − − (1), + + − (2), − + − (3), − + + (4), − − + (5), + − + (6)) are called active states. These produce voltages that lead the power transfer between the DC bus and the machine.

There are different sets of pole voltages for different states. For example, if we consider a state (+ − −), then the corresponding pole voltages are:

$$V_{RO} = +(V_{DC}/2), V_{YO} = -(V_{DC}/2), V_{BO} = -(V_{DC}/2)$$

We can calculate the corresponding line-to-neutral voltages at the motor terminals by using Eqs. (4)–(5). Space vector transformation, which is defined in Eq. (1), can be applied on line-to-neutral voltages corresponding to each inverter state, as shown in Fig. 1. The space vectors that corresponds to six active states are called active vectors whose magnitude is equal to V_{DC}. There are two voltage space vectors that have zero magnitude, and they are called zero vectors. When we join the head points of active vectors, it forms

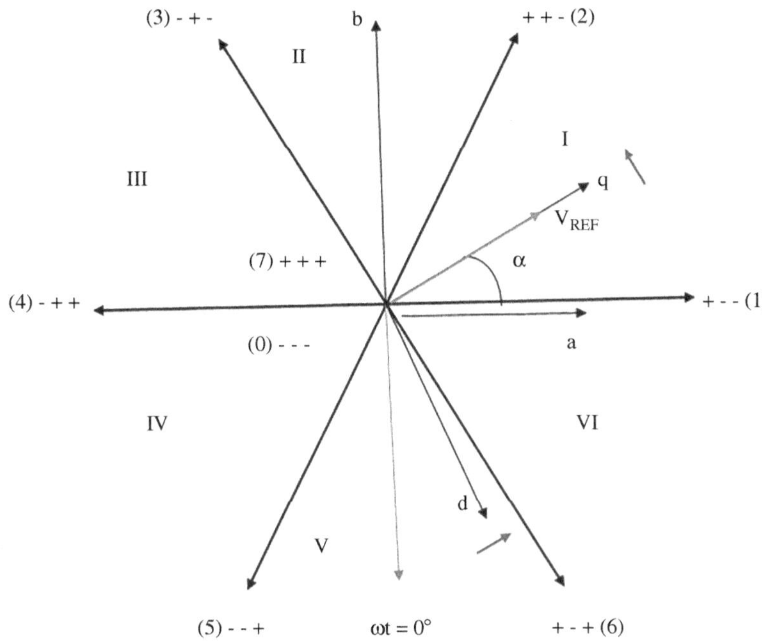

Fig. 1 Voltage space vectors of a voltage source inverter.

a regular hexagon. Zero vectors are located at the center of the hexagon. The space vector plane is divided into six sectors (I, II, III, IV, V, and VI) by the six active vectors, as shown in Fig. 1.

4 Permanent magnet synchronous motor

PMSMs are becoming very popular in industrial drives because of their effectiveness to produce variable-speed drives [14–17]. Several improvements are made in permanent magnet (PM) materials, particularly rare-earth magnets that have substantially advanced new research and applications. Traditional AC analysis and design methods on electrical machines are becoming more sophisticated, but more research is still required to establish an approach for analyzing, designing, and controlling PMSMs with systematic methodology.

A control approach for PMSMs is established in this paper. By investigating PMSMs with different structures of the rotor, the PMSM paradigm was created to operate at a large range of speeds while maintaining steady power. Tests were conducted to ensure the correctness of the optimized design on the constructed prototype motor. It built a technique and a control algorithm.

PMs are used in making electrical equipment for the following reasons [18]:

1. The field excitation system absorbs no electrical energy, hence there are no excitation losses, resulting in a significant boost in efficiency.
2. When compared to electromagnetic excitation, they have better output power per unit volume and better torque.
3. Motors with electromagnetic excitation have superior performance related to dynamics because of the higher magnetic flux density in the air gap.
4. They are simple to construct and easy to maintain.
5. For some equipment, the cost of ownership is reduced over time.

In the 20th century, cage induction motors (IM) were most popular in electric motors. IMs have recently been used in electrical drives as a result of rapid advances in power electronics and control technologies. In comparison to PM synchronous motors, they have less efficiency and a lower power factor (pf).

PM motors have become a more appealing solution for electrical drives than induction motors. Because of advancements in semiconductor drives, controlling PM motors has become easier than before and even more cost-efficient. It has the ability to operate the motor across a wide speed span while developing constant torque, a high power factor and efficiency.

5 Determination of switching time for each switch (S1 to S6) and switching pattern of each sector

When we consider sector 1, the switch (S1) conducts for the complete time period and switch (S4) are completely clamped. Switch (S3) conducts for T0 and T2 and in a similar way the rest of the switch conducts, as shown in Table 1. Determination of T0, T1, and T2 is expressed in Eqs. (6)–(8). The switching pattern of sector 1 is shown below in Fig. 2.

Table 1 Timing sequence of sector 1.

S1 = T0 + T1 + T2	S4 = 0
S3 = T0 + T2	S6 = T1
S5 = T0	S2 = T1 + T2

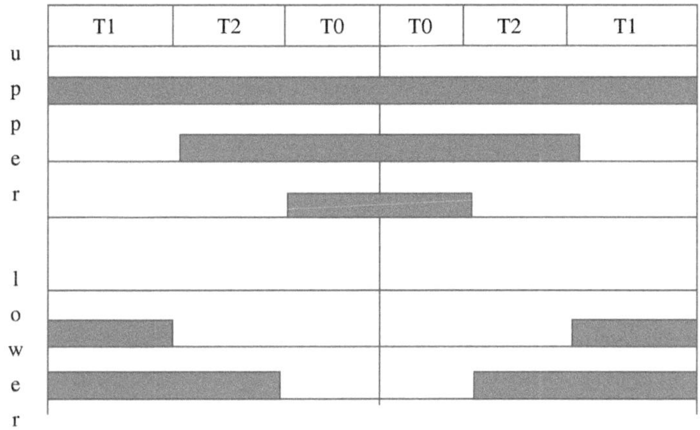

Fig. 2 Pictorial representation of the clamping timing sequence of sector 1.

In sector 2, the switch (S3) is conducting for all time, switch (S6) is clamped for all time, and the rest of the switches are conducting for different times, as shown in Table 2. The switching pattern of sector 2 is shown below in Fig. 3:

Table 2 Timing sequence of sector 2.

S1 = T0 + T1	S4 = T2
S3 = T0 + T1 + T2	S6 = 0
S5 = T0	S2 = T1 + T2

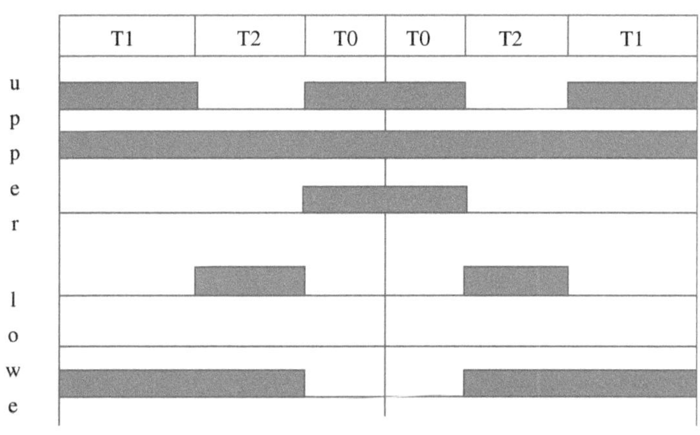

Fig. 3 Pictorial representation of the clamping timing sequence of sector 2.

In sector 3, switch (S3) is conducting for all time, switch (S6) is clamped for all time, and the rest of the switches are conducting for different times, as shown in Table 3. The switching pattern of sector 3 is shown below in Fig. 4:

In sector 4, switch (S5) is conducting for all time, switch (S2) is clamped for all time, and the rest of the switches are conducting for different times, as shown in Table 4. The switching pattern of sector 4 is shown below in Fig. 5:

In sector 5, switch (S5) is conducting for all time, switch (S2) is clamped for all time, and the rest of the switches are conducting for different times, as shown in Table 5. The switching witching pattern of sector 5 is shown below in Fig. 6:

Table 3 Timing sequence of sector 3.

S1 = T0	S4 = T1 + T2
S3 = T0 + T1 + T2	S6 = 0
S5 = T0 + T2	S2 = T1

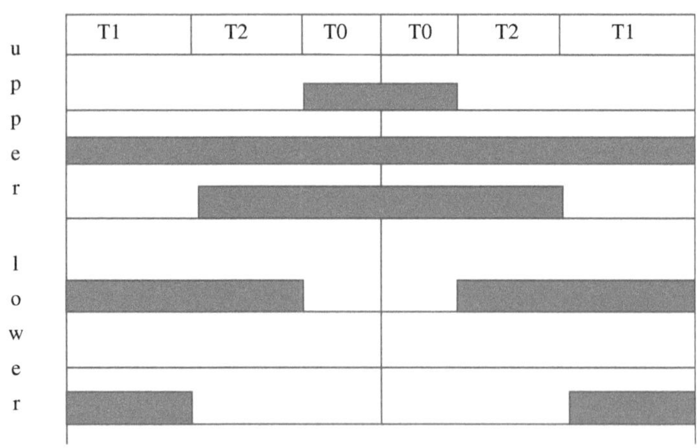

Fig. 4 Pictorial representation of the clamping timing sequence of sector 3.

Table 4 Timing sequence of sector 4.

S1 = T0	S4 = T1 + T2
S3 = T0 + T1	S6 = T2
S5 = T0 + T1 + T2	S2 = 0

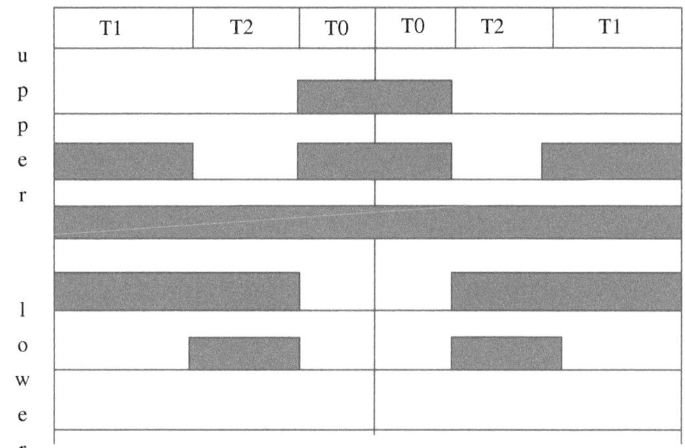

Fig. 5 Pictorial representation of the clamping timing sequence of sector 4.

Table 5 Timing sequence of sector 5.

S1 = T0 + T2	S4 = T1
S3 = T0	S6 = T1 + T2
S5 = T0 + T1 + T2	S2 = 0

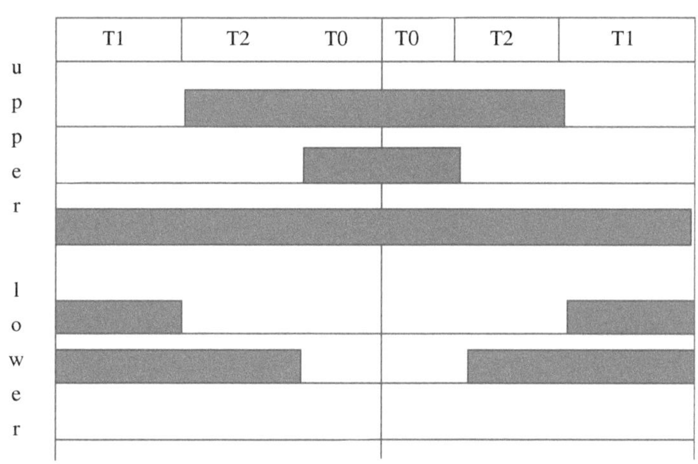

Fig. 6 Pictorial representation of the clamping timing sequence of sector 5.

In sector 6, switch (S1) is conducting for all time, switch (S4) is clamped for all time, and the rest of the switches are conducting for different time, as shown in Table 6. The switching pattern of sector 6 is shown below in Fig. 7:

Table 6 Timing sequence of sector 6.

S1 = T0 + T1 + T2	S4 = 0
S3 = T0	S6 = T1 + T2
S5 = T0 + T1	S2 = T2

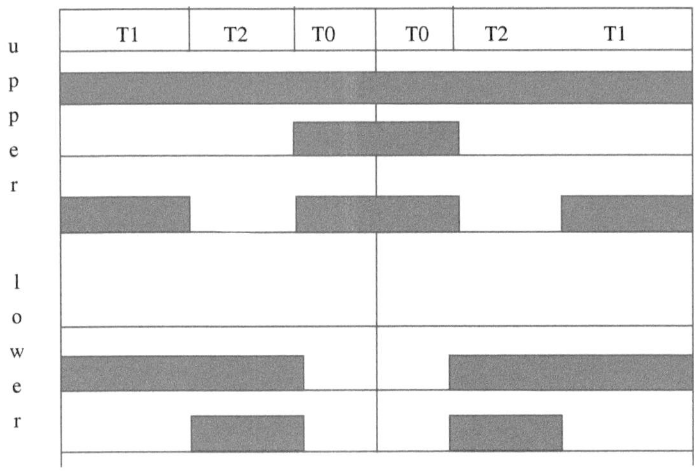

Fig. 7 Pictorial representation of the clamping timing sequence of sector 6.

6 Connection of ports

We obtain T1, T2, and T0 by the equations given below:

$$T_1 = \frac{\sqrt{3}\boldsymbol{T_z}|\boldsymbol{V_{ref}}|}{\boldsymbol{V_{dc}}}\left(\sin\left(\frac{n\Pi}{3}\right)\cos \boldsymbol{a} - \cos\left(\frac{n\Pi}{3}\right)\sin \boldsymbol{a}\right) \tag{6}$$

$$T_2 = \frac{\sqrt{3}\boldsymbol{T_z}|\boldsymbol{V_{ref}}|}{\boldsymbol{V_{dc}}}\left(-\cos \boldsymbol{a}\sin\left(\frac{n-1}{3}\Pi\right) + \sin \boldsymbol{a}\cos\left(\frac{n-1}{3}\boldsymbol{\Pi}\right)\right) \tag{7}$$

$$\mathbf{T_o = T_z - T_1 - T_2} \tag{8}$$

As the first sector is selected by angle identification, then it decides the time duration to operate as the time duration varies for a switch in different sectors. The connection of switches (S1, S2, and S3) is shown in Fig. 8. These are the upper leg switches. Lower leg switches are not considered as they are the reversal of upper leg switches.

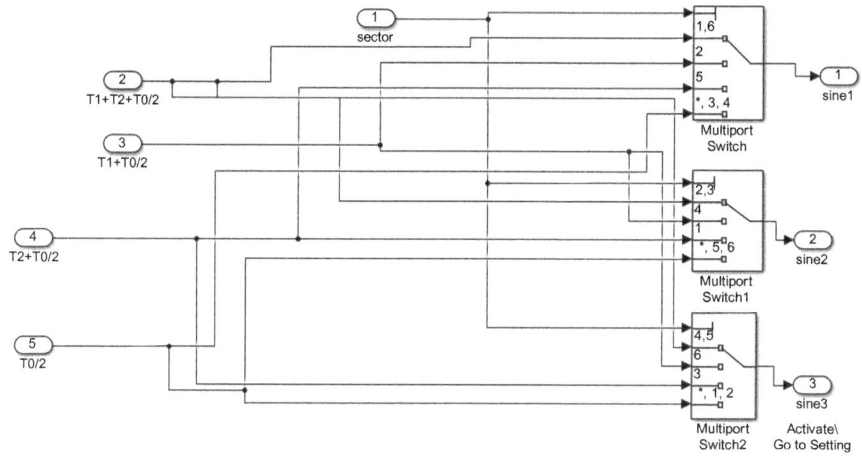

Fig. 8 Connection of ports for S1, S3, S5.

Let us consider a switch (S1). S1 operates for (T0 + T1 + T2) time duration in sector 1 and sector 6, so they are connected to the same port. S1 operates for (T0 + T1) time duration in sector 2, so it is connected to a different port. In the same manner, different sectors and their timing durations are allotted to different ports to produce a SVPWM clamping waveform. (000 – 100 – 110 – 111) (111 – 011 – 001 – 000) is the sequence followed for one sector.

7 Simulation and result

This section describes the output of a PMSM fed by a modified SVPWM inverter for a simulation study, including output voltage, torque, gating signal generation, current, and speed response. Simulation tools may perform dynamic simulations of motor drives in a visual environment, which can help with the development of novel systems. MATLAB Simulink's features include the ability to simulate complex dynamic systems, a graphical environment, and a large number of toolboxes. Simulink's simulation environment is extremely flexible and expandable, allowing for the creation of a collection of techniques for the complete examination of a different electric drive. The PMSM drive is simulated in the MATLAB Simulink environment. The inverter module, PMSM module, SVPWM module, coordinate transformation production

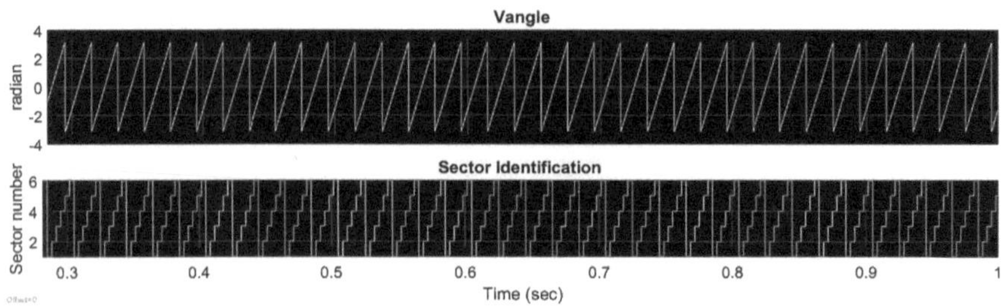

Fig. 9 Angle and sector.

module, PI controller, and many more are all independent functional modules that make up the PMSM control system. These elements are used in a simulation model of a modified SVPWM used in the PMSM control system.

The performance of the speed control of a PMSM drive system using a modified space vector PWM fed inverter has been designed. Clark's transformation is used to transform a balanced three-phase sinusoidal voltage with a 200 V magnitude and a 50 Hz frequency into a, d. and q frame.

V_{ref} has a magnitude of 200 V and the range of angle is from 3.14 to −3.14 rad, which is translated to degrees so that we can select a sector. The corresponding sector will be recognized based on the magnitude and angle, as shown in Fig. 9. Because the VSI has two levels and three legs, it has a total of six sectors, ranging from one to six, as indicated in Fig. 9. Following the identification of the sector, its unique dwell timing is computed, which serves as the foundation for switching duration computation.

Fig. 10 shows the waveforms of the dwell time. For 50 ms, this dwell period aids in the calculation of T1, T2, and T0, as illustrated in Fig. 11. This will assist in determining the duration of the switch operation. The applied time for the switches indicated in Fig. 10 is calculated separately, and they all serve as the reference wave for the space vector PWM approach, with an amplitude of 0.9 units. To generate the gate pulses, this wave is compared to a carrier wave (triangular waveform) with an amplitude of one unit.

Fig. 12 shows the different gate pulses generated by each phase of the reference wave as well as the inverted gate pulses that are also generated and fed into VSI's six power switches.

Fig. 13 shows the line-line output voltage (volts) and the accompanying line current, which is the same as the stator current of the PMSM in amps of

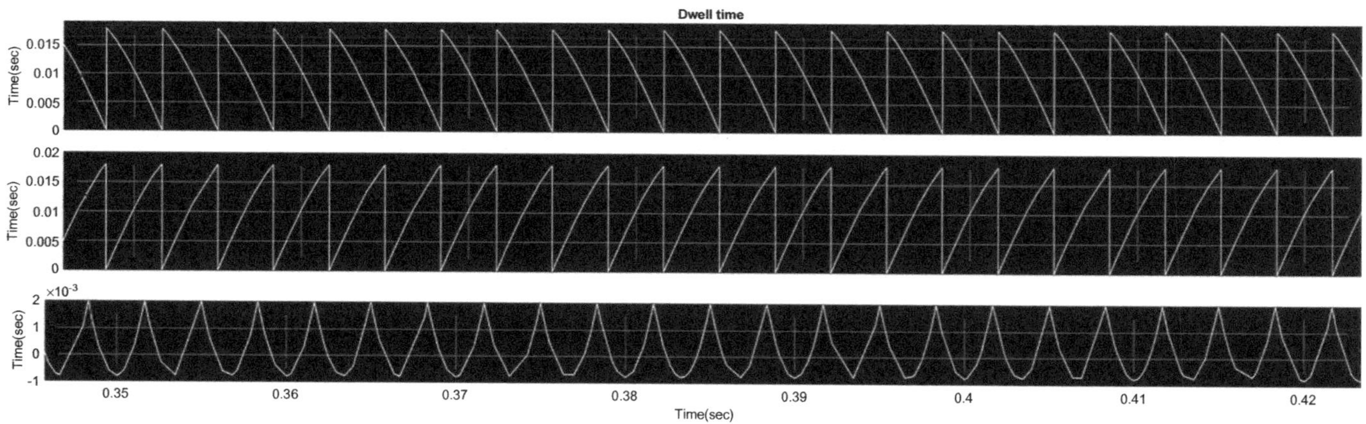

Fig. 10 Dwell time.

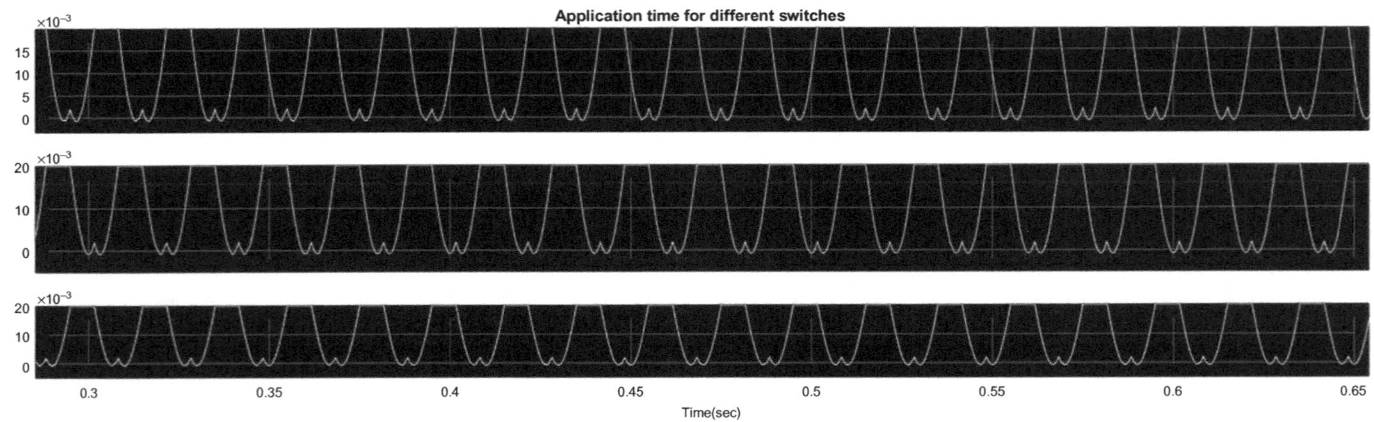

Fig. 11 Clamped SVPWM waveform.

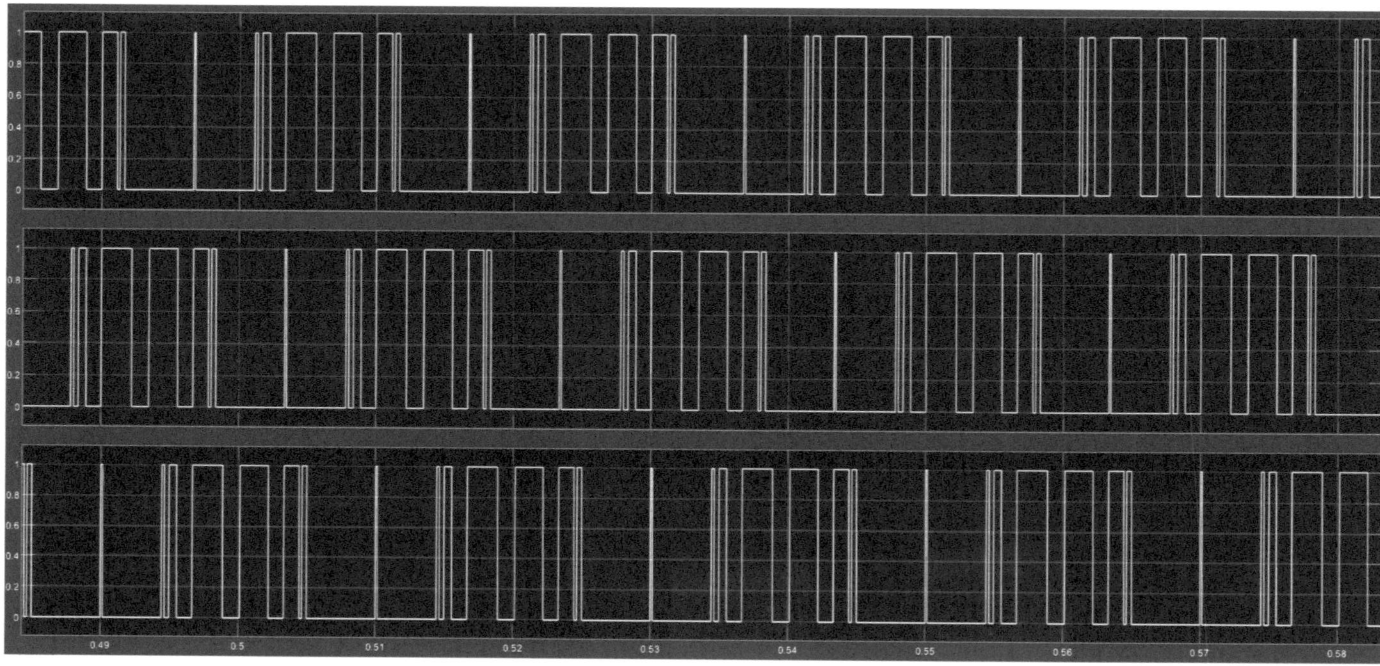

Fig. 12 Gate pulse generated.

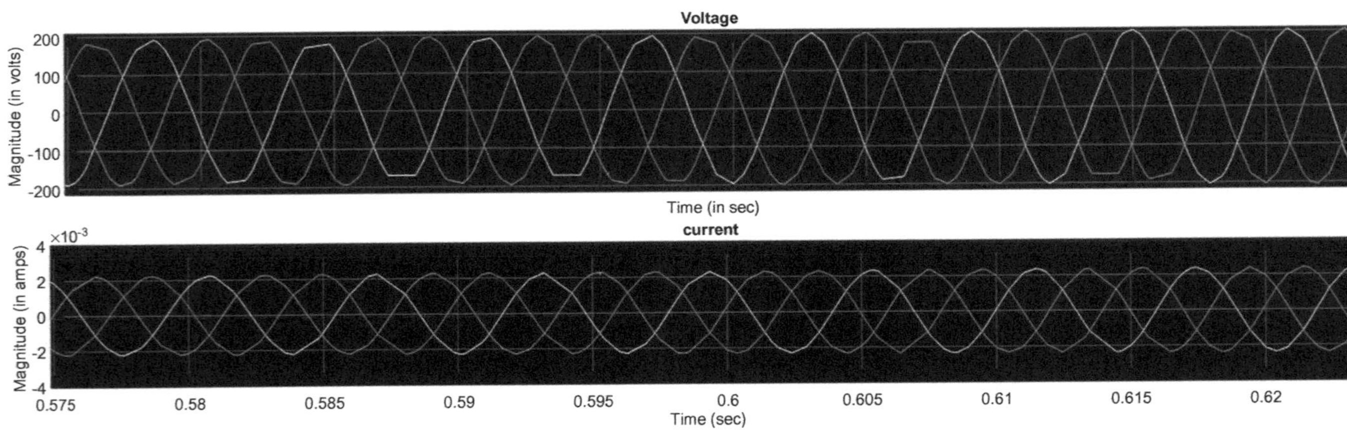

Fig. 13 Output voltage and current.

Table 7 PMSM parameters.

Parameters	Symbol	Value	Unit
Stator resistance	R_s	1.42	Ω
Daxis inductance	L_d	6.61	mH
Qaxis inductance	L_q	5.80	mH
Rotational inertia	J	0.00117	Kgm^2
PM flux	λ_M	0.1154	Wb
Number of poles	p	4	

the SVPWM powered VSI. It is deduced that VSI's output voltage and current are balanced in all three phases, and that they are also free of harmonic content.

The VSI output voltage as shown in Fig. 13 is fed to the stator of a PM synchronous motor (PMSM) as an input supply, which is developed using the parameters shown in Table 7. Figs. 14 and 15 illustrate the motor speed and torque created by the machine, respectively. There are no ripples in the output torque of the machine illustrated in Fig. 14, which boosts the machine's efficiency.

8 Conclusion

In this chapter, waveforms of the angle sector used for different angles are shown. After selection of the sector, the dwell time is assigned. When dwell times (T1, T2, T0) are generated by arranging a different timing sequence, a clamped SVPWM waveform is generated that is compared with the carrier wave, and a gate pulse is produced. The generated waveform is fed to the inverter and then a sinusoidal output waveform of voltage and current is represented. These output waveforms are fed to PMSM such that electromagnetic torque and rotor speed are developed to control the PMSM.

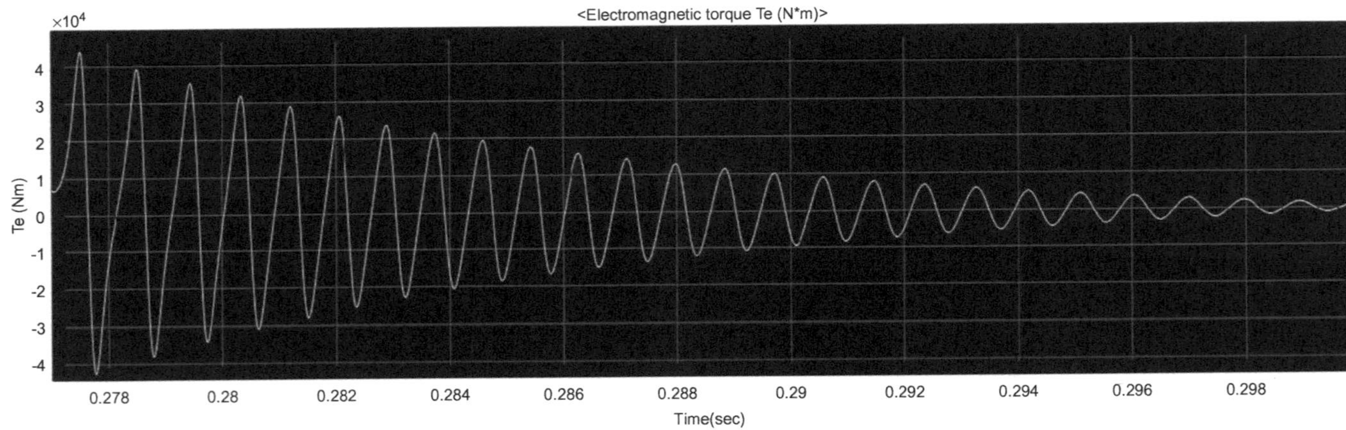

Fig. 14 Electromagnetic torque.

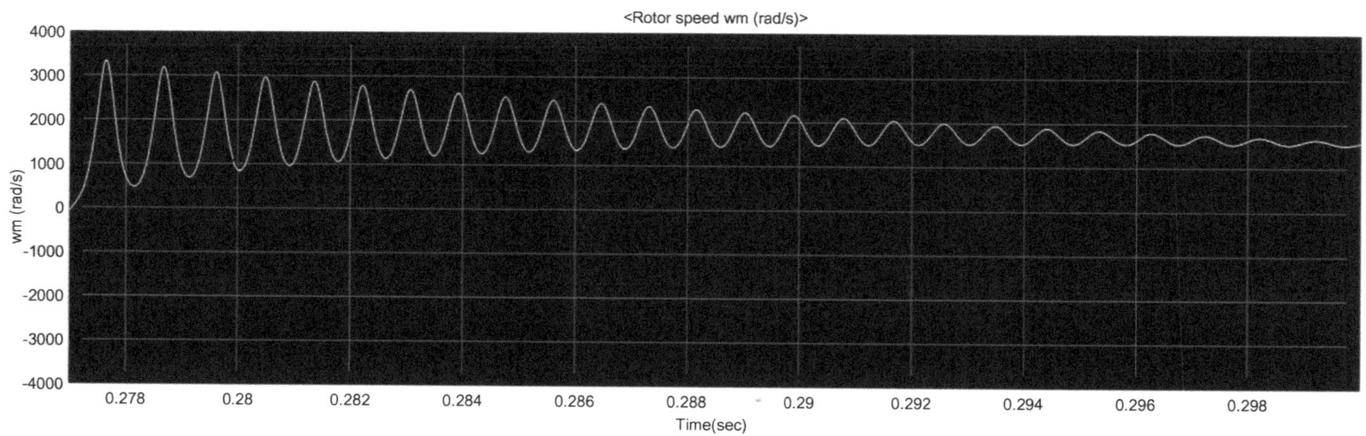

Fig. 15 Rotor speed w_m.

References

[1] M. Satoh, M. Tomita, S. Kanedo, Comparison Between SPM Motors and IPM Motors for Compressors of Air-Conditioners, IPEC, Tokyo, 2000, pp. 1286–1289, 181182.

[2] T.M. Jahns, Interior permanent-magnet synchronous motors for adjustable speed drives, IEEE Trans. Ind. Appl. IA-22 (1986) 738–746.

[3] G.R. Slemon, Electric Machines and Drives, Addison-Wesley, 1992.

[4] T.J.E. Miller, Brushless Permanent-Magnet and Reluctance Motor Drive, Claredon Press, Oxford, London, 1989.

[5] P.C. Sen, Principles of Electric Machines and Power Electronics, 2nd, J. Wiley and Sons, New York, 1997.

[6] R. Krishnan, Electric Motor Drives: Modeling, Analysis, and Control, Prentice Hall, 2001.

[7] H.W. van der Broeck, H.C. Skudenly, G.V. Stanke, Analysis and realization of a pulse width modulator based on voltage space vectors, IEEE Trans. Ind. Appl. 24 (1) (1988) 142–150.

[8] J. Holtz, P. Lammert, W. Lotzkat, High-speed drive system with ultrasonic MOSFET PWM inverter and single-chip microprocessor control, IEEE Trans. Ind. Appl. 23 (6) (1987) 1010–1015.

[9] G. Narayanan, V.T. Ranganathan, Analytical evaluation of harmonic distortion in PWM AC drives using the notion of stator flux ripple, IEEE Trans. Power Electron. 20 (2) (2005) 466–474.

[10] D. Zhao, G. Narayanan, R. Ayyanar, Switching loss characteristics of sequences involving active state division in space vector based PWM, in: Proceedings of IEEEAPEC2004, 2004, pp. 479–485.

[11] G. Narayanan, Synchronized Pulse Width Modulation Strategies Based on Space Vector Approach for Induction Motor Drives (Ph.D. thesis), Department of Electrical Engineering, Indian Institute of Science, Bangalore, August 1999.

[12] U. Senthil, B.G. Fernandes, Hybrid space vector pulse width modulation based direct torque controlled induction motor drive 3 (June 2003) 1112–1117.

[13] D. Zhao, Space Vector Methods for AC Drives to Achieve High Efficiency and Superior Waveform Quality (Ph.D. thesis), Arizona State University, Tempe, AZ, December 2006.

[14] P. Pillay, R. Krisnan, Modeling, simulation, and analysis of permanent magnet motor drives, IEEE Trans. Ind. Appl. IA-25 (1989) 265–279.

[15] T.-S. Low, M.A. Jabbar, M.A. Rahman, Permanent-magnet motors for brushless operation, IEEE Trans. Ind. Appl. 26 (1990) 124–129.

[16] T.M. Jahns, Motion control with permanent-magnet ac machines, Proc. IEEE 82 (1994) 1241–1252.

[17] G.R. Slemon, Electrical machines for variable-frequency drives, Proc. IEEE 82 (1994) 1123–1139.

[18] J.F. Gieras, M. Wing, Permanent Magnet Motor Technology (Design and Application), second ed., Marcel. Dekker, Inc, 2002. revised and expanded ed.

CHAPTER 7.6

Power management in electric vehicles using a hybrid energy storage system

Vishwajeet Kumar Sinha[a], Ram Niwash Mahia[a], and Om Prakash Mahela[b]
[a]Department of Electrical Engineering, NIT Hamirpur, Hamirpur, India
[b]Power System Planning Division, Rajasthan Rajya Vidyut Prasaran Nigam Ltd., Jaipur, Rajasthan, India

Contents

1 Introduction

These days, most conventional internal combustion engine (ICE) vehicles are based on petroleum products such as petrol, diesel, and liquefied petroleum gas (LPG). But these fuels are being continuously depleted day by day, so it becomes very important to minimize the use of petroleum products as much as possible. Also, these vehicles emit very harmful gases such as carbon dioxide, carbon monoxide, and sulfur dioxide, which are responsible for increasing global warming to some extent [1]. So the modern world is moving toward electric vehicles (EVs).

Vehicles fueled by electricity have some advantages over ICE vehicles. As the electric motor responds very fast, EVs generally have very high torque and are also very responsive in nature. Also, EVs are generally more linked with technology compared to ICE vehicles. As a result, many EV charging stations allowing charging to be controlled through a mobile application. It is expected that there will be a large increase in the EV global share from 2% in 2016 to 22% in 2030 [2]. With the aim of erasing pollution from

Active Electrical Distribution Network
https://doi.org/10.1016/B978-0-323-85169-5.00014-9

gasoline-powered engines, various kinds of EVs have been recently designed. One main disadvantage that hinders the use of EVs is the unavailability of a mechanism for the effective identification of the state of charge (SOC) as well as the projected life of the battery. The battery must meet electrical requirements such as operating voltage, current, power and energy densities, and a long working time and life. Different types of batteries are on the market such as lead cell, lithium (Li)-ion, Li-ion polymer, nickel-cadmium cell, nickel metal hydroxide, etc. Of these batteries, the Li-ion battery is generally used in most EVs as it has high energy density, which provides a longer range for EVs in one charging. But an energy storage system (ESS) comprising only the battery faces several issues such as a shorter battery life and thermal effects [3,4]. Also, the battery is less responsive than a supercapacitor due to its lower power density [5,6]. This prompts a search for additional solutions. Although the higher power density battery is available, its price is much higher than the lower power density pack. However, we can overcome this problem by increasing the size of battery, increases the cost of the system. The problem of thermal management of the battery must be addressed to use the battery at its full power capacity by taking safety measures. It is necessary to cool the battery after running at full load. Additionally, it should be capable of starting EVs in cold weather so that it can reach the desired operating power limit. This condition actually becomes dangerous when charging and discharging of the battery are done at a higher rate [7,8].

In order to solve some of these problems, the authors have proposed a hybrid energy storage system (HESS) comprising a battery and supercapacitor. Also, a control module is designed that controls the flow of power from both the battery and the ultracapacitor. This paper is organized as follows: Section 2 describes the hybrid energy storage system in brief, Section 3 details the proposed model design, Section 4 discusses the simulation results, and Section 5 concludes and offers a future scope.

2 Hybrid energy storage system

HESS is a combination of two or more energy sources. This combination of sources may be either a battery and a supercapacitor or it may be an ICE with a battery or flywheel. The supercapacitor and battery both come under the same category of electrochemical device [9]. But the principle of operation of these sources is different, which makes their features highly different from each other [10]. Table 1 shows some of the key features of the different types of battery sources. Table 2 shows some of the key features of supercapacitors. Table 1 shows that the battery has a relatively higher energy density of

Table 1 Key features of different types of battery cells [11].

Type of cell	Nominal cell voltage (V)	Energy density (Wh/kg)	Power density (kW/kg)	Life cycle (Times)
Lead cell	2.1	35–40	0.18	800
Nickel metal hydride cell	1.2	60–120	0.25–1	1000
Lithium-ion	3.8	100–265	0.25–0.34	1200
Lithium polymer	3.7	100–265	1–2.8	1000
Li-iron phosphate	3.2/3.3	90–160	0.2	2000

Table 2 Key features of different types of supercapacitors [11].

Chemistry	Nominal cell voltage (V)	Energy density (Wh/kg)	Power density (kW/kg)	Life cycle (times)
Supercapacitor	2.5/2.7	2–30	4–10	1,000,000

35–265 Wh/kg and it varies with the type of cell due to the different chemistry. Similarly, Table 2 shows that the supercapacitor has a relatively higher power density and a lower energy density. The tables also show that the life of the supercapacitor is above 1 million cycles, which is much larger than that of the battery cell. When compared to batteries, the supercapacitor performs better at low temperatures. These traits enable the best possible combination to obtain a better result. Also from Table 1, we can see the performance comparison of different batteries such as lead cell, nickel metal hydroxide cell, Li-ion, Li-polymer, and Li-iron phosphate. By studying the specification of each battery, we can choose the most efficient one for hybrid combination.

To optimally decide the type of hybrid combination, various research has been done. Fig. 1 shows the relation between various cells and different types of capacitors [12]. It gives the optimum condition in terms of power density (W/kg) and energy density (Wh/kg) to use. Fig. 1 represent the comparison among different batteries.

From Fig. 1, it is observed that a Li-ion battery has a higher energy density among all other cells and the electrolytic capacitor has higher power density among all other capacitors. These two can be used to get the optimum desired results.

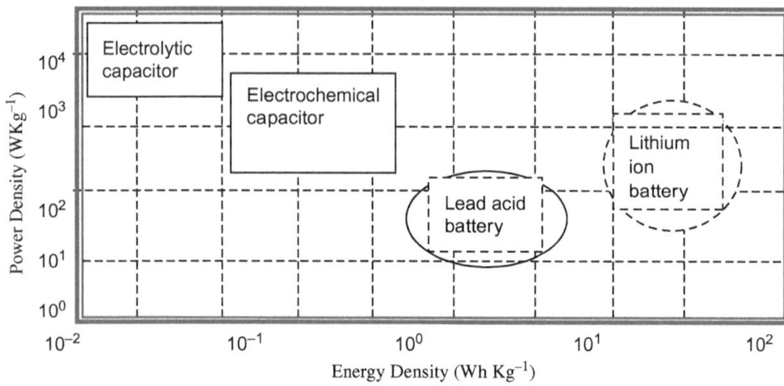

Fig. 1 Relation between power density and energy density for the battery and supercapacitor.

3 Proposed HESS system

Figs. 2 and 3 depict the proposed system's block diagram and schematic circuit diagram, respectively. It consists of a PV panel array, a boost converter, an Li-ion battery, a supercapacitor, an inverter, and a bidirectional DC-DC converter. Different components are described below.

A. *PV panel*

The PV panel is used to provide power to the battery and supercapacitor. Their effectiveness in converting sunlight into electricity is typically less than 17% and the generated electricity varies with weather conditions

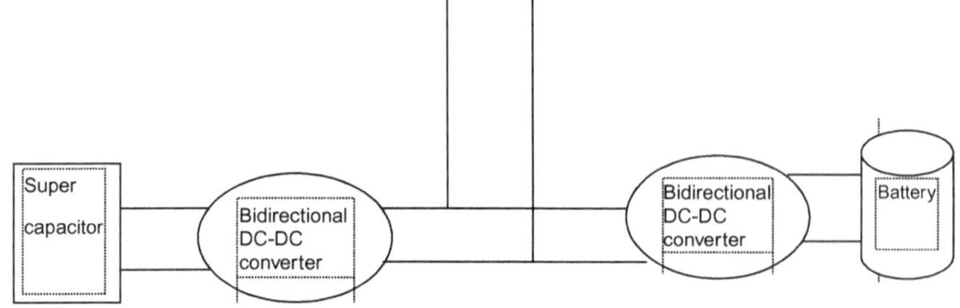

Fig. 2 Proposed system block diagram.

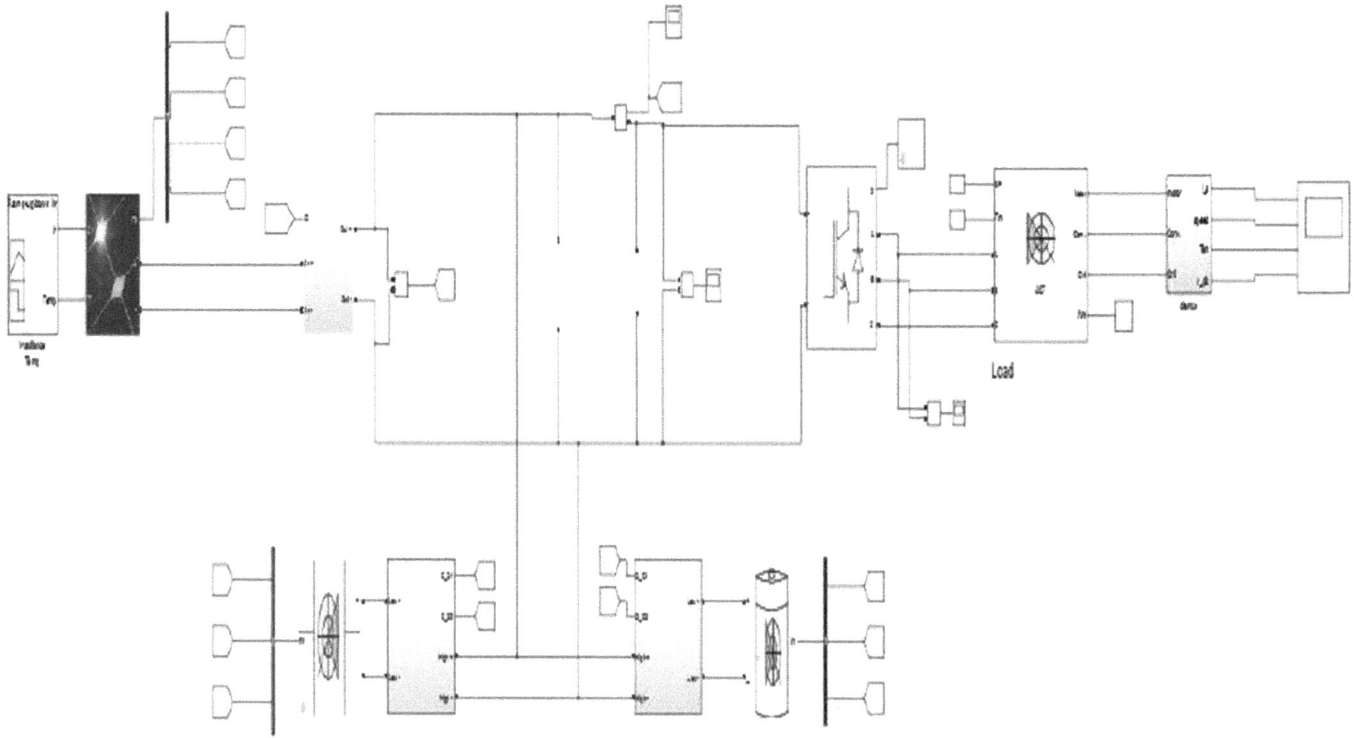

Fig. 3 Schematic circuit diagram of a proposed system.

[13,14]. To increase PV panel efficiency, the authors have used a perturb and observation (P&O) algorithm for maximum power point tracking (MPPT) [15,16]. A signal builder is also used to get different types of interchangeable linear signals.

Fig. 4 shows the various parameters of the PV panel such as voltage, current, and power. The maximum power point tracking occurs after 0.2s and any small change in irradiance or temperature slower then 0.2 is tracked by the MPPT. At 0.2s, the PV voltage is reduced but simultaneously there is an increase in current level so that the PV power remain balanced.

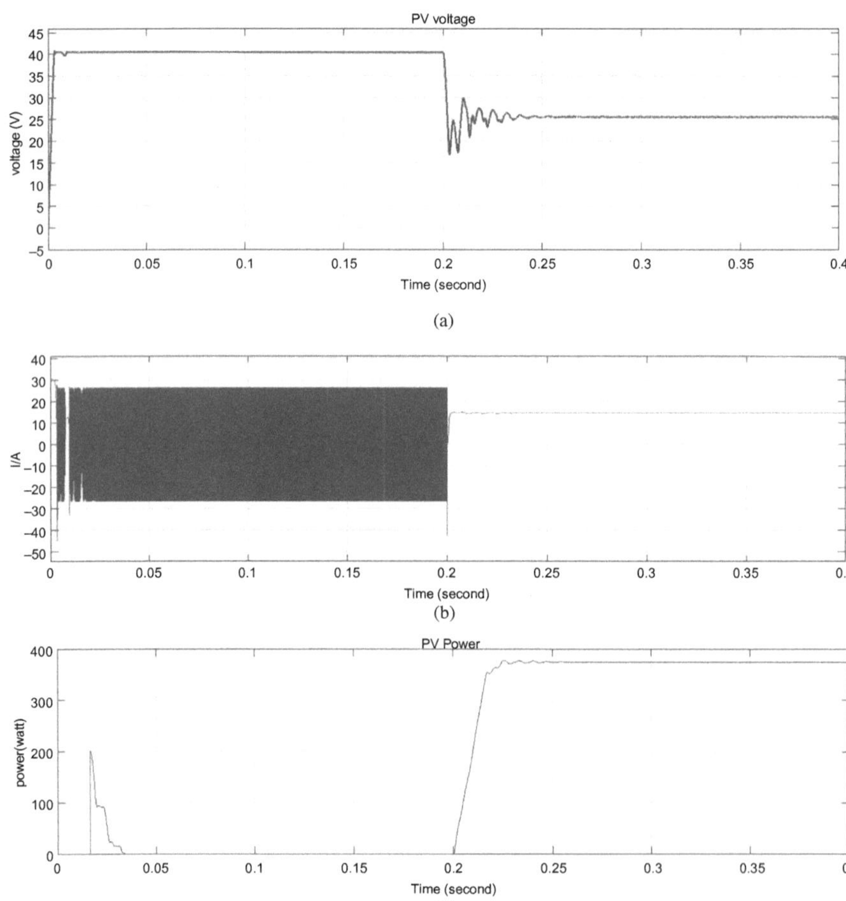

Fig. 4 Solar panel parameters: (A) PV panel output voltage, (B) PV panel output current, and (C) PV. power.

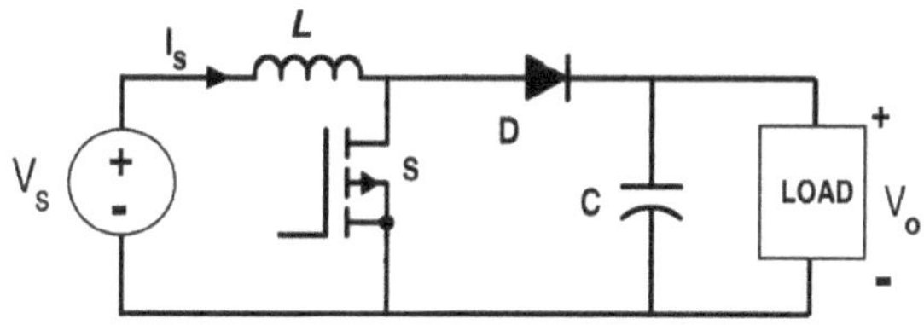

Fig. 5 Boost converter.

B. *Boost converter*

Fig. 5 shows a boost converter circuit that is a step-up converter. It is used for boosting the voltage from a low to a high level. The voltage generated from the solar PV panel is very low. So by using a boost converter, we are increasing the voltage level before giving it as input to the battery and the supercapacitor. It consists of switches that are either of an insulated gate bipolar transistor (IGBT) or a metal oxide semiconductor field effect transistor (MOSFET). They operate in ON/OFF condition according to requirements [17].

C. *Li-ion battery*

The motor current may be supplied in a smooth manner with battery pack management. To eliminate the output current ripples or to avoid big fluctuation, a 50 Hz bessel low pass filter with a second-order cut-off frequency is used [18]. Assuming there is no loss in the converter, the current coming from the battery is equal to the brushless DC motor current. This can be given by the below relation.

$$I_{\text{battery}} = \frac{V_{\text{load}} \times I_{\text{load}}}{V_{\text{battery}}} \tag{1}$$

The battery reference current is given by:

$$I^{*}_{\text{battery}} = \frac{V_{\text{load}} * I_{\text{load}}}{V_{\text{battery}}} B_{\text{LP}}(S) \tag{2}$$

where I^{*}_{battery} represents the battery reference current, V_{battery} and I_{battery} are the voltage and current of the Li-ion battery, and V_{load} and I_{load} are the voltage and current of the brushless DC motor.

$B_{\text{LP}}(S)$ is the bessel low pass filter's transfer function and it can be given as:

$$B_{\text{LP}}(S) = \frac{\theta_{n(0)}}{\theta_n\left(\frac{S}{\omega_0}\right)} = \frac{b(1)S^n + \ldots + b(n+1)}{S^n + a(2)S^{n-1} + \ldots + a(n+1)} \tag{3}$$

where,
ω_0 is the cut-off frequency,

$$\theta_n(s) \text{ is the reverse bessel low pass filter},$$

a_n and b_n are the bessel polynomial coefficient.

During charging the HESS, if the SOC of the supercapacitor rises above the maximum threshold limit of q_{scchar_high} then the power limit of the Li-ion rises to $P_{char\_high\_limit}$

where,

q_{scchar_high} represents the supercapacitor high charging limit.

$P_{char\_high\_limit}$ represents the high charging limit of the Li-ion battery.

The Li-ion power limitation is decreased to $P_{charlow\text{-}limit}$ if the SOC of the supercapacitor is less than the minimum constraint $q_{scchar\text{-}low}$. The dynamic constraint of the term "lithium-ion battery" can be written as:

If

$$q_{sc} - q_{scref} \geq q_{scchar\_high} \tag{4}$$

$$P_{batt\_char\_limit} = P_{char\_high\_limit} \tag{5}$$

where,

q_{sc} represents the supercapacitor charge,

q_{scref} represents the reference supercapacitor charge,

$P_{batt\_char\_limit}$ represents the Li-ion battery charging limit.

If

$$q_{sc} - q_{scref} < q_{scchar\_low} \tag{6}$$

$$P_{batt\_char\_limit} = P_{char\_high\_limit} \tag{7}$$

At the moment HESS starts discharging, if the SOC of the supercapacitor rises above the upper threshold limitation $q_{sc\_disch\_high}$, then the power range of the Li-ion rises to $P_{dish\_high\_limit}$.

where $q_{sc\_disch\_high}$ represents the supercapacitor high discharging limit.

$P_{dish\_high\_limit}$. represents the high discharge limit of the Li-ion battery.

The control parameter used above is obtained by the hybrid algorithm established by the Nelder-Mead simplex approach and particle swarm optimization.

D. *Novel control design of supercapacitor and battery*

The schematic diagram of the combined control of the supercapacitor and the battery is shown in Fig. 6. In Fig. 6 $V_{dc\ ref}$ and V_{dc} are provided to summing block and then the error signal pass through the PI controller so that the error gets reduces. Further, the output of the PI controller is passed through the LPF so that we get only the desired value of the DC voltage. Then, the battery current $I_{battery}$ and the DC voltage obtained are given to the summing block. . The output of the summing block is passed through the PI controller to suppress the error. A saturation level is applied that allows only a certain level of lower and upper output to pass through them. The battery will activate or start providing power to the brushless DC motor drive when the pulse is given. In a similar way, the supercapacitor will also activate or start providing power to the brushless DC motor drive when the corresponding pulse is given or when it requires more power to accelerate. So, in this way both the supercapacitor and the battery work in accordance.

E. *Bidirectional DC-DC converter*

A bidirectional DC-DC converter is illustrated in Fig. 7. It is also termed as half bridge DC-DC converter. When the boost and buck converters are interlinked in antiparallel across each other, the resultant circuit has the same fundamental structure as the boost and buck converters. But with the added feature of bidirectional power flow, it is called a bidirectional DC-DC converter. It is reversible in nature, that is, it works in both directions. The primary advantages of the half-bridge bidirectional converter over the bidirectional Cuk converter are that it only requires one inductor instead of two, the latter's value is half that of the former, and the power switch ratings required for the half-bridge bidirectional converter are significantly lower. Because of the lower inductor current and hence reduced conduction and switching losses, the half-bridge converter has a greater efficiency than the Cuk converter. Here, it works as an interface between the battery and supercapacitor [19,20].

4 Simulation and result

The simulation of the proposed hybrid energy storage system is applied to the brushless DC motor drive in MATLAB/Simulink. The various parameter values used are presented in Table 3.

Fig. 8 shows the various output characteristics of the brushless DC motor drive, which we obtained after implementing the designed module on that

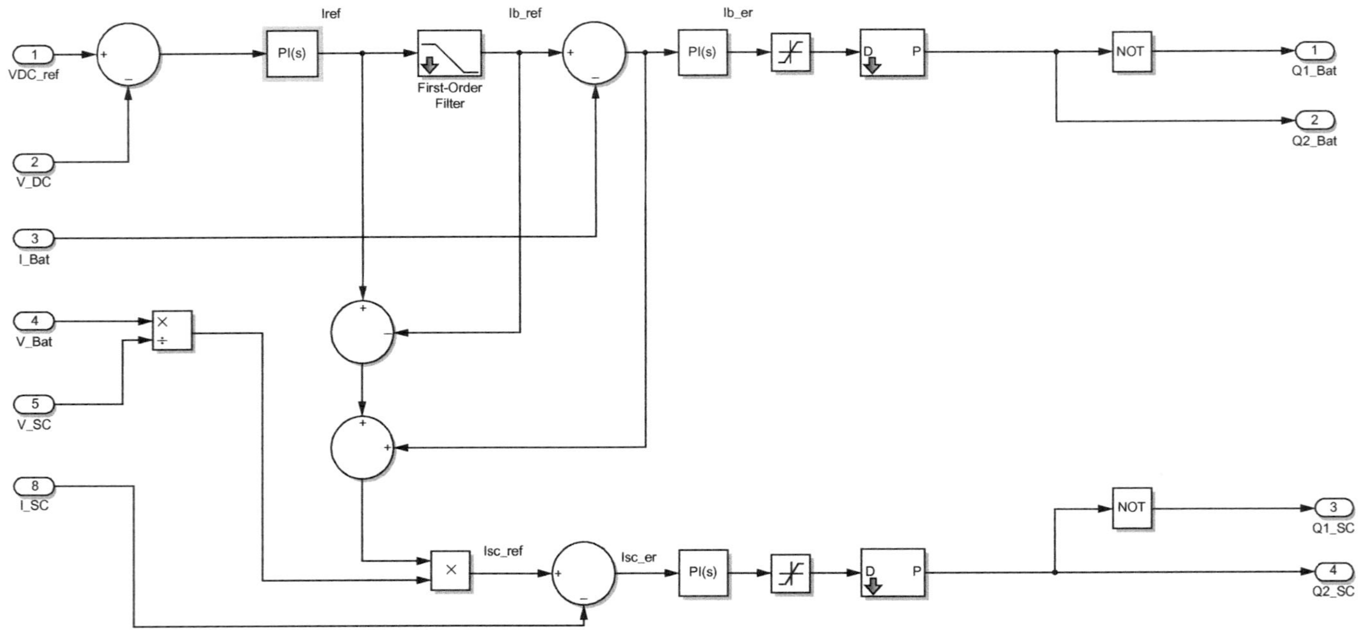

Fig. 6 Combined control battery and supercapacitor.

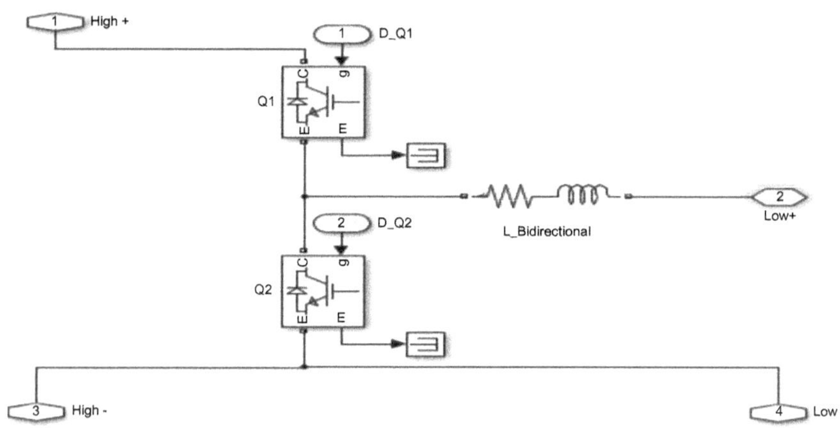

Fig. 7 Bidirectional DC-DC converter.

motor drive. Fig. 8A shows the electromagnetic toque. Here, we observe that after 0.27 s, the electromagnetic torque1 increases compared to electromagnetic torque2, which indicates that the supercapacitor is providing more power to the EVs when it is moving on an uneven road or hilly area due to its higher power density. Fig. 8B depicts the rotor speed of the brushless DC motor drive. Here, we observe that the rotor speed keeps increasing even after 0.27 s, which signifies that power from the HESS is sufficient to meet the EV power demand when moving on an uneven road or hilly area. Fig. 8C depicts the stator current of the brushless DC motor drive. Here, we observe that the stator current is smooth until 0.27 s. After that, it changes due to the change in power demand by EVs due to movement on a hilly area or uneven road.

Table 3 Simulation parameter values.

Parameter	Value
V_{dc} ref	40 V
% SOC of battery	50%
% SOC of supercapacitor	98%
Rated capacitance	29 F
Rated voltage	32 V
Battery nominal voltage	24 V
Battery rated capacity	14 Ah
Inductance	$0.3^{*}e^{-3}$ H

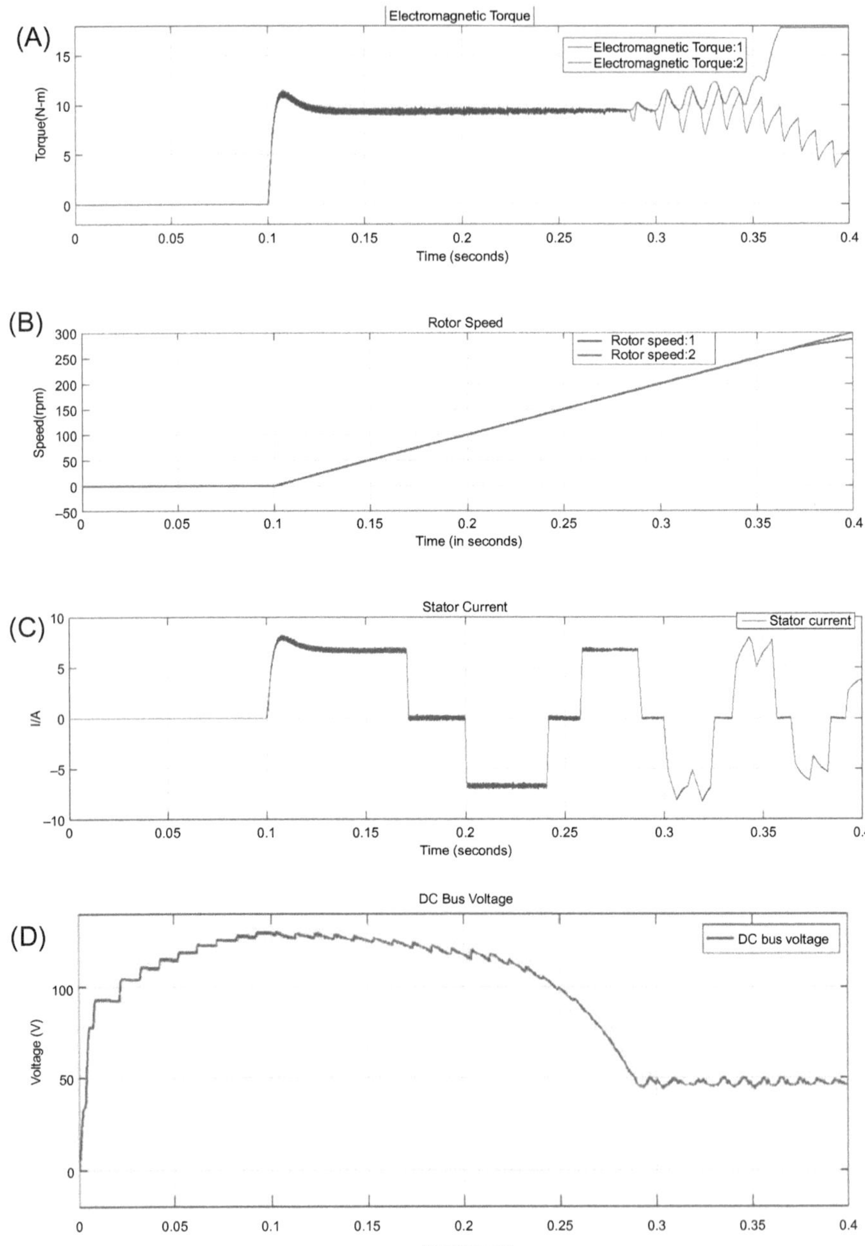

Fig. 8 Simulation result of the proposed HESS when applied to a brushless DC motor drive: (A) stator current, (B) electromagnetic torque, (C) rotor speed, and (D) dc bus voltage.

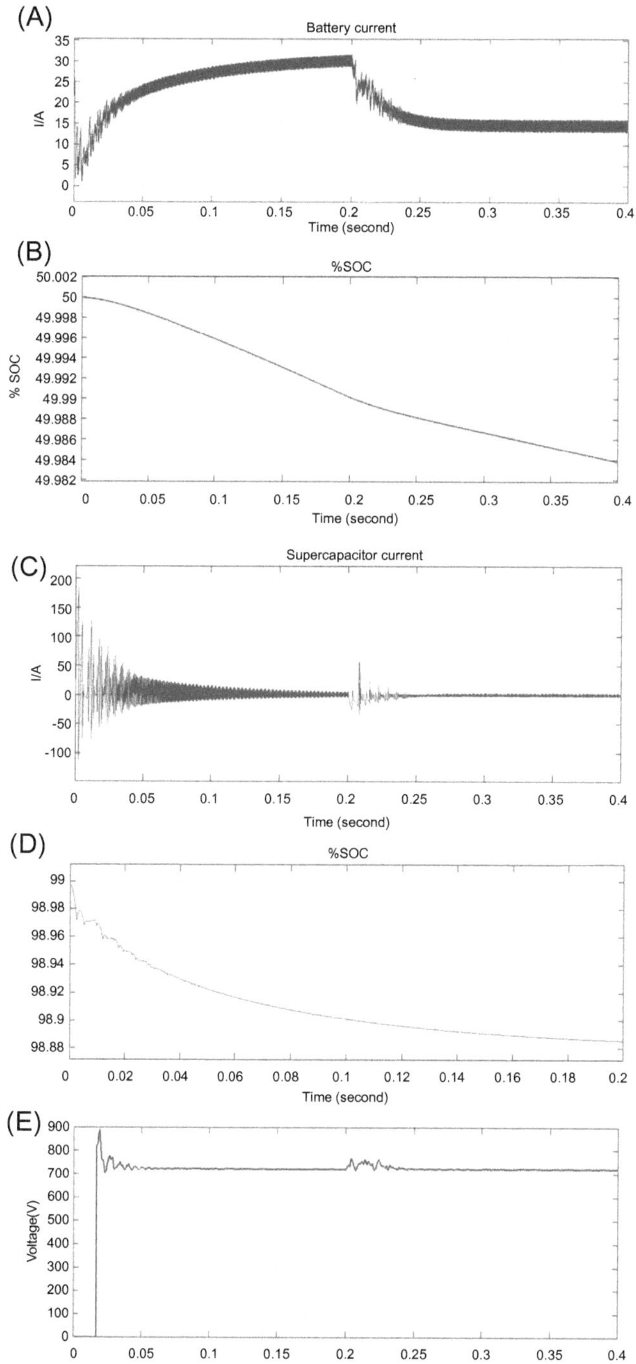

Fig. 9 Simulation result of proposed HESS combination: (A) Battery current, (B) percentage SOC of battery, (C) supercapacitor current, (D) percentage SOC of supercapacitor, and (E) load voltage.

Fig. 9 shows the simulation result of the various HESS parameters when implemented on brushless DC motor drive. Fig. 9A shows the battery current and Fig. 9C shows the supercapacitor current. We observe that changes in battery current are smooth compared to supercapacitor current. As there is less ripple content in the battery, it leads to less stress on the battery and as a result its lifespan increases. Also, the sudden or instant variations in the supercapacitor are normal as this is responsible for the load having a high frequency content. Fig. 9E shows that due to vehicle acceleration or deceleration, a little amount of variation in the load volatage occurs but that voltage is promptly recovered by the supercapacitor. Fig. 9B and D represent how the percentage SOC of the battery and supercapacitor varies as the course of action proceeds in the HESS.

5 Conclusion

In this paper, we have designed a control module for a new HESS consisting of a supercapacitor, a battery and a bidirectional DC-DC converter. A solar PV panel is used to provide power to the battery and supercapacitor. The proposed model is applied to the brushless dc motor in the MATLAB/Simulink model and the various parameters of the brushless dc motor and HESS are analyzed. From the analysis, we can infer that by using the HESS power control module, we can improve the life of the battery as less stress occurs on it due to low or minimum ripple current present when operated as HESS. Also, the sudden or instant variation in the supercapacitor current shows the normal behavior of the supercapacitor as it is responsible for the high frequency variation load.

References

[1] C. Samaras, K. Meisterling, Life Cycle Assessment of Greenhouse Gas Emissions from Plug-in Hybrid Vehicles: Implications for Policy, 2008, pp. 3170–3176.
[2] Global Electric Vehicle Outlook, International Energy Agency, May 2019.
[3] A. Emadi, S.S. Williamson, A. Khaligh, Power electronic intensive solutions for advanced electric, hybrid electric, and fuel cell vehicular power systems, IEEE Trans. Power Electron. 21 (3) (2006) 567–577.
[4] A. Emadi, K. Rajashekara, S.S. Williamson, S.M. Lukic, Topological overview of hybrid electric and fuel cell vehicular power system architectures and configurations, IEEE Trans. Vehicle Technol. 54 (3) (2005) 763–770.
[5] S.M. Lukic, J. Cao, R.C. Bansal, F. Rodriguez, A. Emadi, Energy storage systems for automotive applications, IEEE Trans. Ind. Electron. 55 (6) (2008) 2258–2267.

[6] S.M. Lukic, S.G. Wirasingha, F. Rodriguez, J. Cao, A. Emadi, Power management of an ultra-capacitor/battery hybrid energy storage system in an HEV, in: Proc. IEEE Vehicle Power Propulsion Conf., Windsor, U.K, Sep. 2006, pp. 1–6.
[7] A.C. Baisden, A. Emadi, An ADVISOR based model of a battery and an ultra-capacitor energy source for hybrid electric vehicles, IEEE Trans. Vehicle Technol. 53 (1) (2004) 199–205.
[8] J. Cao, N. Schofield, A. Emadi, Battery balancing methods: a comprehensive review, in: Proc. IEEE Vehicle. Power Propulsion Conf., Harbin, China, Sep. 2008, pp. 1–6.
[9] D. Dubal, O. Ayyad, V. Ruiz, P. Gomez Romero, Hybrid energy storage: the merging of battery and super-capacitor chemistries, Chem. Soc. Rev. 44 (7) (2015) 1777–1790.
[10] M. Liu, F. Minfan, Y. Wang, C. Ma, Battery cell equalization via megahertz multiple-receiver wireless power transfer, IEEE Trans. Power Electron. 33 (5) (2017) 4135–4144.
[11] J. Cao, A. Emadi, A new battery/ultracapacitor hybrid energy storage system for electric, hybrid, and plug-in hybrid electric vehicles, IEEE Trans. Power Electron. 27 (1) (2011) 122–132.
[12] F. Zhang, T. Zhang, Xiang, L. Zhang, A high-performance supercapacitor-battery hybrid energy storage device based on graphene-enhanced electrode materials with ultrahigh energy density, Energy Environ. J. 6 (5) (2013) 1623–1632.
[13] J.H.R. Enslin, et al., Integrated photovoltaic maximum power point tracking converter, IEEE Trans. Ind. Electron. 44 (6) (1997) 769–773.
[14] B. Bekker, H.J. Beukes, Finding an optimal PV panel maximum power point tracking method, in: 2004 *IEEE Africon. 7th Africon Conference in Africa (IEEE Cat. No. 04CH37590)*. Vol. 2. IEEE, 2004.
[15] B. Subudhi, R. Pradhan, A comparative study on maximum power point tracking techniques for photovoltaic power systems, IEEE Trans. Sustain. Energy 4 (1) (2012) 89–98.
[16] A. Chowdhury, S. Samal, T. Rout, M. Maharana, A complete comparison of the two maximum power point tracking technique (MPPT) for photovoltaic systems, in: 2018 4th International Conference on Electrical Energy Systems (ICEES). IEEE, 2018.
[17] K. Chomsuwan, P. Prisuwanna, V. Monyakul, Photovoltaic grid-connected inverter using two-switch buck-boost converter, in: *Conference Record of the Twenty-Ninth IEEE Photovoltaic Specialists Conference*, 2002. IEEE, 2002.
[18] R. Kaszynski, J. Piskorowski, Bessel filters with varying parameters, in: *IEEE Instrumentation and Measurement Technology Conference Proceedings*, 2005.
[19] M. Jain, M. Daniele, P.K. Jain, A bidirectional DC-DC converter topology for low power application, IEEE Trans. Power Electron. 15 (4) (2000) 595–606.
[20] L.-S. Yang, T.-J. Liang, Analysis and implementation of a novel bidirectional DC–DC converter, IEEE Trans. Ind. Electro. 59 (1) (2011) 422–434.

CHAPTER 7.7

Outage management in a smart distribution grid integrated with PEV and its V2B applications

Endeshaw Solomon[a], Issaias Gidey Hagos[a], and Baseem Khan[a,b]
[a]Institute of Technology, Hawassa University, Hawassa, Ethiopia
[b]Department of Electrical and Computer Engineering, Hawassa University, Hawassa, Ethiopia

Contents

1 Introduction

Currently, many car owners use their vehicles for different purposes such as work, transportation. and as a source of energy. Drivers mostly use their vehicles for transportation from their home [1]. Consequently, it is predicted that cars are parked on average for 96% of their lifespans [2]. So, car owners are not utilizing their cars most of the time. However, with the emergence of plug-in electric vehicles (PEVs), this trend can be reversed. A smart distribution grid is a virtual and physical system, which utilizes the physical structure of the electric system to observe, control and maintain the distribution system under different operating conditions. In this review, a brief

Active Electrical Distribution Network
https://doi.org/10.1016/B978-0-323-85169-5.00013-7

discussion of an outage management system (OMS) to the smart microgrid has been presented. The comprehensive OMS is applied to observe and manipulate the smart electric grid and simulation in advance [3]. The main objective of the OMS is to enhance the consistency of the system by identifying outages and by sensing and protecting against instability. The enhanced OMS is realized by different optimization approaches which uses its capacity to perform real time analysis and control of smart grid. PEVs are a more attractive technology for drivers, car companies, and the community in general [4]. Due to different PEV applications, their sales have rapidly increased over the last few years [5].

PEVs have many benefits compared with internal combustion engine vehicles such as decreased greenhouse gas (GHG) emissions, increase efficiency, and decreased pollution [6–8]. Electric vehicles (EVs) are powered by batteries that can store and supply energy to the system when needed; other EVs such as fuel cells can also supply energy. In this case, the PEV can function as distributed generation (DG) and provide different applications when parked while charging the vehicle at the electric outlet. Utilizing these facilities gives additional economical advantage to PEV owners [9].

2 Framework of demand aggregator

An aggregator is the driving force to speed up practical decision making to arrange the charging time for PEVs. When the car reaches the parking place, the owner charges the vehicle and communicates with the aggregator to share the necessary data. These data include the status of the battery and the time the EV was left. A future arrival pattern affects the aggregator's decision making when PEVs are already available in the parking lot. The PEV load forecast (PLF) is an intelligent element inside the aggregator that is responsible for evaluating the loading effect of the future PEVs that enter the parking lot [9].

The following assumptions have been made:

(1) Drivers use their PEVs in the same fashion as the conventional internal combustion engine (ICE) vehicles.

(2) The aggregator is not a market participant, but it is only an agent responsible for the convenience of PEV charging and benefits through demand response (DR) cooperation.

(3) All PEV chargers function at a power factor of one.

3 PEV for demand-side management

(A) Demand-side management (DSM)

For a grid, DSM is the planning, implementing, and manipulating of a distribution system facility designed to affect the consumer use of electricity that produces the intended alterations in the demand side that consist of maximum cutting, deep filling, load shifting, planned management, planned demand growth, and variable load shape [10]. The two main attributes of DSM are energy efficiency (EE) and demand response (DR). EE is planned to decrease electrical energy usage during the whole time; DR is planned to alter the site load for energy intervals [11]. The grid and consumers' taking part in demand-side management programs provide the benefits to consumer, service provider, and community as a whole.

4 Blackouts and outages

Power loss causes harm to people, economic, security, and community levels that is not measurable. Outage causes may be natural, accidents, poor maintenance, or an attack. The distribution grid may be damaged during outages, causing cascading failures in distribution transformers. The utility depends on customer reports to sense a blackout. To build the load supply, the gap measurement could be taken within milliseconds or microseconds.

5 PEV for V2B

There are many types of PEVs for the study of charging and discharging patterns, PEV penetration, and DG connections.

5.1 PEV penetration scenarios

The total PEV depends upon the percentage of PEV penetration. Three different penetration levels discerned from different studies are presented in this paper for the PEV penetration level [12].

Scenario 1: 35% of the total cars are assumed to be PEVs. From this, only hybrid PEV models are included in this paper. Slow (0.2 charging) and fast (1 charging) charging are also assumed.

Scenario 2: Assume 51% of total cars are PEVs. In this case both, battery electric vehicles (BEVs) and plug-in hybrid electric vehicles (PHEVs) are included and charging is measured similar to Scenario 1.

Scenario 3: Assume that 62% of cars are PEVs. BEVs are included in the study. To understand the effect of enhancing PEV penetration on system load is considered to be constant.

6 Types of PEVs and charging points

The type of PEV can be defined based on its capacity to store energy and driving miles. PEV types include the hybrid PEV, and one BEV model, United States Advanced Battery Consortium (USABC) and Electric Power Research Institute (EPRI). In the hybrid PEV, the number shows the driving capacity of the PEV in miles with batteries. Table 1 presents the approaches of supplying energy from EV batteries [13–16].

7 Battery technology

The battery is the main component of PEVs so the feasibility of V2G, V2B, and V2H depends on the status of the battery. Current EV battery technology is based on the lithium-ion battery, but nickel metal hydride batteries are also used for PHEV. The cost of the battery is approximately 20%–40% of

Table 1 Approaches of supplying energy from EV batteries.

Approach	Merit	Demerit
V2G	• Large-scale operation • Give source to auxiliary service • Grid reliability enhancement • Electricity transaction • DG integration	• Operation is complex • Requires more EVs • User voluntarism is necessary • Not common business area
V2B	• Operate at the building • Decrease cost of energy • Gives standby power • Lower asset required	• Willingness of user is needed • Operation is complex • Market penetration is not good
V2H	• Operate at home • Requires one electric vehicle • Decrease energy bill • Gives standby power • Interconnection DG	• It is not sufficient for residence • For only single home

the cost of the PEV. There are two types of aging processes: cyclic aging and calendar aging. The calendar life is loss of capacity due to the passage of time. Temperature and state of charge affect the calendar life while cycle aging is a loss of capacity produced by the repetitive charging and discharging process.

8 Enhancing outage management system (OMS) with PEV

The OMS is very crucial for utilities to set up a grid and restore power during outages. It is planned to decrease the economic effect of power interruptions. So, the utility must increase service quality for customers and increase individual operator safety. The OMS analyzes the position and amount of outage estimation abilities. The way to properly manage the outage of a smart distribution system using PEV is to utilize the battery to home (B2H) application.

Nearly every outage management system has two limitations:

- Shortage of an accurate, up to the second model of the distribution network.
- Shortage of connection to a system that monitors and manipulates the grid.

9 Vehicle to building

Vehicle to building technology has attracted more and more interest in today's electric power system. This system is a transitional step between vehicle to grid and vehicle to home but provides the necessaries for PEV and building owners [17]. Both achieve electricity savings of office buildings because the arrival and departure times of workers are known. This characteristic enables PEV management and allows PEV to be fully charged when workers return home. Hospitals, hotels, universities, office buildings, shopping malls, sport centers, and more can benefit from vehicle to building technology [18]. Additionally, V2B can support the critical loads of buildings such as data servers, computers, emergency lights, elevators, water pumps, etc., in emergency situations if there are power outages or shortages. This way, data loss and service interruptions could be avoided; the diagram shows a V2B system that consists of a PEV, local distributed energy generators, critical loads, and a building energy management system (BEMS) [19]. Fig. 1 shows the vehicle to building scheme.

A building energy management system will perform optimization algorithms and send the calculated set point to many subsystems under its

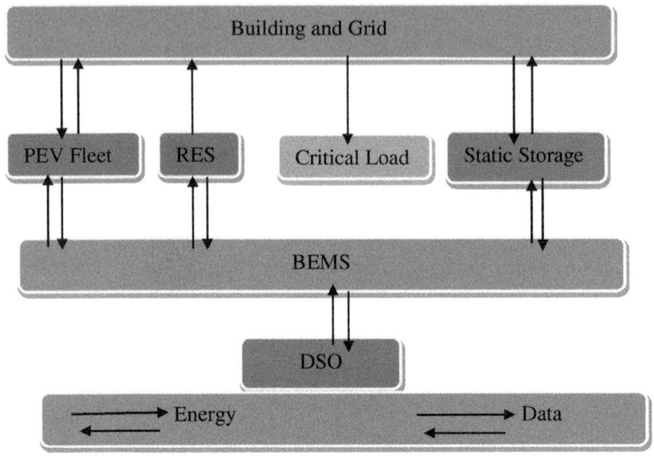

Fig. 1 Vehicle to building scheme.

control, such as a PEV. These algorithms will request economic revenue such as energy trading and peak power reduction.

In North America, power system utilities provide different types of load control and demand-side management to their clients. These provide enhanced power system security and advantages to the participants. Considering that the electricity rate is low when vehicle batteries are charged, when batteries are discharged, the battery storage may be used to balance the high cost during peak load demand. In this review, there are three basic charges for the business rate schedule: customer charge, energy charge, and demand charge. Finally, the monthly total revenue r for a BEV or PHEV based on V2B operation can be:

$$r = E_{ec}\left(r_{rc-onpeak} - R_{rc-midpeak}\right) \times t + r_{tdc}\left(P_{max} - P_{dsm-}\mathrm{max}\right) \tag{1}$$

where E_{ec} is the energy shifted from on-peak time to mid-peak time (kWh); $r_{rc-onpeak}$ is the on-peak time energy charge rate (\$/kWh); t is the number of days in a month; r_{tdc} is the time-related demand charge; P_{max} is the maximum peak power demand (kW); $P_{dsm-max}$ is the maximum peak power demand after DSM.

For practical applications for the given EV, the actual maximum power from the vehicle to building can be calculated as:

$$P_{vehicle} = P_{ideal} \times \eta_{charger} \times \eta_{inv} \times \eta_{other} \tag{2}$$

where $P_{vehicle}$ is the actual maximum power for V2B (kW); P_{ideal} is the ideal maximum power from V2B; $\eta_{charger}$ is the efficiency of the charge; η_{inv} is the electrical conversion efficiency of the DC to AC inverter; η_{other} is other factors such as power and battery self-discharge. The actual revenue will be reduced considering the charging efficiency, conversion efficiency, power loss, battery self-discharge, etc. Assume the efficiency of the charger is

$$\eta_{\text{charger}} = 0.9, \text{ the conversion efficiency}$$

$$\eta_{\text{inv}} = 0.93$$

$$\eta_{\text{other}} = 0.9$$

Vehicle to building has a similar function as V2H in that the EV will operate as a distributed alternative power supply that can discharge its excess energy to nearby buildings without long power transmissions. The buildings may be office towers, supermarkets, etc. [20,21]. Also, V2B is similar with V2G in that V2B in general needs to aggregate the energy from the EV fleet and energy aggregation needs infrastructure support, such as small-scale smart parking lots with bidirectional power flow capacity. A software building energy management system (BEMS) is also needed to supervise the operation. The BEMS will do its task based on its built-in optimization algorithm and the data from smart meters to further manage the whole building's operation [22].

Still more infrastructure and research are required to make V2B applications stand up to building operation when an interruption occurs, by protecting the main system and critical load from shutting down [19,20]. Enterprise buildings are more efficient for vehicle to building execution.

10 Approaches of delivering energy from PEV battery

Vehicle to building is similar to V2H, although it needs more technology support such as an aggregator. The transmission power loss is very small because it does not travel along a long transmission line. Vehicle to grid requires more installation costs of aggregators and power lines that helps bidirectional electrical power transmission between the PEV and the microgrid. Additionally, the power loss with a double transmission line could be significant (Table 2).

Here, vehicle to grid is more dominant because it is transmitting energy at the grid level. For conditions in which a generator fails to operate, V2G can stabilize all systems by injecting power into the grid. When PEV acts as energy storage, the main concern is battery operation convergence.

Table 2 Approaches of delivering energy from PEV batteries.

	V2G	V2H	V2B
Supportive infrastructure	Aggregator Long transmission network Parking lots are large	HEMS Smart energy meter (for home)	BEMS Smart energy meter (for building) Parking lots are small
Features	Large-scale EV Operation is complex Requires more regulation Sophisticated construction Requires income	Small scale (for home) Operation is easy Works automatically Easy construction	For small scale (for building) For local building Needs incentive Complicated construction
Applications	Auxiliary service Compensate reactive power Large-scale operation	Operates at home level, single EV Application is easy	Operates at building level Gives backup power Connect with V2G
Drawbacks	Complex operation Complex prediction of EV demand Battery degradation Absence of regulatory framework User preference to be considered	Medium difficulty operation Market penetration is poor user preference to be considered Battery degradation	Well-matched with a small family home More loss Battery degradation

11 Conclusion

In the future, PEVs will not only be used for transportation but also for providing other facilities like energy storage during power outages. In this chapter, outage management in a smart distribution grid integrated with PEV and its V2B applications is presented and compared with V2G and V2H. From the above approaches, the vehicle to grid idea is the more motivated one, but it is more complicated to use on a large scale. Vehicle to grid requires a crucial role for the development of PEV market share and a favorable and stable regulation framework. Vehicle to building and vehicle to home are recently introduced ideas that are concentrated on buildings and homes, respectively. As compared to vehicle to grid, this idea is easy to apply, doesn't need external articles to be completely functional, and gives more tangible function for the user. Lastly, of those three discussed systems, only the vehicle-to-home system is called "leaf to home," which has commercial availability, and the vehicle-to-building version is on the way to being marketed.

References

[1] E. Burger, Flexible views for view-based model driven development, in: Proceedings of the 18th international doctoral symposium on Components and architecture, ser. WCOP'13, Vancouver, British Columbia, Canada: ACM, 2013, pp. 25–30. [Online] Available: https://doi.org/10.1145/2465498.2465501.

[2] M.E. Kramer, E. Burger, M. Langhammer, View centric engineering with synchronized heterogeneous models, in: Proceedings of the 1st Workshop on View Based, Aspect-Oriented and Orthographic Software Modelling, ser. VAO'13, Montpellier, France: ACM, 5: 1–5: 6. [Online], 2013.

[3] J. Krumm, How People Use Their Vehicles: Statistics from the 2009 National Household Travel Survey, Apr. 2012.

[4] W. Kempton, J. Tomic, Vehicle-to-grid implementation: from stabilizing the grid to supporting large scale renewable energy, J. Power Sources 144 (1) (2005) 280–294.

[5] E.V. Landscape, Global EV Outlook for 2020, 2013. [Online] Available.

[6] B.K. Sovacool, R.F. Hirsh, Beyond batteries: an examination of the benefits and barriers to plug-in hybrid electric vehicles (PHEVs) and a vehicle-to-grid (V2G) transition, Energy Policy 37 (3) (2009) 1095–1103.

[7] J. Tomić, W. Kempton, Using fleets of electric-drive vehicles for grid support, J. Power Sources 168 (2) (2007) 459–468.

[8] B. Lunz, Z. Yan, J.B. Gerschler, D.U. Sauer, Influence of plug-in hybrid electric vehicle charging strategies on charging and battery degradation costs, Energy Policy 46 (2012) 511–519.

[9] J.D.K. Bishop, C.J. Axon, D. Bonilla, M. Tran, D. Banister, M.D. McCulloch, Evaluating the impact of V2G services on the degradation of batteries in PHEV and EV, Appl. Energy 111 (2013) 206–218.

[10] C.W. Gellings, The concept of demand-side management for electric utilities, Proc. IEEE 73 (10) (1985) 1468–1470.

[11] NERC, Data Collection for Demand-Side Management for Quantifying its Influence on Reliability: Results and Recommendations, North American electric Reliability Corporation, Princeton, NJ, Dec. 2007.
[12] L.P. Fernandez, T.G.S. Roman, R. Cossent, C.M. Domingo, P. Frias, Assessment of the impact of plug-in electric vehicles on distribution networks, IEEE Trans. Power Syst. 26 (1) (2011) 206–213.
[13] E.V. Everywhere, Road to Success, 2014, [Online] Available: http://energy.gov/sites/prod/files/2014/02/f8/eveverywhere_road_to_success.pdf. (Accessed 14 January 2015).
[14] C. Liu, K.T. Chau, D. Wu, S. Gao, Opportunities and challenges of vehicle-to-home vehicle-to-vehicle and vehicle-to-grid technologies, Proc. IEEE 101 (11) (2013) 2409–2427.
[15] L. Hua, J. Wang, C. Zhou, Adaptive electric vehicle charging coordination on distribution network, IEEE Trans. Smart Grid 5 (6) (2014) 2666–2675.
[16] U. Chukwu, S. Mahajan, V2G electric power capacity estimation and ancillary service market evaluation, in: 2011 IEEE Power and Energy Society General Meeting: 1–8, 2011.
[17] J. Donoghue, A. Cruden, Whole system modelling of V2G power network control, communications and management, in: EVS27 Symposium Barcelona, Spain, November 17–20, 2013, pp. 1–9.
[18] L. Zhao, V. Aravinthan, Strategies of residential peak shaving with integration of demand response and V2H, in: 2013 IEEE PES Asia-Pacific Power and Energy Engineering Conference (APPEEC): 1–5, 2013.
[19] E. Saberbaril, H. Sabooril, S. Sabood, Utilizing PREYs for peak-shaving, loss reduction and voltage profile improvement via V2B mode, in: The 19th Electrical Power Distribution Conference (EPDC2014), 6–7 May, Niroo Research Institute: 59–65, 2014.
[20] C. Pang, P. Dutta, S. Kim, et al., PHEVs as dynamically configurable dispersed energy storage for V2B uses in the smart grid, in: 7th Mediterranean Conference and Exhibition on Power Generation, Transmission, Distribution and Energy Conversion 7–10 November 2010, Agia Napa, Cyprus (Paper No. MED10/174): 1–6, 2010.
[21] C. Pang, P. Dutta, M. Kezunovic, BEVs/PHEVs as dispersed energy storage for V2B uses in the smart grid, IEEE Trans. Smart Grid 3 (2012) 473–482.
[22] S. Shao, M. Pipattanasompom, S. Rahman, Demand response as a load shaping tool in an intelligent grid with electric vehicles, IEEE Trans. Smart Grid 2 (4) (2011) 624–631.

SECTION 8

Power quality issues mitigation in smart distribution network

CHAPTER 8.1

Mitigation of nonlinear load influence on the power quality of the low-voltage distribution network using DSTATCOM

Gajendra Singh Chawda[a] **and Pankaj Kumar**[b]
[a]Department of Electrical Engineering, IIT Jodhpur, Jodhpur, India
[b]Department of Electrical Engineering, RTU Kota, Kota, India

Contents

1 Introduction

Smart electrical appliances for utility networks are increasing daily, and this introduces power quality issues as an impact on the nonlinear nature of the loads. Therefore, the demand for quality electrical power is increased, and conventional energy sources are not fulfilling these demands. This has led to a move away from conventional large linear loads toward modern nonlinear loads (NLs). In general, NLs are subcategorized into dynamic NLs and static loads. These loads cause the increased harmonic distortion of voltage and currents. It may increase when a single phase or two phases of unbalancing occur in the system, deteriorating the distribution equipment's proper functioning. An unbalance in the voltage affects the execution of other loads, mainly the induction motor (IM) load [1].

A suitable power quality (PQ) mitigation technique can be initiated by using the characteristics of power quality disturbances. Different machine

Active Electrical Distribution Network
https://doi.org/10.1016/B978-0-323-85169-5.00010-1

learning/signal processing techniques can help find the types of power quality issues [2–7].

Custom power device-based possible solutions are reported in the literature for mitigating these problems. A distribution static compensator (DSTATCOM) is a vital solution for improving PQ in the distribution network. A three-leg voltage source converter (VSC) topology is presented in [8] to reduce the system's complexity and cost in the three-phase three-wire (3P3W) distribution network during PQ improvement. However, a three-leg DSTATCOM would not be effective in the case of a neutral current compensation issue. Thus, a three-phase four-wire (3P4W) four-leg voltage source converter DSTATCOM topology for neutral current compensation and PQ improvement is presented in [9]. This system is reported in [10] for effective PQ disturbance mitigation. The effect of sources and loads on PQ mitigation using DSTATCOM is investigated in [11]. This research reported that the performance of the PQ improvement DSTATCOM has been dependent on the reference current generation control techniques.

In the past, multiple control techniques have been reported for estimating the reference signals for controlling DSTATCOM. The synchronous reference frame theory in [12] and its modified version in [13] have been presented for controlling DSTATCOM in the presence of NL and IM loads. The performance of DSTATCOM using ADALINE-LMF in [14] and its modified version in [15] has been investigated in the presence of NLs. The second-order generalized integrator-quadrature control technique in [16] and the least mean fourth-based control technique in [17] have been presented for PQ disturbance mitigation in the presence of NLs. Comparisons of different control techniques have been presented in [12]. However, the performances of the control techniques discussed above have not been investigated with combined loads. An investigation of combined distribution NLs is necessary for effective PQ mitigation.

This chapter presents the role of DSTATCOM for PQ improvement in distribution networks in the presence of combined NLs using instantaneous reactive power theory (IRPT). The system under investigation is simulated in MATLAB 2020 and would be validated with the help of hardware prototypes in the future.

2 System under consideration

The sample distribution system is presented in Fig. 1 [9]. The 3P4W distribution network is realized by using three voltage sources connected in a star

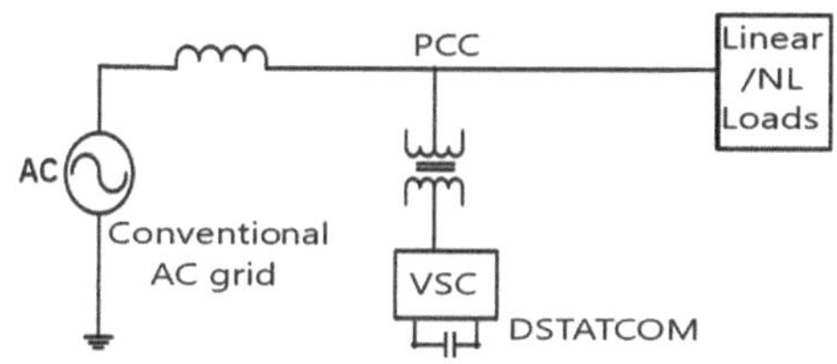

Fig. 1 System under consideration.

and a ground point. The combinations of linear, nonlinear and unbalanced loads are connected at the point of common coupling (PCC). The four-wire DSTATCOM is connected at the PCC for PQ improvement by mitigating the effect of connected loads. The switching transients of the DSTATCOM are controlled with the help of a ripple filter. The system parameters are listed in the Appendix.

3 Distribution static compensator and its control technique

1. **DSTATCOM**: It is a shunt-connected DFACTS device utilized for reactive power compensation and maintaining the power quality at the PCC. The basic diagram of the DSTATCOM is illustrated in Fig. 2. The basic idea to connect the DSTATCOM is to mitigate harmonics in the current as well as voltage signals by injecting compensating current [18–20]. Hence, the shapes of the current and voltage

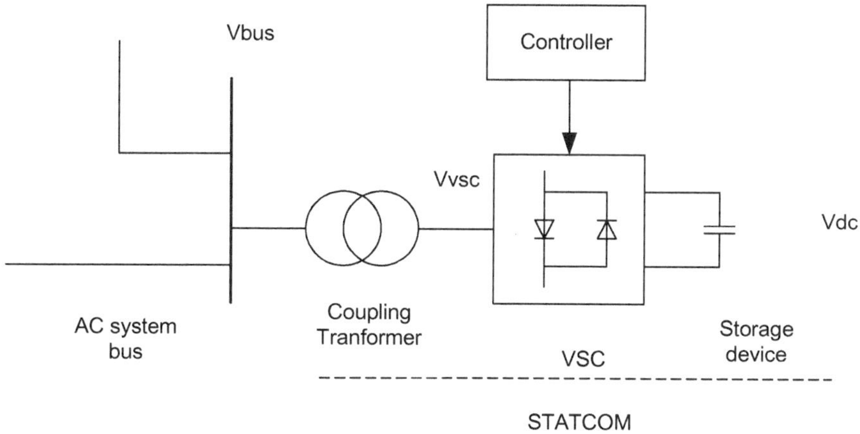

Fig. 2 DSTATCOM.

signals are sinusoidal. DSTATCOM is working on three operational modes:

- Capacitive mode, $V_D > V_S$, the DSTATCOM injects reactive power to the grid.
- Floating mode, $V_D = V_C$.
- Inductive mode, $V_D < V_S$, the DSTATCOM absorbs reactive power from the grid.

2. **Control scheme for DSTATCOM**: An IRPT-based control technique is presented for controlling the DSTATCOM for PQ improvement. The basic idea is to control the changes associated with the grid, load, and DC-link voltage signals, which generates the gate pulses for the DSTATCOM. This system further injects the compensating current at the PCC to maintain the voltage and harmonics at the PCC.

This theory is also known as the reactive power theory (RPT). The process includes the estimation of three-phase active and reactive instantaneous voltages and currents and represents their quantities in the instantaneous space vectors with excluding the zero-phase sequence component [21,22] to minimize the computational complexity. In this process, the $2\pi/3$ difference has been maintained during the calculation of the *a-b-c* coordinates. These coordinate axes are fixed on the same plane, as illustrated in Fig. 3. Also, the amplitudes and directions of the instantaneous space vectors (e_a and i_a) are varied with time on the a-axis.

Similarly, e_b, i_b, and e_c, i_c are set on the b and c-axes [1,22–25]. The conversions of the aforementioned space-vectors into the orthogonal coordinates (α, β) are given as:

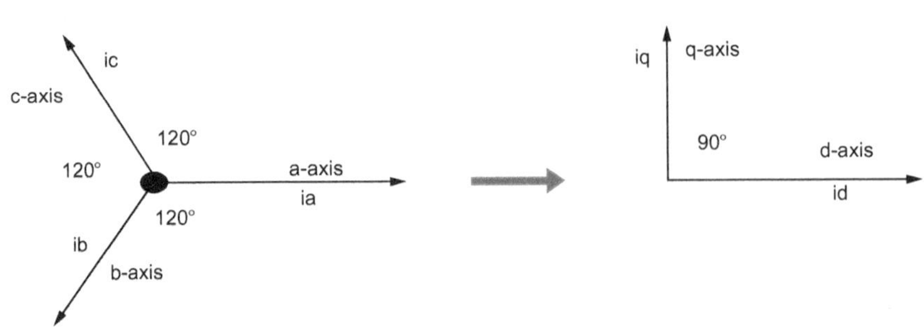

Fig. 3 $\alpha-\beta$ coordinate transformation.

$$\begin{bmatrix} e_\alpha \\ e_\beta \end{bmatrix} = \sqrt{\frac{2}{3}} \begin{bmatrix} 1 & -\frac{1}{2} & -\frac{1}{2} \\ 0 & \sqrt{\frac{3}{2}} & -\sqrt{\frac{3}{2}} \end{bmatrix} \begin{bmatrix} e_a \\ e_b \\ e_c \end{bmatrix} \tag{1}$$

$$\begin{bmatrix} i_\alpha \\ i_\beta \end{bmatrix} = \sqrt{\frac{2}{3}} \begin{bmatrix} 1 & -\frac{1}{2} & -\frac{1}{2} \\ 0 & \sqrt{\frac{3}{2}} & -\sqrt{\frac{3}{2}} \end{bmatrix} \begin{bmatrix} i_a \\ i_b \\ i_c \end{bmatrix} \tag{2}$$

However, the desired output of the α-axis is obtained by the estimated values of e_α and i_α and the β-axis is obtained by the estimated values of e_β and i_β.

The three-phase instantaneous power is given as:

$$P = e_\alpha{}^* i_\alpha + e_\beta{}^* i_\beta \tag{3}$$

$$P = e_a\, i_a + e_b\, i_b + e_c\, i_c \tag{4}$$

The three-phase instantaneous reactive power is given as:

$$q = e_\alpha{}^* i_\beta + e_\beta{}^* i_\alpha \tag{5}$$

$$\begin{bmatrix} p \\ q \end{bmatrix} = \begin{bmatrix} e_\alpha & e_\beta \\ -e_\beta & e_\alpha \end{bmatrix} \begin{bmatrix} i_\alpha \\ i_\beta \end{bmatrix} \tag{6}$$

The instantaneous power is obtained by combining the $e_\alpha{}^* i_\alpha$ and $e_\beta{}^* i_\beta$ in Eq. (6). Similarly the reactive power can be estimated. The block diagram of IRPT is depicted in Fig. 4 to illustrate the procedure of active and reactive power calculation for reference signal generation of DSTATCOM.

The current error is calculated as

$$\begin{aligned} i_{sn}^* &= 0, \\ i_{sn} &= -(i_{sa} + i_{sb} + i_{sc}) \end{aligned} \tag{7}$$

where i_{sn} is the sensed neutral current and i_{sn}^* is the reference neutral current of the fourth leg switches of DSTATCOM.

4 Simulation results and discussion

The IRPT-controlled DSTATCOM-BESS configuration has been examined under balanced and unbalanced loading conditions for load balancing. Power quality disturbances are also mitigated using the proposed approach.

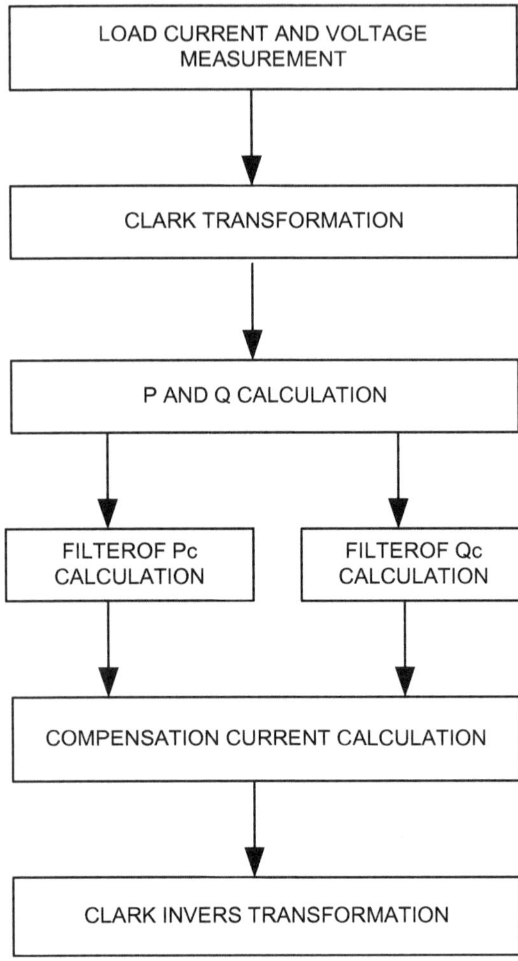

Fig. 4 Flow diagram of proposed IRPT of DSTATCOM.

A. *Response of proposed method with combined NLs*

The performance of the proposed method with combined unbalanced NL and IM load is as depicted in Fig. 5. The unbalancing in the load is created through the switch OFF phase-a of the NL load (I_{NL}) for 0.2–0.25 s and phase-a and phase-b of the NL load for 0.2–0.3 s. During this time, the three-phase load is transformed to a two-phase load, and finally, the two-phase load is converted to a single-phase load. This is illustrated in Fig. 5, which shows the practical loading condition of distribution loads. It has been perceived that the source voltage (V_s) is sinusoidal, balanced, and in-phase with the source current (I_s) under the nonlinear unbalancing and balanced induction motor (IM_L) loading

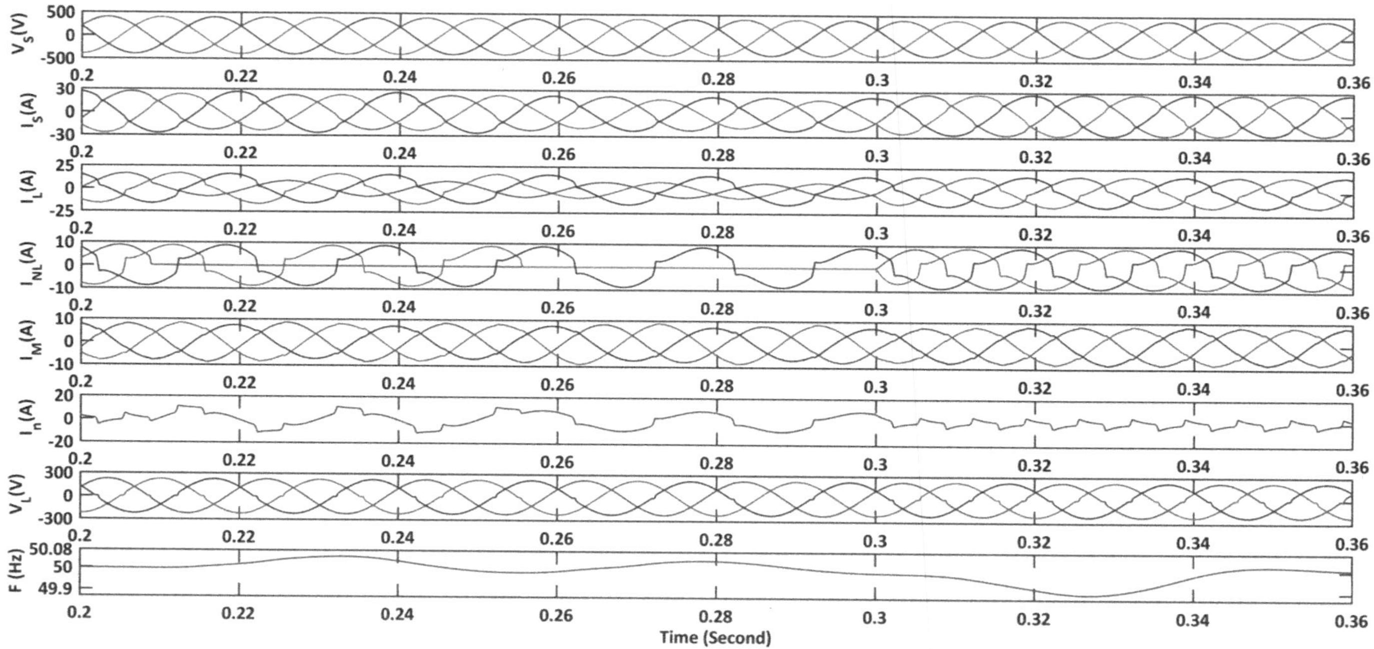

Fig. 5 Response of proposed method with combined NLs.

conditions. During switching of the NL load, the phase difference between the source current and voltage is slightly increased; therefore, few oscillations are found in the frequency (F_{Hz}). Hence, it is a significant cause for the out of balance current to flow through the neutral wire (I_n). It has been noticed that the total I_n is the vector sum of the obtained three-phase line currents. However, under balanced loading conditions, the vector sum is found to be zero. Hence, the obtained I_n is zero. It has been observed that unbalancing in one load of the 3P4W system has shown some variations in total load (I_L, V_L) current and voltage of the distribution system. And during this period, the performance of other loads such as the induction motor is not affected.

The response of the induction motor parameter (IM_P) and rotor speed (N_R) of the IM during nonlinear load unbalancing has is shown in Fig. 6. The power factor of the source (PF_S) and load (PF_L) during the unbalancing period have been found as 0.63 and 0.88, respectively. From 0.25–0.30 s, minor fluctuations have been observed in the stator current and torque of the induction motor. This shows that the rotor current, stator current and torque of the induction motor are not influenced by the NL load unbalancing; the system is shown in efficient working condition in Fig. 6. The source current and source voltage, nonlinear load current, induction motor load current and total load current harmonics as per the IEEE-519-2014 standard [26] are observed as 0.041, 2.53, 8.37, 16.44, and 2.72 respectively, which is shown in Fig. 7A–E.

5 Conclusions

This chapter presents the instantaneous reactive power theory-based control scheme for DSTATCOM for improving power quality with the proliferation of combined loads such as unbalanced linear, nonlinear, and induction motor loads. It has been observed that the fourth wire's neutral current helps maintain the stability of the system. The performance of the IM load remains unchanged during the examination period of unbalancing. The proposed method ensures that the unbalancing in one load does not affect the performance of other loads connected at the PCC. The defined operation boundaries of the DC-link voltage in the IRPT control ensured the reliable execution of the DSTATCOM, which protects the converter throughout large load perturbation. The proposed system has solved the combined NL loading challenges of the distribution network. The outcomes reveal

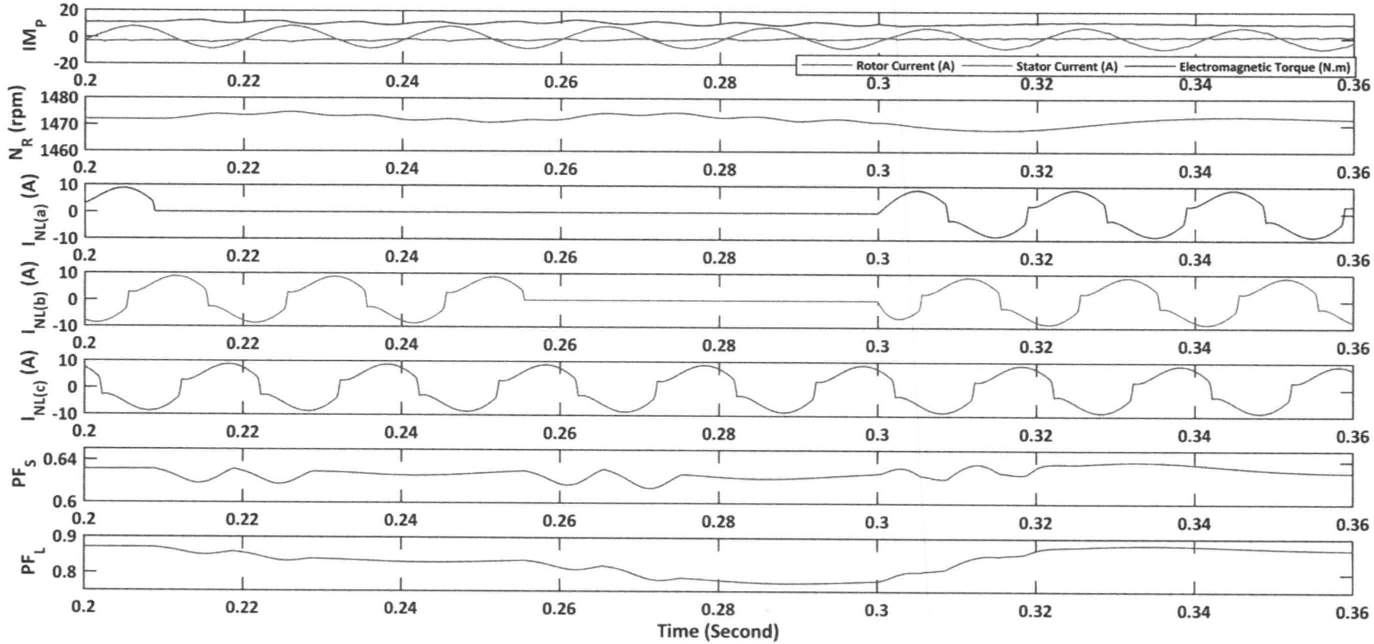

Fig. 6 Response of proposed method with combined NLs.

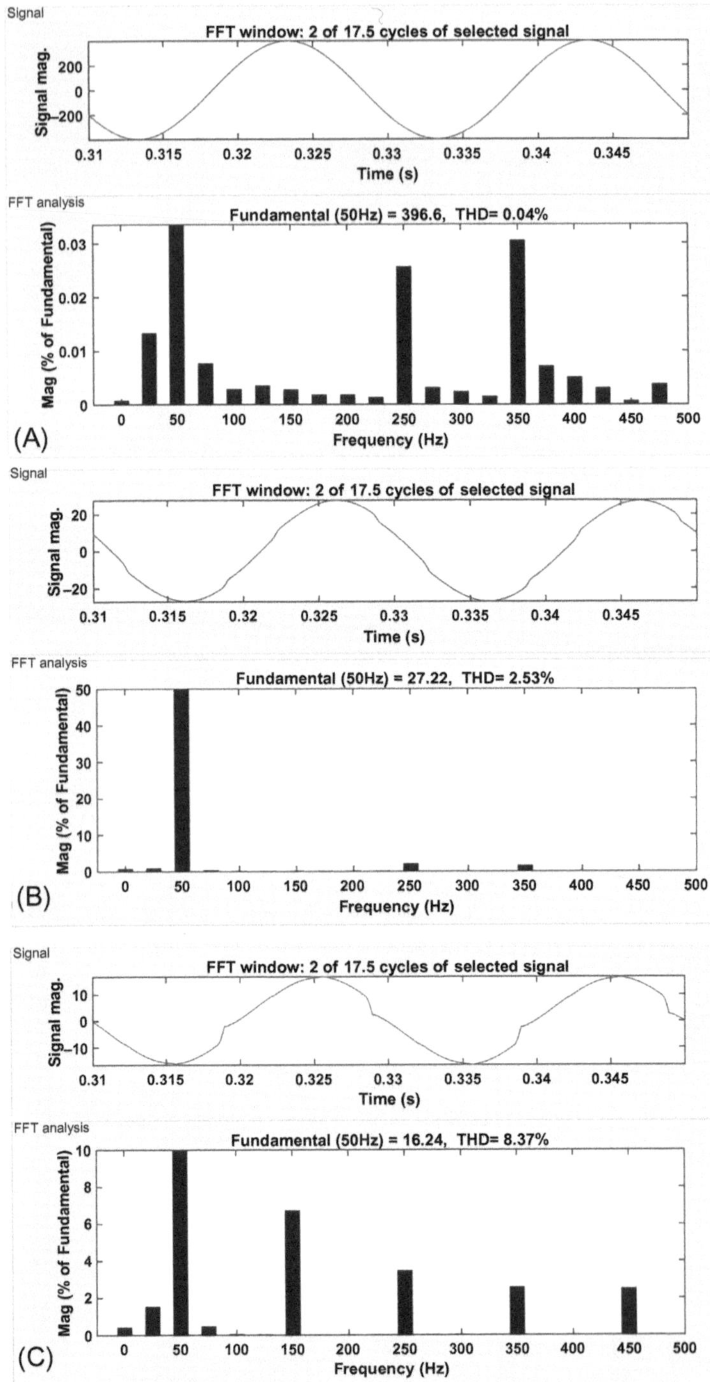

Fig. 7 Harmonics analysis of proposed IRPT controlled DSTATCOM method. (A) Grid voltage harmonics spectrum, (B) grid current harmonics spectrum, (C) NL load current harmonics spectrum,

Continued

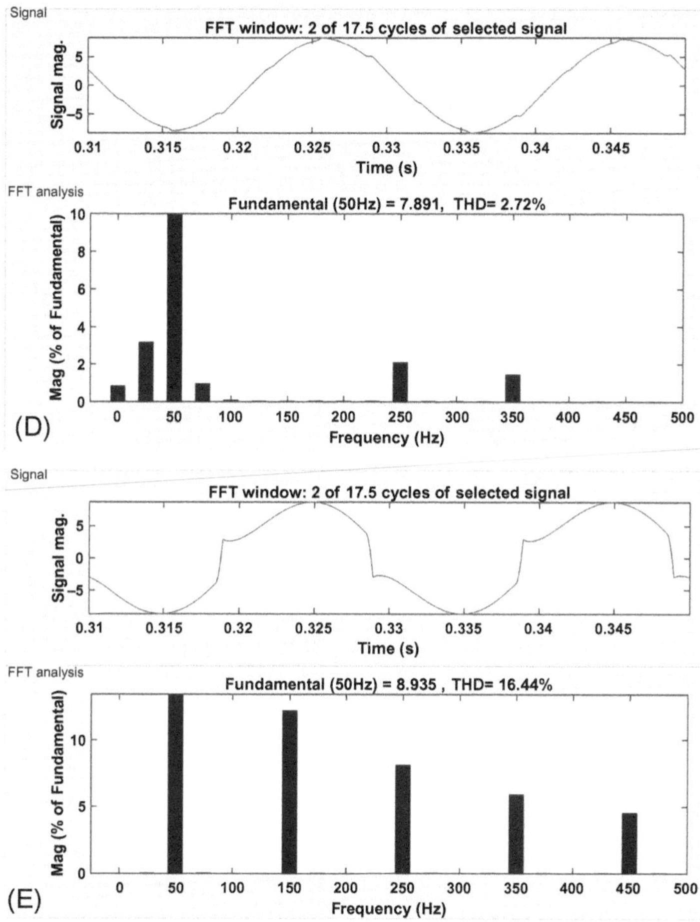

Fig. 7—Cont'd (D) IM load current harmonics spectrum, and (E) combined load current harmonics spectrum.

that the system PQ is maintained, and harmonics are mitigated within the range of voltage stability according to the IEEE-519-2014 international standard.

Appendix

Line-line three-phase grid voltage = 415 V, 50 Hz, Interfacing inductor $L_f = 2\,\text{mH}$, ripple filter $R_f = 5\,\Omega$, $C_f = 10\,\mu\text{F}$, $C_{dc} = 8000\,\mu\text{F}$, Rectifier based nonlinear load $R_{\text{load}} = 40\,\Omega$, $L_{\text{load}} = 100\,\text{mH}$, IM load, low pass filter = 20 Hz (cut off frequency), switching frequency (f_s) = 10 kHz.

References

[1] B. Singh, A. Chandra, K. Al-Haddad, Power Quality: Problems and Mitigation Techniques, John Wiley & Sons, 2014.
[2] O.P. Mahela, A.G. Shaik, N. Gupta, A critical review of detection and classification of power quality events, Renew. Sust. Energ. Rev. 41 (2015) 495–505.
[3] G.S. Chawda, A.G. Shaik, M. Shaik, S. Padmanaban, J.B. Holm-Nielsen, O.P. Mahela, P. Kaliannan, Comprehensive review on detection and classification of power quality disturbances in utility grid with renewable energy penetration, IEEE Access 8 (2020) 146807–146830.
[4] B. Singh, K. Al-Haddad, A. Chandra, A review of active filters for power quality improvement, IEEE Trans. Ind. Electron. 46 (5) (1999) 960–971.
[5] O.P. Mahela, A.G. Shaik, Topological aspects of power quality improvement techniques: a comprehensive overview, Renew. Sust. Energ. Rev. 58 (2016) 1129–1142.
[6] J. Gong, D. Li, T. Wang, W. Pan, X. Ding, A comprehensive review of improving power quality using active power filters, Electr. Power Syst. Res. 199 (2021), 107389.
[7] G.S. Chawda, A.G. Shaik, O.P. Mahela, S. Padmanaban, J.B. Holm-Nielsen, Comprehensive review of distributed FACTS control algorithms for power quality enhancement in utility grid with renewable energy penetration, *IEEE* Access 8 (2020) 107614–107634.
[8] G.S. Chawda, A.G. Shaik, Power quality mitigation in weak AC grid with low X/R ratios using distribution static compensator controlled by LMF algorithm, in: *2020 IEEE Region 10 Symposium (TENSYMP)*, IEEE, 2020, pp. 44–47.
[9] B. Singh, P. Jayaprakash, D.P. Kothari, New control approach for capacitor supported DSTATCOM in three-phase four wire distribution system under non-ideal supply voltage conditions based on synchronous reference frame theory, Int. J. Electr. Power Energy Syst. 33 (5) (2011) 1109–1117.
[10] P. Jayaprakash, B. Singh, D.P. Kothari, DSP based implementation of a three-phase four-wire DSTATCOM for voltage regulation and power quality improvement, in: *2009 35th Annual Conference of IEEE Industrial Electronics*, IEEE, 2009, pp. 3660–3665.
[11] T. Zaveri, B. Bhalja, N. Zaveri, Comparison of control strategies for DSTATCOM in three-phase, four-wire distribution system for power quality improvement under various source voltage and load conditions, Int. J. Electr. Power Energy Syst. 43 (1) (2012) 582–594.
[12] B. Singh, J. Solanki, A comparison of control algorithms for DSTATCOM, IEEE Trans. Ind. Electron. 56 (7) (2009) 2738–2745.
[13] G.S. Chawda, A.G. Shaik, Fuzzy logic based control algorithm for DSTATCOM connected to weak AC grid, in: *2018 2nd International Conference on Power, Energy and Environment: Towards Smart Technology (ICEPE)*, IEEE, 2018, pp. 1–6.
[14] G.S. Chawda, A.G. Shaik, Performance evaluation of adaline controlled DSTATCOM for multifarious load in weak AC grid, in: *2019 IEEE PES GTD Grand International Conference and Exposition Asia (GTD Asia)*, IEEE, 2019, pp. 356–361.
[15] G.S. Chawda, A.G. Shaik, Smooth grid synchronization in weak ac grid with high wind energy penetration using distribution static compensator, in: *2019 2nd International Conference on Smart Grid and Renewable Energy (SGRE)*, IEEE, 2019, pp. 1–6.
[16] A.S. Al-Sumaiti, Performance evaluation of second order generalized integrator-quadrature algorithm for DSTATCOM in non-ideal grid, in: G.S. Chawda, O.P. Mahela, B. Khan, H.H. Alhelou, E. Heydarian-Forushani (Eds.), *IECON 2020 The 46th Annual Conference of the IEEE Industrial Electronics Society*, IEEE, 2020, pp. 2551–2556.
[17] G.S. Chawda, A.G. Shaik, Adaptive reactive power control of DSTATCOM in weak ac grid with high wind energy penetration, in: 2019 IEEE 16th India Council International Conference (INDICON), IEEE, 2019, pp. 1–4.

[18] O.P. Mahela, A.G. Shaik, Power quality improvement in distribution network using DSTATCOM with battery energy storage system, Int. J. Electr. Power Energy Syst. 83 (2016) 229–240.
[19] G.S. Chawda, A.G. Shaik, Enhancement of wind energy penetration levels in rural grid using ADALINE-LMS controlled distribution static compensator, IEEE Trans. Sustain. Energy (2022), https://doi.org/10.1109/TSTE.2021.3105423.
[20] E. Hossain, M.R. Tür, S. Padmanaban, S. Ay, I. Khan, Analysis and mitigation of power quality issues in distributed generation systems using custom power devices, IEEE Access 6 (2018) 16816–16833.
[21] F.Z. Peng, J.-S. Lai, Generalized instantaneous reactive power theory for three-phase power systems, IEEE Trans. Instrum. Meas. 45 (1) (1996) 293–297.
[22] H. Akagi, Y. Kanazawa, A. Nabae, Instantaneous reactive power compensators comprising switching devices without energy storage components, IEEE Trans. Ind. Appl. 3 (1984) 625–630.
[23] S. Rahmani, A. Hamadi, K. Al-Haddad, A.I. Alolah, A DSP-based implementation of an instantaneous current control for a three-phase shunt hybrid power filter, Math. Comput. Simul. 91 (2013) 229–248.
[24] R.S. Herrera, P. Salmerón, H. Kim, Instantaneous reactive power theory applied to active power filter compensation: different approaches, assessment, and experimental results, IEEE Trans. Ind. Electron. 55 (1) (2008) 184–196.
[25] R.S. Herrera, P. Salmerón, Instantaneous reactive power theory: A reference in the nonlinear loads compensation, IEEE Trans. Ind. Electron. 56 (6) (2009) 2015–2022.
[26] IEEE, IEEE Recommended Practice and Requirements for Harmonic Control in Electric Power Systems, IEEE Std 519-2014 (Revision of IEEE Std 519-1992), pp. 1-29, 11 June, 2014, https://doi.org/10.1109/IEEESTD.2014.6826459.

CHAPTER 8.2

Power quality state estimation of power distribution systems: opportunities and challenges in observability enhancement

Abbas Marini, Morteza Shabanzadeh, and Ahmad Esmaeili
Power Systems Operation and Planning Research Department, Niroo Research Institute, Tehran, Iran

Contents

1 Introduction

Power electronic-based devices have been widely employed in various applications in power distribution systems. The nonlinear currents of these devices impose great harmonic distortions on the electrical networks. Besides, the low flexibility of t]he networks, especially distribution grids, makes them vulnerable to various faults and associated risks. As a result, different power system disturbances could occur in the network for a variety of

Active Electrical Distribution Network
https://doi.org/10.1016/B978-0-323-85169-5.00018-6

reasons such as single- and multiphase faults, lightning, weather conditions, human faults, and energizing and switching large loads and capacitor banks. In the viewpoint of power quality (PQ) issues, some PQ challenges such as harmonic distortions, voltage sags and swells, and transient phenomena may emerge in the network.

Power quality is necessary for the continuous and reliable operation of most electrical devices. Electrical and electronic equipment is sensitive to the changes in the magnitude and angle of the supplied power. Also, many electrical appliances and control processors could be tripped due to the voltage sag in several consecutive cycles. Harmonics as the most important PQ issue can lead to short-term and long-term problems on electrical apparatus. Harmonic resonance; reduction in the efficiency of generation, transmission, and consumption of the electric power; aging of devices; and insulation of various components due to the extra heat and malfunctioning of systems or devices are the main harmonic effects in the electrical system. Despite PQ problems, it is required to monitor power quality levels in the network as the first step for compensation actions against various PQ issues. In this scope, PQ measurements play a great role in the identification of power quality levels throughout the network.

From a technical viewpoint, measurement technologies can be classified in a broad range including the sampling frequency, measured quantities, accuracy, and installation costs. Smart meters, phasor measurement units (PMU), micro-PMUs, power sensors, and power quality monitors (PQM) are some conventional PQ measurement devices. The measured quantities of meters will be sent to the distribution control center using wired or wireless communication technologies. Power line carriers, optical fibers, and telephone networks are some examples of wired technologies. On the other hand, radio links, mobile networks, and Wi-Fi networks are some wireless technologies. The received data will be then analyzed in the control center to determine the PQ level of the network. The calculated PQ indices will be also stored for future purposes. In some schemes, the indices will be shown to the system operator using human-machine interfaces that lead to a better representation of the system performance, and as a result, provide a quicker decision capability. Such systems are known as power quality monitoring systems (PQMS) and are being utilized in some countries throughout the world. A typical architecture of a PQMS has been represented in Fig. 1.

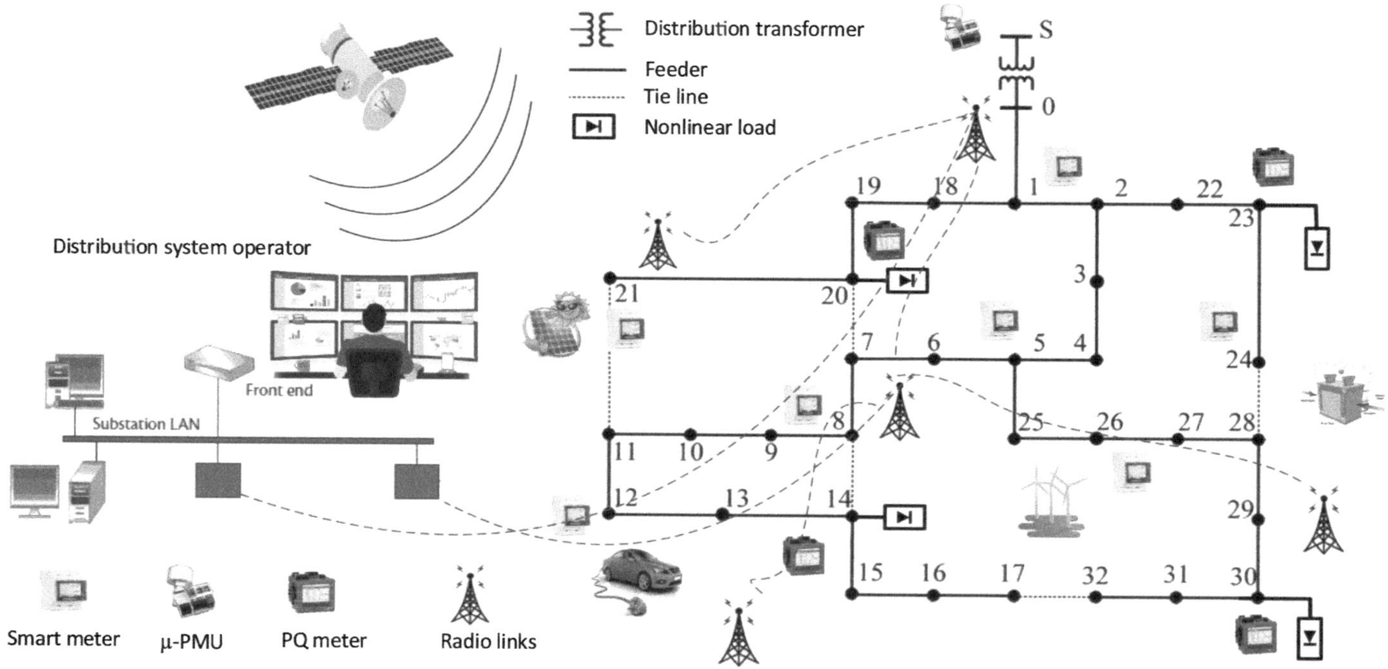

Fig. 1 The architecture of a wide area power quality monitoring system.

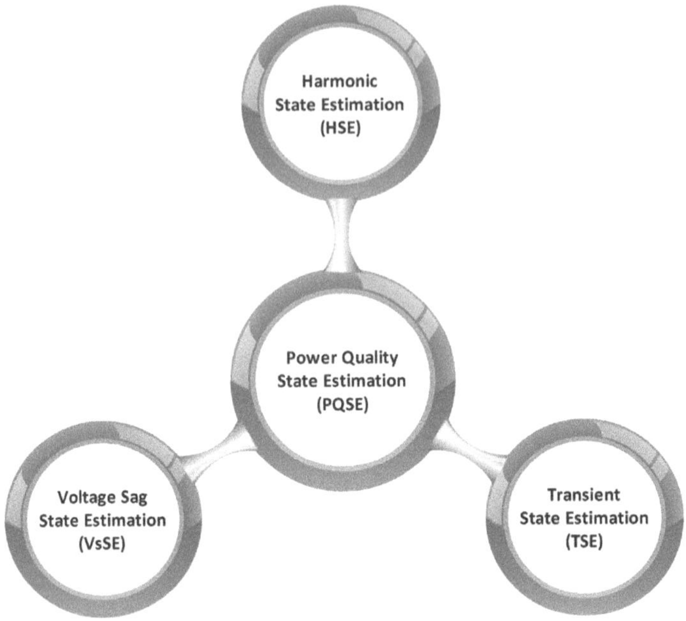

Fig. 2 Power quality state estimation categories.

As already mentioned, it is economically inefficient to install PQ meters at each node/bus of the electrical network. This issue is more important in distribution networks in which the number of load points is generally much higher than transmission systems. Hence, it is required to employ computer techniques and state estimation algorithms to obtain the PQ indices of the unmonitored areas of the network with the help of available measurements. This process is known as power quality state estimation (PQSE). State estimation and in particular, PQSE algorithms, are the main calculations that should be performed in control centers. In the PQ process, it is tried to estimate desirable PQ indices using all available metered data for those areas of the network that lack PQ meters. In general, the PQSE techniques are classified into three main categories named harmonic state estimation (HSE), voltage sag state estimation (VsSE), and transient state estimation (TSE), as shown in Fig. 2.

Harmonic distortions can be easily increased due to the propagations of currents of nonlinear loads. In HSE, the steady-state behavior of the network against harmonic distortions is analyzed. In VsSE, short-term voltage issues such as voltage sags and swells are determined. As already stated, these disturbances are mainly the direct consequences of system faults, motor starting, severe weather, and large load energizing. The number of occurred voltage sags at unmonitored areas could be determined using the recorded voltage sags of the meters. In TSE, the transient behavior resulted from various disturbances such as lightning, electrostatic discharge, and inductive and capacitive switching is determined. This analysis allows determining transient behaviors at unmonitored areas. In this chapter, all three PQSE categories will be further explained and modeled in detail. The main topics of this chapter are as follows.

2 Harmonic state estimation

In a harmonic power flow (HPF) algorithm, the harmonic voltages and currents of the network are typically calculated using known harmonic current sources. To do so, harmonic models of nonlinear loads as well as network characteristics should be utilized in harmonic power flow calculations. A reasonable question is that if the harmonic current sources are unknown, how the harmonic voltages of the network could be calculated (or how the sources of harmonic loads could be identified). In the section of harmonic state estimation (HSE), we are going to answer this question. Needless to say, the HSE is the inverse process of HPF. The goal of the HSE is to determine the harmonic content of harmonic sources by means of all appropriate measurements of the network voltages and currents. Therefore, HSE can be defined as follows:

HSE tries to localize harmonic dominant sources and/or determine harmonic contents of the network in unmonitored nodes based on the historical data and harmonic measurements obtained from other monitored nodes of the network.

It could be inferred that the HSE is of greater practical importance than HPF. For example, some common incidents in real cases such as failure of capacitor banks, communication, and phone interference, or high zero/negative component levels that have been reported in a specific region, could be a sign of violated harmonic distortion levels [1]. Hence, a few incident investigations about the harmonic content of an electrical region and its source would apply to this situation.

A big difference between conventional transmission/distribution SE and HSE is in the observability level of the system. In the fundamental frequency

($f_0=50$ or 60 Hz), the number of available measurements is usually more than the number of state variables (i.e., magnitude and phase angles of bus voltages), hence the number of measurement equations is more than the number of state variables, especially in the transmission system. Therefore, a conventional SE is usually an overdetermined problem. In this case (i.e., full observability), the weighted least square (WLS) is a common method that could result in a unique solution for the estimation of the system state variables. However, the observability level of the system might be degraded as the determined SE problems in which the number of measurement equations is equal to the number of state variables. In this case, WLS is also applicable. But in real power distribution systems, the number of harmonic measurement equations is fewer than the required harmonic state variables. This is due to the fact that the number of harmonic measurements is limited mostly because of financial constraints. Also, the number of state variables for all harmonics of interest is much more than the number of state variables at the fundamental frequency (f_0). Hence, the HSE problem is usually an underdetermined system of equations. Due to partial observability, the conventional WLS method could not be used and alternative solution methods need to be employed.

In order to establish the desired HSE infrastructure, a proper measurement system is definitely required to be installed throughout the distribution network. The monitoring system could have an architecture such as shown in Fig. 1, including PMUs, PQ meters, local processors, and communication infrastructure. Recent advancements in phasor measurement technology, which uses GPS sampled data, bring an opportunity to get the whole harmonic snapshot from the power distribution system. PMUs, as the main representatives of phasor measurements, could be synchronized in the range of 1 μsec. This feature would help local processors to receive the sample measurement data in a very short time to provide an opportunity for the mainframe of the control center to analyze the aggregated measurement data in near real-time intervals [2,3].

At the transmission level, the great number of synchronized measurements allows the harmonic state estimation of the power system in a conventional manner. However, directly applying these methods to the distribution level is not applicable because the number of measurements is lower. Hence, the use of historical data is essential in the management of distribution systems [4]. Historical data are known as pseudomeasurements. However, because the pseudomeasurements have generally less accuracy,

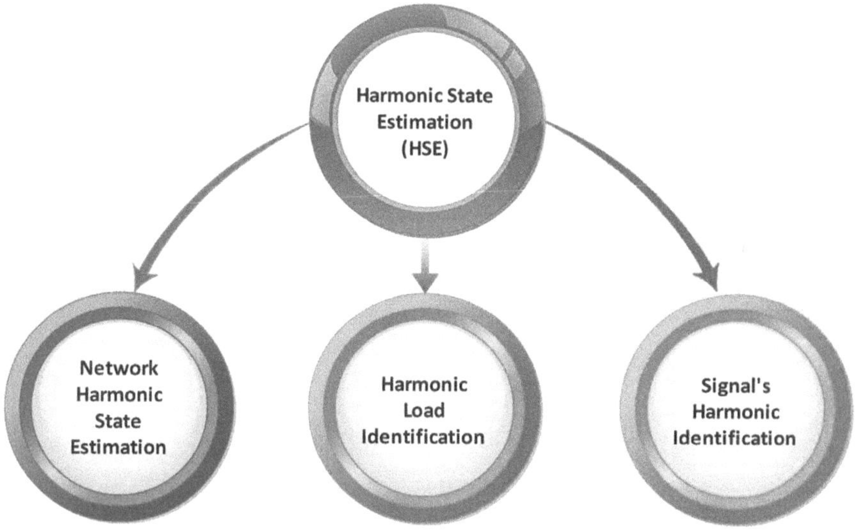

Fig. 3 Harmonic state estimation categories.

their corresponding weights are usually considered low. This assumption makes their effects less in the calculation process of the harmonic state of the network. Furthermore, virtual measurements such as the zero injection points and/or the nodes with known loads are usually employed besides the other measurement data to improve system observability. The reliability of the estimated harmonic state highly depends on the accuracy of pseudo- and virtual measurements and thus, their utilization should be carefully considered.

The HSE problem could be investigated in three general categories, as shown in Fig. 3: (I) the estimation of the harmonic state variables of the whole network, (II) harmonic load identification, and (III) extracting harmonic components from a given signal. In the first category, harmonic voltages and currents of the whole network (or at least its observable areas) are estimated using harmonic measurements from all installed meters in the network. Harmonic load identification, as the second category, deals with the localization of harmonic current sources in a network from harmonic measurements in single or various metering points. This process is employed to attribute harmonic distortions, either for applying penalty factors or for compensation purposes. In the rest of this section, various network HSE modeling methods and solution procedures are described.

2.1 Modeling HSE problem

As previously noted, the harmonic state of the system could be calculated from some harmonic measurements and historical information. Usually, in HSE and especially in distribution systems, the number of state variables to be estimated is more than the number of available measurements. For solving HSE problems, all the solution methods have been developed based on minimizing measurement residuals (i.e., the difference between measured quantities and the real values of state variables). Let x be the vector of network harmonic variables, then the general model to assign harmonic variables to the harmonic measurements could be represented as Eq. (1).

$$Z_h = h_h(x_h) + e_h \tag{1}$$

where subscript h stands for the harmonic component number. Also, the main notation used in Eq. (1) and throughout this chapter is stated as follows:

x is the vector of n harmonic state variables of each harmonic of interest. Harmonic voltages of nodes, harmonic injections, and harmonic currents of lines (i.e., feeders and lateral) are some common state variables.

Z is the vector of m harmonic measurements including a broad range of harmonic measured quantities such as harmonic voltages, harmonic injections, harmonic currents of the lines, and harmonic powers.

h is the function of the measurement matrix with the size of $m \times n$, which relates the harmonic measurements to the harmonic state variables. Concerning the harmonic measurements, this function could be either linear or nonlinear. If the system has been completely observed using PMU measurements, this function would be linear. Otherwise, the mere use of conventional harmonic measurements makes Eq. (1) nonlinear. Note that the h letter is also employed for each harmonic of interest, which, in this case, has been represented as an index to a variable.

e is the vector of m measurement error that shows measurement errors and uncertainties in the process of harmonic measurement and transmission.

Usually, the error is considered as the white noise, that is, the average and covariance of errors are equal to zero ($E(e) = 0$ and $E(ee^T) = \sigma^2$).

Due to the lack of a reliable method for the correct calculation of active and reactive harmonic powers in the presence of waveform distortions, real and reactive power measurements are not usually employed in Eq. (1). Hence, it is common to use voltage and current measurements as the harmonic input data of the model. If the harmonic phasors of voltages and currents are considered, then Eq. (1) can be decomposed into Eqs. (2), (3) and hence, each equation could be calculated separately.

$$Z_h^{\mathrm{v}} = H^{\mathrm{v}} x_h + e_h \tag{2}$$

$$Z_h^{\mathrm{c}} = H^{\mathrm{c}} x_h + e_h \tag{3}$$

In the equations above, the voltage and current measurements and matrices have been represented by the relevant superscripts. The H^{v} matrix is binary, that is, if a harmonic state variable is directly measured, then the corresponding array in the H^{v} will be equal to "1." On the other hand, the H^{c} matrix is of the admittance type, which should be calculated concerning the set of current measurements and harmonic state variables. If any state variable is directly measured using one of the installed meters, then the corresponding array will be equal to "1" in the H^{c}.

Because the measurements have been considered here as phasor quantities, the developed model in Eqs. (2), (3) is linear. The common method for solving such equations is the WLS method. However, some HSE methods have been developed based on conventional measurements (especially in the load identification category). These measurements make the HSE model nonlinear and in turn, nonlinear solution methods need to be employed. In the following section, these solution methods are introduced.

2.2 HSE solution methods

Various analytical and artificial solution methods have so far been implemented to solve the HSE problem. Comprehensive sensing, Bayesian learning, echo state network, fast kernel entropy optimization, hybrid firefly algorithm, seeker optimization, and particle swarm optimization are among the various methods used for the HSE problem. The Kalman filter is also a very useful method in the time domain for HSE modeling and solution. This method is a stochastic optimal estimator that, by using a least-square technique, develops an iterative procedure for the calculation of the next estimations using previous estimations and existing measurements. Our focus here is on the introduction of the HSE, thus, only two common analytical methods are introduced in this section and the interested reader is referred to the literature for further details Eq. (4).

2.2.1 Weighted least square

As already mentioned, the most common solution technique for Eq. (1) is the weighted least square (WLS) method. This method has been developed to precisely estimate all state variables by minimizing the square values of measurement residuals and implicitly considering the measurement accuracies. Let the measurement residual be indicated as r_i, then the minimization of all measurement residuals can be represented by Eq. (4) in each harmonic frequency. To solve this problem, the first-order optimal conditions must be applied by equating partial derivatives to zero, as shown in Eq. (5). This minimization should be employed individually for each harmonic of interest. The W is the measurement weights matrix, which shows the importance of each measurement in the total harmonic state estimation process. Usually, the measurement weights (representing their relative precisions) are determined using measurement variances, such as in Eq. (6). In addition, the *Jacobian* matrix measurement, which is typically represented by H, is shown in Eq. (7).

$$\min \sum_{i=1} \left(\frac{1}{\sigma_i^2} (Z_{ih} - h_{ih}(x_h))^2 \right) = \min \sum_{i=1} \left(\frac{1}{\sigma_i^2} (r_{ih}(x_h))^2 \right) \tag{4}$$

$$\begin{aligned} J_h(x) &= \frac{1}{\partial x} \partial \left(\sum_{i=1} \left(\frac{1}{\sigma_i^2} (Z_{ih} - h_{ih}(x_h))^2 \right) \right) \\ &= (Z_h - H_h x_h)^T W_h (Z_h - H_h x_h) = 0 \end{aligned} \tag{5}$$

$$W_h^{-1} = R_h = \text{diag}\left[\left\{\sigma_{1h}^2, \sigma_{2h}^2, \ldots, \sigma_{mh}^2\right\}\right] \tag{6}$$

$$H_h = \frac{\partial h_h(x_h)}{\partial x_h} \tag{7}$$

To solve the nonlinear problem of Eq. (4), the iterative Newton-Raphson (NR) method is a common approach. Ignoring high-order derivatives and setting the *Jacobian* matrix to zero for each harmonic frequency leads to an iterative solution scheme, as in Eq. (8). In this equation, the symmetric matrix G is called the harmonic *gain* or *information* matrix and calculated as Eq. (9). Starting from an initial guess, the algorithm iteratively updates the state variables until the objective function Eq. (4) is minimized.

$$J_h(x) = 0 \Rightarrow \widehat{x}_h^{k+1} = \widehat{x}_h^k - G_h^{-1}(x_h) H_h^T(x_h) W_h Z_h \tag{8}$$

$$G_h(x_h) = H_h^T(x_h) W_h H_h(x_h) \tag{9}$$

If the measurement harmonic matrix is linear, the solution method requires no iteration and could be determined in one single step. This could be possible if all measurements are taken from PMU or μPMU devices. Such a linear system was previously represented in Eqs. (2), (3) and the direct solution of Eq. (10) could be applied to this HSE problem. Here, the measurement matrix and input measurements could be calculated using Eqs. (11) and (12), respectively.

$$x_h = \left[H_h^{\mathrm{T}} W_h H_h\right]^{-1} \left[H_h^{\mathrm{T}}\right] Z_h \tag{10}$$

$$H_h = \left[H_h^{\mathrm{v}} \mid H_h^{\mathrm{c}}\right] \tag{11}$$

$$Z_h = \left[Z_h^{\mathrm{v}} \mid Z_h^{\mathrm{c}}\right] \tag{12}$$

It should be emphasized that the most important criterion for Eq. (1) to make it solvable by the WLS method is the full observability of the HSE equations. For this reason, the network must be fully observable by considering pseudo and virtual measurements besides all installed meters. Otherwise, the WLS method could not solve the HSE problem. In unobservable systems, some other solution methods such as singular value decomposition (SVD) could be applied. This method is explained in the next subsection.

2.2.2 Singular value decomposition

If the system equations of HSE are underdetermined due to the limited number of proper measurements, the WLS method could not solve this problem. Clearly, this is due to the singularity of WLS equations and the noninvertibility of the Jacobian matrix measurement. Furthermore, if the gain matrix is ill-conditioned, the solution method cannot estimate the state variables and as a result, the problem will mathematically become unsolvable. Even more, the common solutions based on WLS may result in a round-off error. Indeed, if one or some of the measurement equations are linear combinations of other equations (or near to linear approximations) or a zero pivot is created during the solution process, then the WLS will not be converged. This problem could be observed in a set of linear and nonlinear equations [4–6]. Because in real measurement data, some measurement equations may not be independent, singular matrices could have simply emerged in establishing HSE equations.

In underdetermined systems, the $H^{\mathrm{T}}H$ is often singular (i.e., with an infinite condition number) or near singular (i.e., with a large condition number). In such a situation, Gaussian elimination and lower-upper factorization (two common procedures employed in the WLS solution technique) may not provide satisfactory results. In this case, the singular value decomposition

(SVD) technique has been employed. The SVD method is generally slower than the WLS method; however, it could handle singular and near-singular matrices. In an underdetermined system, the SVD method provides a solution that is the best approximation in the sense of weighted least squares. Although singularity in the underdetermined systems means there is no unique solution for the problem, the SVD method could determine a set of solutions that makes the system observable even with the available measurements [7].

To solve the general HSE problem of Eq. (1) using the SVD method, it is necessary to employ a linear variant of the problem. The linear HSE problem could be represented as Eq. (13) for a case when the wide-area measurement system is available with the assistance of μPMU devices.

$$Z_h = H_h x_h + e_h \tag{13}$$

Roughly speaking, the noise effect can be ignored in Eq. (13). The SVD method is based on a property of linear algebra in which it is possible to represent each matrix with a dimension of $m \times n$ with the multiplication of three different matrices; that is, an $m \times n$ perpendicular vector (U); an $n \times n$ diagonal matrix (Λ) with singular values greater or equal to zero; and an $n \times n$ perpendicular matrix (V) [5,8]. Hence, it is possible to rewrite the measurement matrix of H in Eq. (13) as shown in Eq. (14) in which the h index is ignored for the sake of simplicity.

$$H = U\Lambda V^T \tag{14}$$

The matrix Λ includes all singular values of the measurement matrix H that have been sorted descending, as shown in Eq. (15). If $m \neq n$, then only the number of the main diameter would be nonzero, which is equal to the minimum value of m and n. The arrays of *the* U matrix are eigenvectors of HH^T and their columns are called *left singular vectors*. The columns of matrix V are normalized eigenvectors of $H^T H$ that are called *right singular vectors* of H. The SVD of a matrix always exists even if the matrix is singular or unique. In the SVD method, the summation of measurement residual squares would be minimum and hence, as shown in Eq. (16), it is statistically equivalent to the *Chi-square* distribution function. The solution of the SVD method could be calculated as Eq. (17), which is a set of columns of V that have been weighted with singular values of H and the internal product of Z and columns of U.

$$\Lambda = \text{diag}\left\{ \left[\lambda_1^{-2}, \lambda_2^{-2}, \ldots, \lambda_m^{-2}\right] \right\}; \; \lambda_1^2 \geq \lambda_2^2 \geq \ldots \geq \lambda_m^2 \tag{15}$$

$$\min_{x}\chi^2 = |Hx - Z|^2 \tag{16}$$

$$x = \sum_{i=1}^{m} \frac{1}{\lambda_i^2}\left(U_{(i)} \cdot Z\right) V_{(i)} \tag{17}$$

In the matrix form, the solution can be represented as Eq. (18).

$$x = V\Lambda^{-1}U^T Z \tag{18}$$

If the system is observable, the final solution of the SVD method is equivalent to the solution of the WLS method. However, if the system is not fully observable, the SVD method would find a set of state variables that is observable with the current set of input measurements. Hence, it is possible for the SVD method to find some state variables without any need for the full observability of the system while the WLS method is not able to solve this problem in such cases.

After SVD calculations, the null space vector of the measurement matrix would be examined for the observability analysis of the system. The null space vectors could be any constant, then add the result to the final solution of the problem. This could make a new solution to the underdetermined problem. Thus, after applying the SVD method, an infinite number of possible solutions (that can be inferred as the unobservability of the whole system) is obtained. The state variables that are zero in all null space vectors do not change in the process and could be estimated with current input measurements. Therefore, it could be accounted as observable state variables of the network. From a power system perspective, these variables are related to the observable areas of the network and thus, the SVD method could be employed as an observability tool for network analysis [7].

2.3 An illustrative example of the harmonic state estimation problem

In this section, we evaluate numerically the HSE problem using the IEEE 33-node distribution test feeder. In this feeder, four three-phase rectifiers are considered as nonlinear loads and added at nodes #14, #20, #23, and #30. The meters are considered PMUs and μPMUs able to measure harmonic phasors up to the 25th harmonic at various locations. The modified network is shown in Fig. 4. The harmonic condition of this network was already calculated using a harmonic power flow algorithm. For taking uncertainties into account, the calculated values are deviated by some

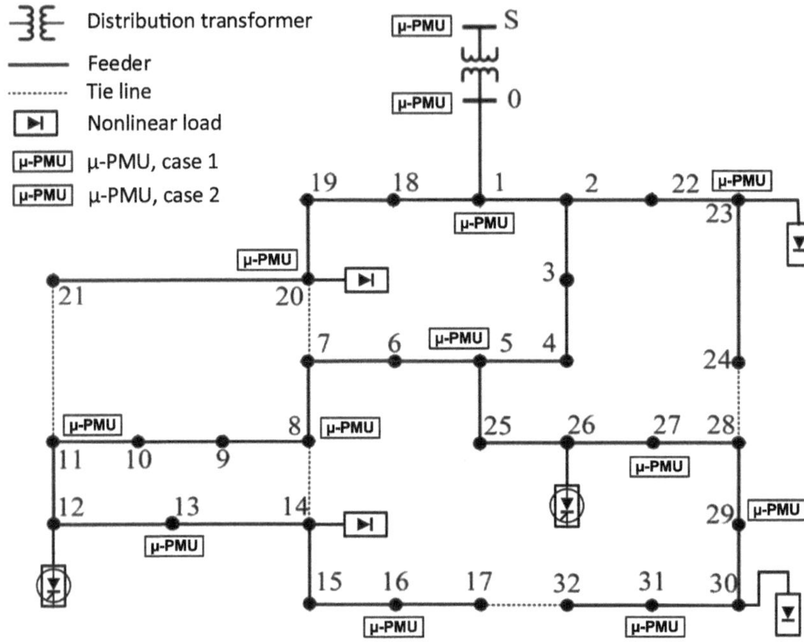

Fig. 4 Modified IEEE 33-node distribution test feeder for HSE evaluation.

Gaussian noises and considered phasor measurements. The measurement variances are also assumed to be 1% for all devices.

Two general cases are considered here as follows:

- **Case 1 (observable network)**: In this case, all μPMUs are assumed to be included in the network; their locations are also shown in Fig. 4.
- **Case 2 (unobservable network)**: In this case, some μPMUs indicated in red in Fig. 4 are ignored on purpose. It means the HSE problem is implemented in an underdetermined system.

In case 1, the measurement reading data of all μPMU devices are considered and imported into the HSE model. Because the model is linear, the equations developed in Eqs. (11), (12) are implemented here. In this case, the HSE problem is an overdetermined set of equations and hence, both WLS and SVD methods could be applied. Hence, the harmonic phasors of voltage at unmonitored nodes could be calculated using Eqs. (10), (18) for WLS and SVD methods, respectively. In light of the results, the WLD method is faster than the SVD method. Both methods have successfully calculated the harmonic voltages at unmonitored nodes. For instance, the estimated voltage magnitudes of the 7th harmonic for all feeder nodes are demonstrated in Fig. 5.

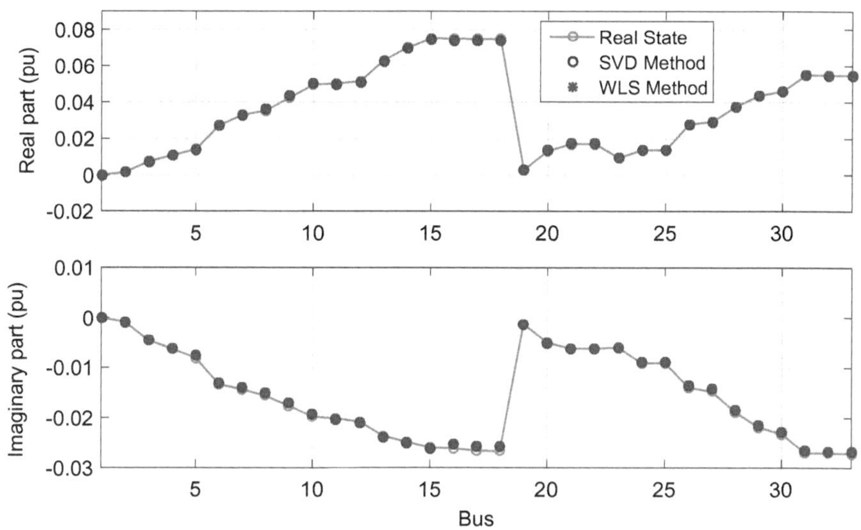

Fig. 5 The estimated voltage magnitude for 7th harmonic, case 1.

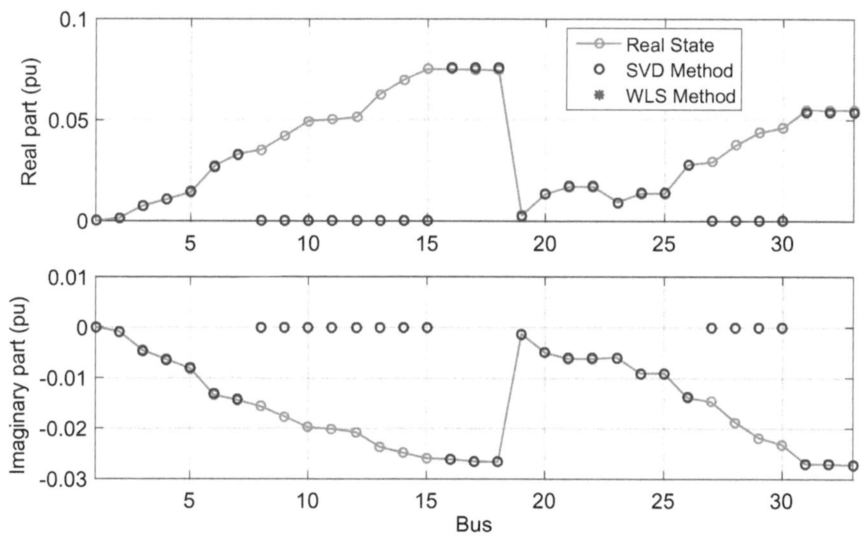

Fig. 6 The estimated voltage magnitude for 7th harmonic, case 2.

In case 2, as previously pointed out, some μPMU devices are ignored in the analyses. The considered meters are shown in Fig. 4 in black. This assumption leads the whole distribution system to become unobservable. Hence, only the SVD method could be applied in this case. The SVD can successfully determine the state variables that are observable by the

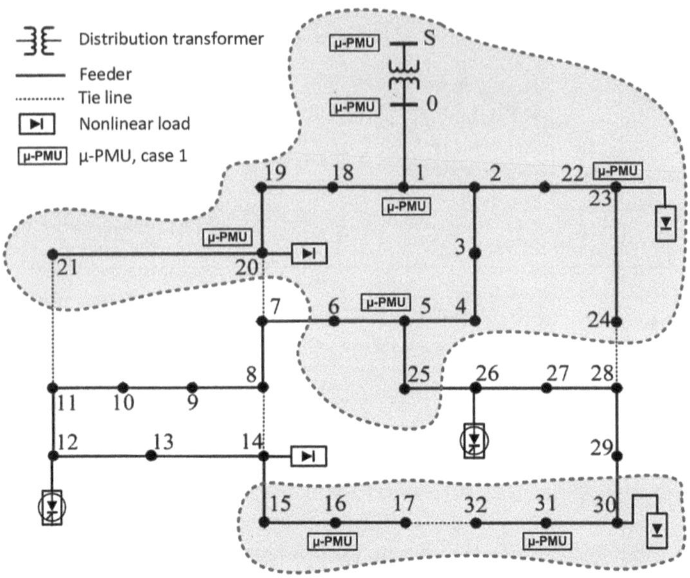

Fig. 7 Observable areas of IEEE 33-node distribution test network, case 2.

current set of input measurements. Also, the model in Eq. (18) could be applied to find the harmonic voltages at unmonitored nodes. For example, the estimated harmonic phasors of all nodes are shown in Fig. 6 for case 2. As is obvious, the SVD method estimates the observable quantities with proper accuracy. In this figure, the results of applying the WLS method are also demonstrated, indicating that the WLS method could not converge and as a result could not estimate any state variables.

As previously stated, the SVD method could be employed as a good alternative for finding the observable areas of the distribution network. In case 2, the system is not completely observable and hence, some observable islands could be found by analyzing the null space of the SVD method. Such an analysis has been carried out for case 2 and its result has been also graphically illustrated in Fig. 7. Two observable islands can be seen in this figure, presuming the set of μPMUs in case 2. There is a general rule of thumb for μPMUs that says any installed μPMU device makes its adjacent nodes observable. This rule could be validated in this figure for the installed μPMUs and as can be easily seen, it is exactly consistent with the results calculated by the null space analysis.

3 Voltage sag state estimation

The most probable and important power quality disturbances are voltage sags and long/short-term power interruptions. In this vein, voltage sags are more important because, on the one hand, events related to the voltage phenomena occur much more frequently and, on the other hand, this phenomenon even may trip an industrial factory due to its destructive effect on the control systems and sensitive electrical apparatus. From an economic perspective, the voltage sags phenomenon may impose high costs on the utility, estimated between $5000 and $250,000 [9]. From a technical viewpoint, voltage sags are generally stochastic and mainly are the results of short circuits in the network lines. In this regard, some uncertain parameters such as fault location, fault type, fault resistance, and prefault voltage would affect the resulting characteristics of voltage sags. With this aim in mind, voltage sag state estimation (VsSE) is proposed for the identification of this PQ phenomenon and also the analysis of its characteristics [10,11]. The VsSE could be generally defined as:

Estimation of profile and/or characteristics of voltage sag in the entire network or its monitored area using historical data and/or provided measurements in the monitored areas.

According to the above definition, two general objectives could be expected from VsSE, that is, the estimation of the voltage sag profile and the estimation of voltage sag performance characteristics. These objectives are explained as follows:

- *Estimation of the voltage sag profile*: In this approach, the profile of the voltage sag is estimated in the unmonitored nodes of the distribution network. Indeed, this approach carefully analyzes the network model through the calculation of the voltage sag profile based on current and previous information. The final results depend on the proper modeling of the network and fault conditions. The primary principles of this method were first proposed in [10], specifically for radial distribution networks; later on, they were developed in [11,12].
- *Estimation of voltage sag performance characteristics*: In this approach, the number of voltage sags with specified magnitude and duration are estimated using historical records and measurements data in monitored nodes [13,14]. The most common index for representing voltage sag performance is the $SARFI_x$[a] index. It is determined as the number of

[a] System Average RMS Frequency Index.

voltage sags (N_i) that violates the limit of x pu (either for voltage sags or swells, increasing or decreasing x pu in the magnitude) for the total feeder customers (N_T). This index is mathematically stated in Eq. (19). For example, $SARFI_{0.9}$ shows the number of voltage sag incidents with magnitudes lower than 0.9 pu for the total feeder customers.

$$SARFI_x = \frac{\sum N_i}{N_T} \tag{19}$$

In the following subsection, the modeling and solution methods of VsSE are further explained.

3.1 VsSE modeling methods

So far, different methods have been presented in the literature for VsSE modeling. They could be generally categorized as the model-based VsSE (stochastic approach) and the measurement-based VsSE (monitoring approach). These categories are briefly defined as follows:

- *Model-based VsSE*: In this approach, the voltage sag profile or its characteristics is merely estimated by the network modeling and fault rate of the network equipment. Hence, no measurement is employed in such a stochastic approach. Indeed, a probabilistic analysis is provided in this approach using the recorded historical data of the distribution system and equipment fault rates.
- *Measurement-based VsSE*: In this approach, various measurements together with the network model are employed to determine the voltage sag performance of the distribution grid.

So far, various modeling methods have been proposed for these two categories, but due to the importance of monitoring the electrical networks, here we only focus on the measurement-based VsSE models. In addition, more often than not, there is no validated historical data that may restrict the application of model-based approaches. Based on the above explanation and before introducing the measurement-based VsSE models, the two most important model-based VsSE methods are briefly explained:

- *Fault position method*: The fundamental principle of this method is based on dividing the distribution network into some sections in which the fault conditions are similar. These sections should be selected in such a way that fault conditions (depth of voltage sag, duration, clearing time of the fault, network topology, etc.) would be the same at each point of them. In each section, a fault location is selected as its representative. Therefore, the whole network can be divided into some fault positions.

Then, each fault position would be analyzed in various simulations with different fault properties. The results are then combined with the fault rates of the equipment for extracting voltage sag performance [15–18].

- *Critical distance method:* In this method, a voltage divider model is employed to determine the critical distances (distance from the PCC or any specific node in which if a certain fault occurs, a voltage sag with specified properties arises). In this method, by using known voltage sag characteristics, critical distances that may result in such voltage sags are determined. This approach requires less information compared to the fault position methods; however, it also provides less accuracy. This method was first proposed in [19] and later developed in [15,20,21].

3.1.1 The principle of measurement-based VsSE models

Short circuit analysis has a great role in voltage sag modeling and meter placement. There are four basic principles for VsSE modeling shown in Fig. 8 that are described as follows:

- *Principle 1 (the concept of fault positions)*: This concept has a great application in VsSE modeling. In this step, all areas of the network are divided into fault positions based on types of lines, fault consequences, topology, etc. Each fault position represents an area of the network in which faults have the same effects. For example, it is possible to divide each line every 2 or 5 km. Hence, fault positions represent the network with some discrete points.
- *Principle 2*: According to the network topology and characteristics, different faults with different properties (prefault voltage, fault impedance, and fault type) could affect any point in the network, whether it is close or far to the fault location. To this end, various VsSE modeling methods are presented based on analyzing the effects of different faults on the steady-state voltages of nodes (as candidates for installing indicators and meters). Moreover, in this step, various faults are intentionally applied in fault positions to analyze their effects on the network nodes in terms of residual voltages.

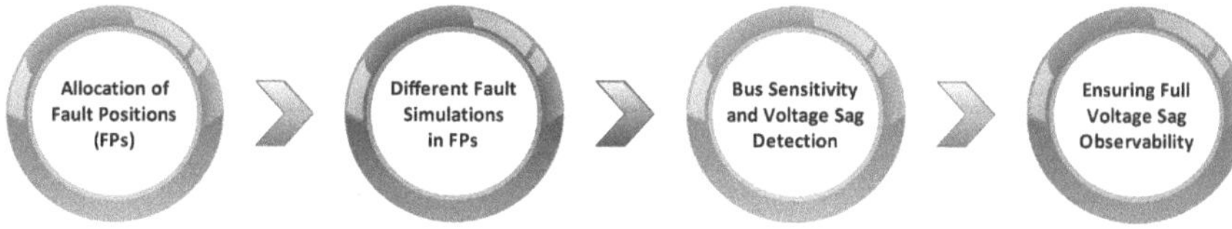

Fig. 8 Four basic principles of measurement-based modeling of the VsSE problem.

- *Principle 3*: After evaluation of the effects of the faults, residual voltages can be determined. In this step, considering a threshold for the identification of voltage sags (e.g., 0.9 pu), the sensitivity of each node to the corresponding fault type can be simply determined. If the residual voltage at the selected node is lower than the voltage sag threshold, then that node could provide an opportunity for its meter to successfully detect the incident fault at the relevant fault position. In other words, that fault type at that fault position is observable by the installed meter at that node.
- *Principle 4*: Principle 3 describes the concept of fault observability. However, the main goal in VsSE modeling is full observability of fault positions with a minimum number of required meters. Accordingly, an optimization problem could be defined for the optimal selection of metering points considering various economic and technical limitations.

The four basic principles mentioned above establish the main blocks for measurement-based VsSE modeling and fault observability. These principles are shown in Fig. 8. In accordance with this figure, one could follow the main steps in the definition of the VsSE problem. Various approaches based on these four principles are proposed in the literature. Of them, we adopt the most usual modeling method, which is based on the concept of monitor reach area (MRA).

3.2 Modeling based on monitor reach area

As previously stated, the voltage sags are mainly the results of different faults in the distribution system. Faults on lines and nodes exhibit noticeable effects. Line faults are very prevalent while faults on nodes are less frequent, although they produce much more voltage drop effects. The monitor reach area (MRA) is a concept for identifying affected areas of faults highly beneficial for meter placement purposes. The MRA could be defined as follows:

The MRA for node k is defined as an area of the distribution network in which the installed meter at this node can detect the occurrence of a voltage sag based on the threshold specified in that area.

To develop the MRA model, it is necessary to define fault positions. At first, we should only consider the fault positions of nodes and then extend the model for lines. To do so, let us suppose that the total number of fault positions is equal to N^{fp}. According to *principle 2* in Fig. 8, it is necessary to simulate various faults at each fault position. Conventional short circuit analysis using an impedance matrix could be conducted. In this method, the during-fault voltage of a node (not faulted node) that is due to the fault that occurred at an arbitrary node can be calculated as Eq. (20).

$$\left[V^{\mathrm{df}}\right] = \left[V^{\mathrm{pf}}\right] + [\Delta V] \tag{20}$$

In Eq. (20), the superscripts "pf" and "df" stand for prefault and during-fault conditions of variables and parameters, respectively. Here, they have been only used for the node voltages. For example, for a typical symmetric three-phase to ground fault, the during-fault node voltages can be represented as Eq. (21).

$$\left[V^{\mathrm{df}}\right] = \left[V^{\mathrm{pf}}\right] - [Z](\mathrm{diag}[Z])^{-1}\left[V^{\mathrm{pf}}\right] \tag{21}$$

where Z is the impedance matrix of the network and diag $[Z]$ is the diagonal matrix of the self-impedance elements of the Z. If we consider the prefault voltages as 1 pu (as is common), then Eq. (21) can be simplified and equivalently expressed in matrix form as Eq. (22). In the same vein, the during-fault voltage of node m due to the occurring fault at node k could be calculated using Eq. (23).

$$\left[V^{\mathrm{df}}\right] = [1] - [Z](\mathrm{diag}[Z])^{-1} \tag{22}$$

$$v_m^{\mathrm{df}} = v_m^{\mathrm{pf}} - \frac{Z_{mk}}{Z_{kk}} v_k^{\mathrm{pf}} \simeq 1 - \frac{Z_{mk}}{Z_{kk}} \tag{23}$$

The $\boldsymbol{V}^{df}$ is an $n \times n$ dimension matrix in which the voltage residuals of nodes resulting from the faults that occur at any node, that is, generally fault positions, represent its rows and columns, respectively. The during-fault voltage on a typical node i and with a fault at that node is represented in the (i, i) element of $\boldsymbol{V}^{\mathbf{df}}$; this element equals zero for a symmetrical three-phase to ground fault (fault impedance is assumed to be zero). Off-diagonal arrays of $\boldsymbol{V}^{\mathbf{df}}$ represent the residual voltages of node i during a fault at fault position j. Hence, the effects of fault in the form of the residual voltage or magnitude of voltage sags due to the fault at position j could be found in the column j of the matrix $\boldsymbol{V}^{\mathbf{df}}$. Therefore, this matrix is called the *voltage sag matrix*. Each row of the voltage sag matrix also can represent the behavior of a typical node to the various faults that occurred at all fault positions.

Likewise, if a fault occurs at a network line, it is also possible to extend an analogous equation (23) for the calculation of residual voltages. To do so, suppose that a fault occurs at location p in the line from bus k to j, as can

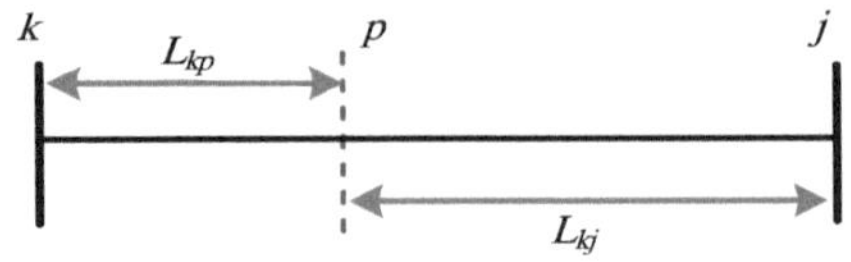

Fig. 9 The schematic of a typical distribution line for the calculation of during-fault voltage.

be seen in Fig. 9. In this case, the residual voltage would be simply calculated using Eqs. (24)–(26).

$$0 < p \triangleq \frac{L_{kp}}{L_{kj}} < 1 \tag{24}$$

$$Z_{mp} = (1-p)Z_{mk} + pZ_{mj} \tag{25}$$

$$Z_{pp} = (1-p)^2 Z_{kk} + p^2 Z_{jj} + 2p(1-p)Z_{kj} + p(1-p)z_{kj} \tag{26}$$

In the equations above, it is possible to equalize Z_{mp} and Z_{pp} with Eq. (23) and use the same equations for the calculation of residual voltages. Accordingly, the voltage sag matrix ($\boldsymbol{V}^{\mathbf{df}}$) will be completely computable for all fault positions using Eqs. (23)–(26). Note that in Eq. (26), Z_{kj} is the transfer impedance of node k and j in the impedance matrix, and z_{kj} is the impedance of line $k-j$.

According to *principle 3* in Fig. 8, the sensitivity of network nodes to the different faults needs to be determined. To this end, it is possible to define two other important concepts from the voltage sag matrix, namely the *affected area* and *exposed area*. These two new concepts are described below.

- ***Affected area***: This is an area of the network in which the during-fault voltages are lower than the voltage sag threshold when a fault occurs at a certain fault position. The columns of the voltage sag matrix ($\boldsymbol{V}^{\mathbf{df}}$) are associated with this area.

In Fig. 10A, the affected area is graphically illustrated for a typical system when a three-phase fault occurs at node #11. The larger area is for the 0.9 pu threshold, which shows that more loads experience a during-fault voltage lower than 0.9 pu while the smaller area is for the 0.7 pu threshold. In general, the lower the threshold value, the smaller the affected area; this is due to the fact that the lower loads mostly experience a deeper voltage sag, especially in power distribution networks. Finally, it should be noted that the affected area in Fig. 10A is plotted assuming a three-phase to ground fault

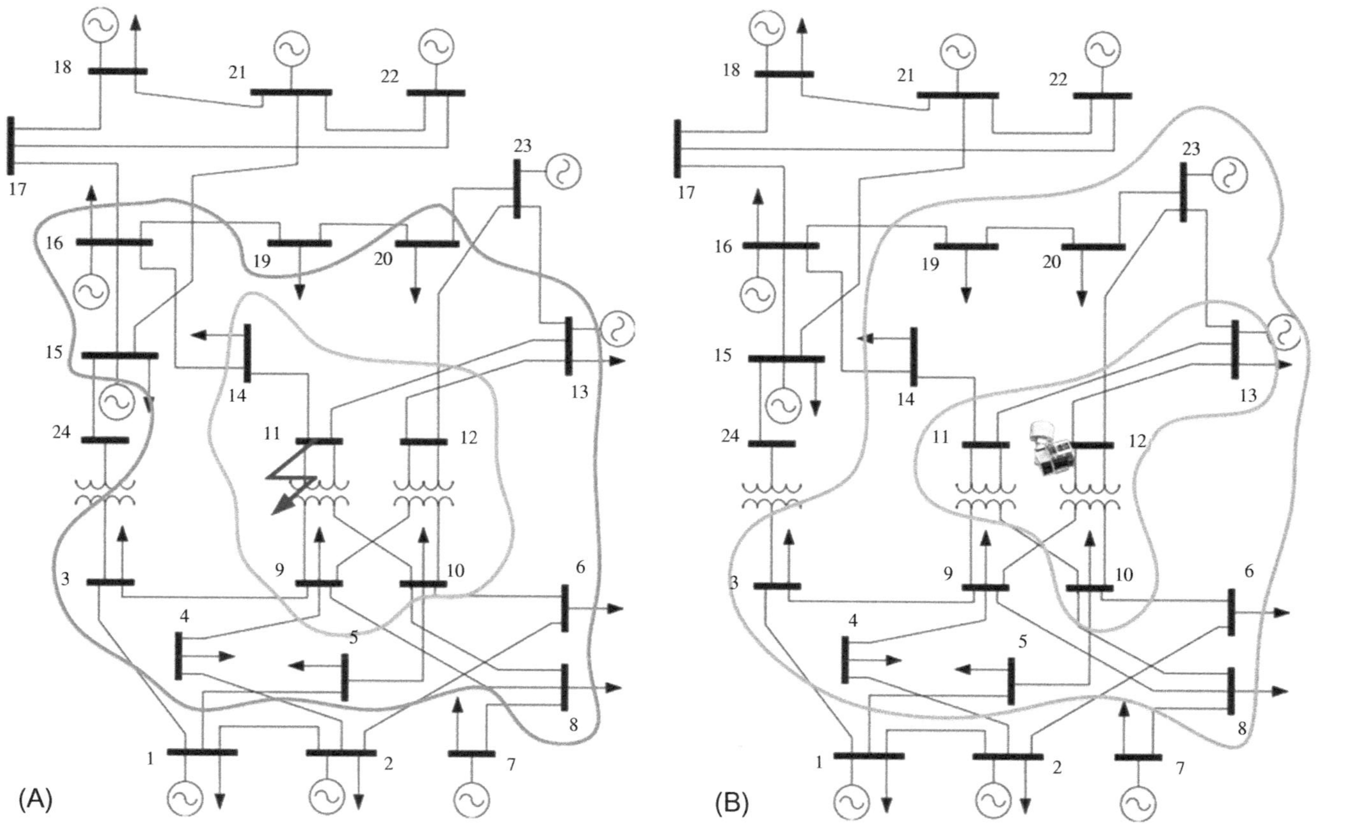

Fig. 10 (A) The affected area for a fault occurred at node #11 and for 0.9 and 0.7 pu thresholds; (B) the exposed area for a monitor installed at node #12 and for 0.9 and 0.5 pu thresholds.

for positive sequence. Thus, for a comprehensive analysis, different faults with various fault parameters should be considered.

- ***Exposed area (or vulnerable area):*** This is the set of network sections including some nodes and lines (or even only some segments of a line) that from the point of view of a typical node *k*, the voltage sag below the considered threshold could be observed. The rows of the voltage sag matrix (V^{df}) are associated with this area.

In Fig. 10B, the exposed area is also graphically illustrated for node #12. In this figure, the exposed areas for two voltage sag thresholds 0.5 and 0.9 pu are depicted. For the lower threshold (i.e., 0.5 pu), nodes #10, #11, and #13 and some parts of the network are located in the exposed area of node #12. It means that all equipment within this area can be easily observed from node #12 if a fault occurs and the voltage goes below 0.5 pu.

According to *principle 4*, an optimization problem needs to be developed for modeling the VsSE problem. To this end, it is necessary to develop an effective meter placement problem for the full observability of the network. Toward this aim, proper measurements could also be employed for the calibration of stochastic VsSE models. Anyway, it is evident that the monitoring program has a great role in the estimation of voltage sag profile and characteristics. Because each point in the network must be observable at least once, such a monitoring program is expected to be able to determine the number and location of measuring devices according to the intended voltage threshold. Due to the key role of meter location, in the following, a meter placement algorithm based on the MRA concept is introduced.

3.2.1 Meter placement based on MRA

For voltage sag observability purposes, a meter placement procedure with the aim of finding the minimum number of measurements should be employed. As discussed previously, the number of recorded voltage sags directly depends on the considered voltage sag threshold. Thus, for a meter installed at a node, the concept of MRA that was previously introduced can be redefined here as follows:

The MRA of a node is equivalent to its exposed area.

According to the MRA definition, if a meter is installed at a node, it could simply detect the voltage sag incidents at some fault positions (corresponding to the rows of the voltage sag matrix) by means of the residual voltage of this node. The MRA of a node contains some fault positions (nodes and lines) of the distribution network where a fault occurring on those locations will definitely result in a voltage sag on that node below the voltage threshold.

For convenience, we denote the threshold as t and in turn, the MRA of node k as MRA_k^t. The index of fault position j can also be considered in MRA_k^t if the residual voltage of node k due to the fault on j is lower than t. This is mathematically stated in Eq. (27). Because the binary matrices are often employed in the placement algorithms, it is also possible to use the definition of Eq. (28) instead of Eq. (27) for representing the MRA of all nodes of the distribution network.

$$\mathrm{MRA}_k^t = \left\{k, j | v_{kj} \in \left[V^{df}\right], v_{kj} \leq t\right\}, j = 1, 2, \ldots, N^{fp}. \tag{27}$$

$$\mathrm{MRA}^t(i,j) = \begin{cases} 1 \text{ if } v_{ij} \leq t \\ 0 \text{ if } v_{ij} > t \end{cases} \tag{28}$$

The MRA matrix is of $N \times N^{fp}$ dimension for each candidate of meter installation points (i.e., nodes) and fault positions. The binary MRA matrix shows the sensitivity of each meter location to the voltage sag of each fault position; that is, if the array (i, j) equals "1," then the meter at node i is sensitive to the fault that occurred at the fault position j with reference to the threshold of t.

Here, the decision variable vector of X with the length of N is also defined as the availability of a meter at a network. It is mathematically expressed in Eq. (29).

$$X(i) = \begin{cases} 1 & \text{if} \quad \text{monitor installed in node } i \\ 0 & \text{if} \quad \text{o.w.} \end{cases} \tag{29}$$

Hence, if X multiplies with a column j of MRA, the result is the total number of monitor reach areas of all meters that includes fault position j. This shows the total observability of fault position j. Accordingly, the meter placement problem can be expressed linearly within an optimization problem as Eq. (30) and shows good mathematical properties.

$$\begin{aligned} &\mathrm{OF}: \ \min \sum_{i=1}^{N} X(i) \\ &s.t. \ \sum_{i=1}^{N} X(i) \times \mathrm{MRA}(i,j) \geq b_j, \ j = 1, 2, \ldots, N^{fp} \\ &X(i) \text{ binary} \end{aligned} \tag{30}$$

where b is a vector with the size of N^{fp} and is equal to "1" as the minimum level of fault observability. It is also possible to reach various observability levels applying a change in this parameter.

3.2.2 Modeling of VsSE based on MRA and measurements

As we already know, the installed meters for full fault observability can now be employed to deal with the VsSE problem. Because the number of measurements is usually fewer than the state variables, this problem is usually underdetermined. The VsSE model is mathematically expressed in Eq. (31) [22].

$$H = MX + E \tag{31}$$

where,

H is the measurement vector representing the number of voltage sags with a magnitude lower than t. This vector is obtained directly from input measurements.

X is the vector of state variables and defined as the number of faults in each fault position within the study area.

M is the binary measurement matrix relating measurements to the state variables.

E is the vector of measurement errors associated with the input measurements.

It should be emphasized that the VsSE model stated in Eq. (31) is basically designed to yield the best estimation of the number of voltage sags at the unmonitored nodes using the number of recorded voltage sags of the meters. Note also that the state variable X is defined here as the number of faults different from the number of voltage sags. Hence, it is required to relate the number of faults with the number of voltage sags. To do so, the measurement matrix M is defined as binary with the aim of relating the state variables to the measurements. Each element of M represents a line segment and a node. Thus, all network lines should be divided into some long enough segments in which fault incidents will result in voltage sags. If the fault in segment j results in a voltage sag below the threshold value of node i, then the element (i, j) of matrix M must be equal to "1."

In summary, to solve the VsSE problem, the items below need to be taken into account:

- Line segment calculations.
- Calculation of the measurement matrix.
- VsSE constraints and modeling.

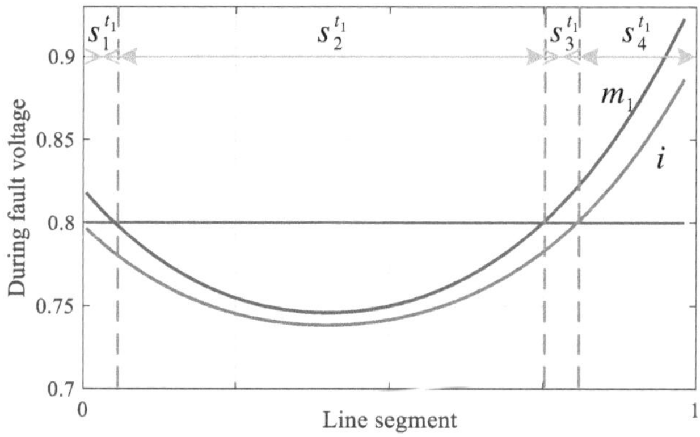

Fig. 11 The segmentizing procedure for the residual voltage calculation of a line.

- Estimation of the number of voltage sags.

For implementation purposes, the essential calculations for VsSE modeling and solving are elaborated in the following.

Line segment calculations

The first step in VsSE modeling is properly defining line segments and their boundaries. For this purpose, the short circuit analysis would be a meritorious alternative. Let t_1 be the voltage sag threshold. The measurement matrix calculated for t_1 is represented as H^1. Similarly, the state variables and the total number of line segments are considered as X^1 and S^1, respectively. All network lines should be divided in such a way that the calculated residual voltages for nodes of interest are exactly higher or lower than the voltage threshold of t_1. The interesting nodes are candidate locations for meter installation (m_1, m_2, ..., m_M) and/or nonmonitored points for which we are interested in determining their PQ indices. To this end, the procedure for assigning line segments can be followed as below:

- Eqs. (23)–(26) should be employed for the analytical calculation of the residual voltage at nodes of interest according to the fault in a typical line of l_1.
- For each typical line of $l1$, the intersection points and the beginning and end of each segment could be determined using the calculated residual voltage curves. This is graphically shown in Fig. 11 for two nodes, that is, the node of interest i and the measurement node m_1. Hence, segments of line l_1 could be easily determined using the calculated residual voltages of these two nodes for faults in line l_1. Here, for voltage threshold t_1, the line l_1 has been divided into four segments as $s_1^{t_1}$, $s_2^{t_1}$, $s_3^{t_1}$, and $s_4^{t_1}$.

- This procedure should be applied to all other network lines and nodes of interest. Alternatively, all calculated line segments will constitute the matrix S^1 for the threshold of t_1.

After determining all line segments, the state variables will be simply defined according to the size of all segments. Here, the vector of X^1 is considered as the state variable that corresponds to the threshold of t_1. Each state variable represents the number of faults that occurred at each line segment in the study horizon.

Measurement binary matrix calculation

After the step above, the measurement matrix related to the t_1 that is denoted by M^1 must be calculated. The matrix M^1 has the dimension of $M \times S^1$ according to the total measurement nodes (M) and the total line segments (S^1). This matrix could be calculated using Eq. (32).

$$M^1 : (m_i s_i) = \begin{cases} 1 & \text{if} \quad v_{m_i} \leq t_1 \text{ (fault in } s_i) \\ 0 & \text{if} \quad v_{m_i} > t_1 \text{ (fault in } s_i) \end{cases} \tag{32}$$

For example, according to Fig. 11, the measurement matrix for line l_1 would be determined as Eq. (33). The element of $m_1(1, 2)$ should be equal to "1" because the fault on segment $s_2^{t_1}$ has led to a residual voltage lower than the threshold of t_1 at node m_1. Hence, it should be considered a voltage sag for this fault at the related segments. All other elements will be determined similarly for all nodes of interest and network lines. Alternatively, each network line adds a set of new columns (according to its total segments) to the measurement matrix.

$$M^1 = \begin{array}{c} \\ m_1 \\ \\ \\ \\ m_M \end{array} \begin{array}{c} \overbrace{s_1^{t_1} \; s_2^{t_1} \; s_3^{t_1} \; s_4^{t_1}}^{S_{l_1}^{t_1}} \quad \ldots \; S_L^{t_1} \\ \begin{bmatrix} 0 & 1 & 0 & 0 & \cdot & \cdot \\ \cdot & \cdot & \cdot & \cdot & \cdot & \cdot \\ \cdot & \cdot & \cdot & \cdot & \cdot & \cdot \\ \cdot & \cdot & \cdot & \cdot & \cdot & \cdot \\ \cdot & \cdot & \cdot & \cdot & \cdot & \cdot \end{bmatrix} \end{array} \tag{33}$$

Because some voltage sag monitors have multiple voltage thresholds, various voltage thresholds might be considered. An example of three voltage sag thresholds is represented in Fig. 12. Similarly, measurement matrices M^i, input measurements H^i, and state variables X^i should be determined for each voltage threshold t_i. Such a general matrix for various voltage thresholds could be described as Eq. (34).

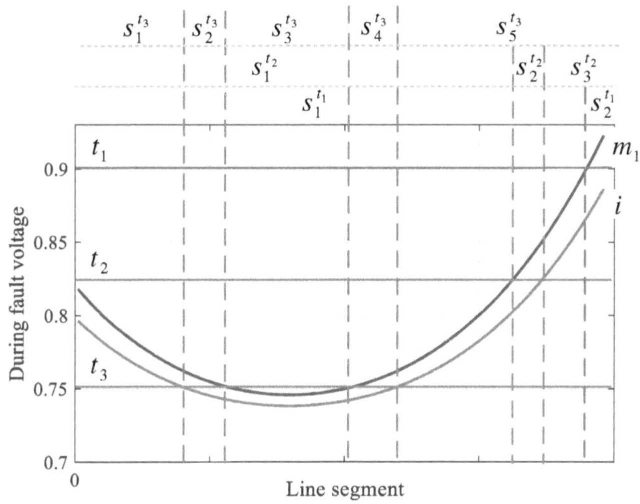

Fig. 12 Calculated residual voltage for a line and three voltage thresholds.

$$\begin{bmatrix} H^1 \\ H^2 \\ H^3 \end{bmatrix} = \begin{bmatrix} M^1 & 0 & 0 \\ 0 & M^2 & 0 \\ 0 & 0 & M^3 \end{bmatrix} \begin{bmatrix} X^1 \\ X^2 \\ X^3 \end{bmatrix} \tag{34}$$

In a matrix form, for each distribution system assuming T voltage thresholds, the general equation of the VsSE problem could be represented as Eq. (35). Also, according to the total number of voltage thresholds, the total number of line segments (S) is equal to the summation of all segments in each voltage threshold.

$$[H] = [M][X] \tag{35}$$

Moreover, the total number of state variables is equal to the total number of line segments (S). From a mathematical point of view, the compact form Eq. (35) is underdetermined because the number of measurements is limited and the voltage thresholds are more than the state variables as well. However, the new state variables are not completely independent and it is possible to define some relations for enhancing the solvability chance of the problem, as will be further explained in the next step.

Constraints and formulation of VsSE problem

From the solvability point of view, the problem of Eq. (35) could be mathematically improved using the following constraints:

- *Constraint* 1: The summation of all state variables for a typical line and for all voltage thresholds should be equal; this is because when a fault occurs at a line, it will be experienced in all segments. This constraint could be mathematically expressed in Eq. (36) for variables associated with the segments in Fig. 12.

$$x_1^{t_1} + x_2^{t_1} = x_1^{t_2} + x_2^{t_2} + x_3^{t_2} = x_1^{t_3} + x_2^{t_3} + x_3^{t_3} + x_4^{t_3} + x_5^{t_3} \tag{36}$$

- *Constraint* 2: For the faults that either do not occur (zero) or occur (positive), let $x_i \geq 0$.
- *Constraint* 3: According to the line segments, some inequality constraints could be extended for the segments that are entirely within other segments. For example, according to Fig. 12, because the segment $s_1^{t_2}$ is completely positioned in the segment $s_1^{t_1}$, then it could be inferred that $x_1^{t_2} \leq x_1^{t_1}$.

According to the constraints above, the VsSE problem can be formulated as a linear programming (LP) problem Eq. (37) that is efficiently solved using commercially available software.

$$\begin{aligned}
&\text{OF}: \ \min \left| x_k^t - \gamma_k^t \sum_{k=1}^{S_j^t} x_k^t \right|, \ \forall k, j, t \\
&s.t. \ \text{C0}: \ [H] = [M][X], \ x_i^t \in \mathbb{N} \ \forall i \in \left\{1, 2, \ldots, \sum_{t=1}^{T} \sum_{j=1}^{L} S_j^t \right\}, \ t \in \{1, 2, \ldots, T\} \\
&\quad \text{C1}: \ \sum_{k=1}^{S_j^t} x_k^t - \sum_{k=1}^{S_j^{t+1}} x_k^{t+1} = 0, \ \forall j \in \{1, 2, \ldots, L\}, \forall t \\
&\quad \text{C2}: \ x_k^t \geq 0 \\
&\quad \text{C3}: \ \begin{cases} x_k^t - x_{k'}^{t+1} \geq 0 \text{ if } a_k^t \leq a_{k'}^{t+1} \text{ and } b_k^t \geq b_{k'}^{t+1} \\ x_k^t - x_{k'}^{t+1} \leq 0 \text{ if } a_k^t \geq a_{k'}^{t+1} \text{ and } b_k^t \leq b_{k'}^{t+1} \end{cases}, \ x_k^t \in \left[a_k^t b_k^t\right], \ x_k^{t+1} \in \left[a_{k'}^{t+1} b_{k'}^{t+1}\right]
\end{aligned} \tag{37}$$

In the above equation;

j is the index of network lines.

t is the index of the considered voltage threshold.

i is the total segments of all system lines in all considered voltage thresholds.

$1 \leq k, \ k' \leq S_j^t$ are the segments in which line j has been divided.

γ_t^k is the parameter equal to the quotient obtained from dividing the length of the segment of x_k^t by the total length of line j containing the segment.

The objective function in Eq. (37) has the most compatibility with fault probability distribution in the network lines. This is very useful because the uniform distribution for faults in the line should distribute the number of faults according to the length of each segment.

Estimation of the number of voltage sags

The optimization problem in Eq. (37) yields an optimal number of faults, X. On the other hand, it is necessary to determine the number of voltage sags at the nodes of interest. To this end, the expected voltage sags at node i would be calculated according to the binary measurement matrix of M_i. The M_i matrix is calculated similarly to the matrix M for the measurement nodes. For example, for node i represented in Fig. 11, the M_i matrix could be computed as Eq. (38).

$$M_i^1 = [\overbrace{\overset{s_1^{t_1}}{1}\ \overset{s_2^{t_1}}{1}\ \overset{s_3^{t_1}}{1}\ \overset{s_4^{t_1}}{0}}^{l_j} \cdot\ \cdot\ \cdot] \tag{38}$$

Hence, the number of voltage sag h_i for node i of interest could be calculated as Eq. (39). The method developed here could be easily extended to various fault types [15,23].

$$h_i = M_i^1 X^1 \tag{39}$$

Finally, it should be noted that the VsSE models based on the MRA concept do not utilize fault location methods. However, some researchers have proposed the use of fault location methods in the VsSE models to enhance performance [13,14] because, in this way, the solution method is not affected by fault properties such as the fault impedance and the prefault voltage. Interested readers are invited to check out these methods as well.

3.3 An illustrative example of the voltage sag state estimation problem

For both teaching and learning purposes, in this section, an illustrative example of the VsSE problem is described. The method is also applied to the typical test system as previously shown in Fig. 10. The monitor locations are calculated by the MRA method for a voltage sag threshold of 0.7 pu. Hence, any fault that resulted in a voltage below 0.7 pu would be detected by at least one monitor. As a consequence, nodes #3, #6, #8, and #17 are selected as the installation points, and thereby, the number of voltage sags detected at the monitored nodes forms the measurement vector.

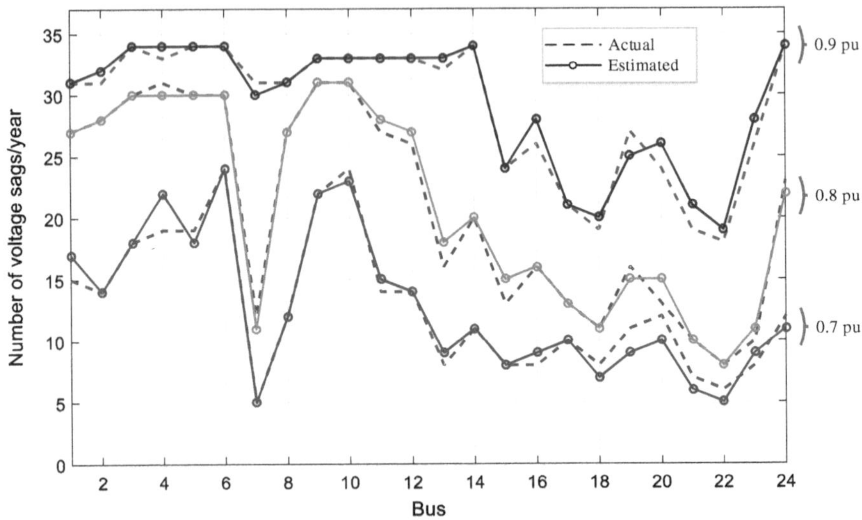

Fig. 13 Voltage sag state estimation results for IEEE 24-bus RTS.

Different faults are assumed to have occurred concerning a predefined fault distribution. Their consequences are analyzed using the short circuit calculation by which the residual voltages at the network nodes are simply calculated. Then, according to the voltage sag thresholds, the voltage sags are detected at various locations. These values are considered as actual numbers of voltage sags. On the other hand, the numbers of voltage sags recorded by the installed meters are employed to implement the VsSE model of Eq. (37), and thereby, to calculate the number of incident faults. After that, the number of estimated voltage sags for all other network nodes is determined using Eq. (39). These values are finally considered as the estimated numbers of voltage sags.

The results for different voltage sag thresholds are plotted in Fig. 13. The actual and estimated values are represented by the dashed and bold lines, respectively. As can be seen, the VsSE model has successfully estimated the number of voltage sags in almost all cases. For monitored nodes, the estimated values are the same as the actual values. The maximum errors are obtained as 5.23%, 8.42%, and 11.36% for 0.9 pu, 0.8 pu, and 0.7 pu thresholds, respectively. The reason why the error for lower-voltage thresholds has been increased is that the current set of meters guarantees that each voltage sag with a magnitude lower than 0.7 pu has been certainly detected by at least one monitor. Therefore, voltage sags with a magnitude greater than 0.7 pu could be recorded by more monitors, resulting in better and more accurate estimates.

4 Transient state estimation

In power transmission and distribution system analysis, fault location, identification, and detection of the disturbance sources have crucial importance. Therefore, transient state estimation (TSE) is basically proposed for the identification of transient phenomena due to various disturbances in the electrical grids. The usual procedure for transient analysis is to simulate some probable scenarios and investigate their impacts and consequences on system variables [24,25]. Transient state simulation software such as EMTP and PSCAD is mostly employed in simulations as well. However, this procedure is often time-consuming and requires a large number of scenarios to be simulated. Another solution is using TSE for analyzing transient effects in the electrical system. TSE in both transmission and (modern) distribution systems could be defined as follows.

Transient state estimation is a process in which by using synchronized high-resolution measurements in monitored buses/nodes, a variety of transient phenomena due to power system disturbances can be estimated in the unmonitored buses/nodes.

The concept of TSE is simply represented in Fig. 14. The transient simulation is mainly employed for analyzing the impacts of transient disturbances on the state variables of the system while TSE is utilized for identifying the reasons for transient disturbances in the system. Hence, if the TSE is completely implemented, it is possible to quickly locate and identify various system disturbances. It is interesting to underline that TSE could be considered the reverse process of the transient simulation [24,26].

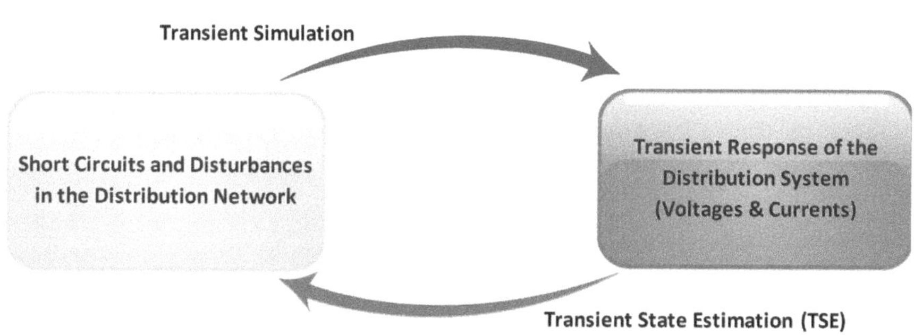

Fig. 14 Transient state estimation vs transient simulation.

Given the TSE definition, two general requirements must be taken into account in the TSE algorithm:

- Synchronized measurements are required for the proper estimation of the dynamics of the network. Meanwhile, the data should be trackable using GPS tags.
- The dynamic model of various system components is required. Note that the network dynamic model is built using all individual models.

On the other hand, because the formulations of conventional SE problems are mostly developed in the phasor domain for steady-state variables, new formulations should also be employed for TSE modeling. In this context, two general models have been proposed for modeling TSE: the state-space (SS) and numerical integral substitution (NIS) models.

- *State-space model*: This model is formulated using the state-space model of all network elements. The elements are modeled by differential equations and these equations are then rebuilt to establish a recursive algorithm between the state-space variables. In this model, inductance currents and nodal voltages are considered state variables [24,27,28].

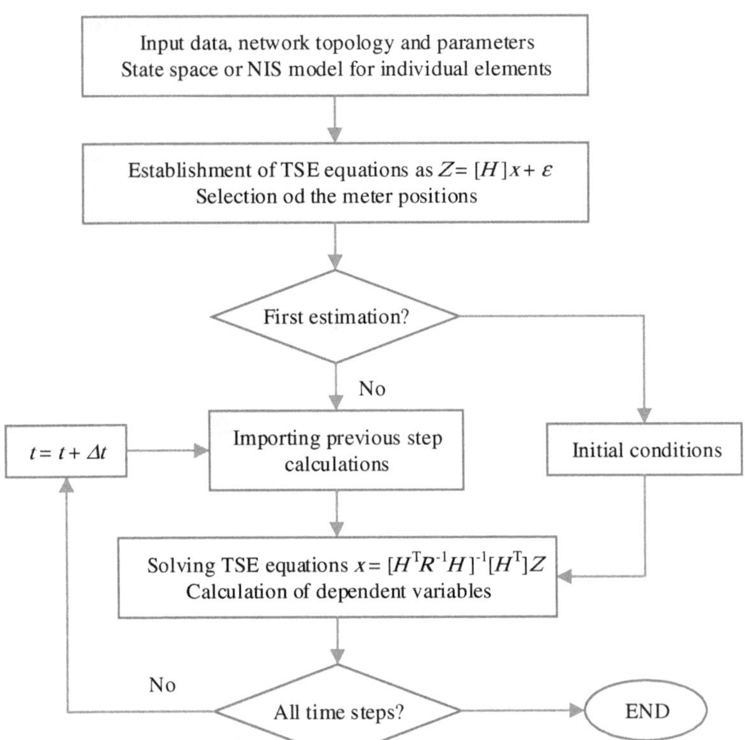

Fig. 15 The general algorithm for the TSE problem

- *Numerical integral substitution*: In this model, an equivalent dynamic model using *Dommel's method* is employed [29]. The general method is based on converting various elements into their *Norton* equivalents. In the NIS method, the state variables are only nodal voltages [25,26,30–33].

The general algorithm implemented for the TSE solution procedure is illustrated through the flow chart depicted in Fig. 15 for both SS and NIS models. Needless to say, the instantaneous measurements of voltages and currents must be utilized for the step-by-step calculation of the state variables from current inputs and previous results.

For TSE simulation purposes, a time step of 50 μs is considered in almost all studies. The 50 μs time step is considered a standard criterion for transient simulation and state estimation because it could properly represent and record transients in the power transmission and distribution systems. Moreover, this time constant is the same as the default simulation time step in EMTP and PSCAD. Hence, this criterion should be considered as the minimum required resolution for the PQ meters. On the other hand, smart measurement devices with a high sampling rate (at least 20 kHz) are rarely installed in the power distribution systems at this time. Therefore, the TSE algorithms may require more advanced measurement infrastructures to be applicable. However, recent advancements in the field of microphasor measurements have created expectations for the possibility of TSE algorithms in the near future [33,34].

Another difference between TSE and other SE algorithms is the different solution methods for solving it. Because only instantaneous voltages and currents are imported into the TSE algorithm, this model is linear. Therefore, not only is it fast, but also both WLD and SVD methods could be easily applicable. Furthermore, the observability level of the network could be simply analyzed by evaluating the measurement matrix.

It has been noted in the literature that the establishment of the SS model is time-consuming and nonoptimal, even for small networks [28]. Besides, this model could not handle inductive cutsets and capacitive loops. Hence, the NIS method has attracted more attention in recent years. For this reason, the NIS model is only explained in this chapter. The reader could follow both models in detail using the mentioned references.

4.1 TSE based on numerical integral substitution

The NIS method is currently the most generally accepted method for solving electromagnetic transients. As already mentioned, this method is faster and more optimal compared to the SS method. In the NIS method, the

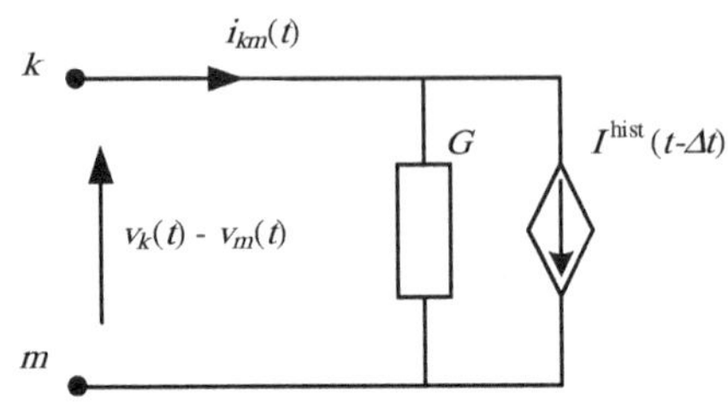

Fig. 16 Dynamic *Norton* equivalent model used by the NIS method.

system is modeled using a series of differential equations for various elements to establish the dynamic model of the network. These models are then combined using the numerical integral rule, which leads to a discrete dynamic model. In the NIS method, each element is shown as an equivalent model shown in Fig. 16 [25,26,30]. This model is the dynamic *Norton* equivalent model of the element.

As previously stated, the modeling approach in the NIS method is based on *Dommel*'s method [29]. This method is the basis of available commercial software specifically developed for transient studies (e.g., EMTP). Considering an element between two nodes of k and m, as shown in Fig. 16, the element model in the NIS method could be represented as Eq. (40).

$$[G]v(t) = i_s(t) - I^{hist} \tag{40}$$

where,

$[G]$ is the conductance matrix representing the internal conductance of the element.

$i_s(t)$ is the input current of the element.

I^{hist} is the injection history for each memorial element (inductor and capacitor).

In fact, the NIS method is equivalent to the nodal analysis method in the electrical circuit analysis theory. In this method, the instantaneous voltages of network buses/nodes are considered state variables.

The RLC elements have various models in *Dommel*'s approach. With the help of the circuit model in Fig. 16, the descriptive equations of each element could be represented as follows in the NIS method:

- *For resistor:*

$$i_{km}(t) = G(v_k(t) - v_m(t)) \tag{41}$$

$$I^{hist}(t - \Delta t) = 0 \tag{42}$$

$$G = {}^{1}/_{R} \tag{43}$$

- *For inductor:*

$$i_{km}(t) = G(v_k(t) - v_m(t)) + I^{\text{hist}}(t - \Delta t) \tag{44}$$

$$I^{\text{hist}}(t - \Delta t) = -G_{eff}(v_k(t - \Delta t) - v_m(t - \Delta t)) - i_{km}(t - \Delta t) \tag{45}$$

$$G_{eff} = {}^{\Delta t}\!/_{2L} \tag{46}$$

- *For capacitor:*

$$i_{km}(t) = G(v_k(t) - v_m(t)) + I^{\text{hist}}(t - \Delta t) \tag{47}$$

$$I^{\text{hist}}(t - \Delta t) = G_{eff}(v_k(t - \Delta t) - v_m(t - \Delta t)) + i_{km}(t - \Delta t) \tag{48}$$

$$G_{eff} = {}^{2C}\!/_{\Delta t} \tag{49}$$

It can be inferred that the dynamic model of the power distribution system equipment is developed according to its internal modeling elements. For example, for a small-scale generator (as distribution generation) that can be modeled using an inductor and resistor, Eqs. (41)–(46) should be combined to build its model. Dynamic models of all other equipment could also be developed in the same way. Lastly, the complete dynamic model of the distribution network can be easily obtained by means of the dynamic models of all network equipment. Note finally that the network dynamic model is actually similar to Eq. (40) for unknown instantaneous voltages of all nodes.

To develop a standard state estimation model, it is necessary to insert various measurements into the model and build a measurement matrix. The rows of matrix H would be determined concerning the type of input measurements. As already mentioned, the state variables in the NIS method are the instantaneous nodal voltages. Hence, according to the individual model of elements shown in Fig. 16 and the type of input measurements, the rth row of the measurement matrix can be simply calculated as follows:

- *For a nodal voltage measurement at node k:*

$$h_{rk} = 1 \tag{50}$$

$$z_r = v_k^{\text{meas}}(t) \tag{51}$$

- *For voltage measurements from two ends of an RLC element:*

$$h_{rk} = 1, \quad h_{rm} = -1 \tag{52}$$

$$z_r = v_k^{\text{meas}}(t) - v_m^{\text{meas}}(t) \tag{53}$$

- *For a current measurement of an RLC element:*

$$h_{rk} = {}^{1}\!/_{G_{eff}}, \quad h_{rm} = -{}^{1}\!/_{G_{eff}} \tag{54}$$

$$z_r = i_{km}^{\text{meas}}(t) - I_{km}^{\text{hist}}(t) \tag{55}$$

For each measurement, all rows of the measurement matrix can be calculated using Eqs. (50)–(55). Analogously, the standard state estimation model of TSE could be mathematically expressed as Eq. (56). In this model, the nodal voltages are considered state variables.

$$z = [H]x + \varepsilon \tag{56}$$

The TSE model in Eq. (56) is linear, and therefore efficient estimation techniques such as WLS (for fully observable systems) and SVD (for partially observable systems) could be easily applied to find its state variables. Considering measurement variances, the solution of state variables can be obtained using Eq. (57); this is exactly analogous to the approach already mentioned. Because the state variables are considered as nodal voltages, other network instantaneous variables such as line currents and node injections would be directly calculated from nodal instantaneous voltages.

$$x = \left[H^T R^{-1} H\right]^{-1} \left[H^T\right] z \tag{57}$$

The reader should keep in mind that the measurement models provided in Eqs. (50)–(55) could also be extended for three-phase networks [25]. Further, the transmission lines could be modeled using the PI model and lumped elements. However, for this purpose, the following considerations should be met:

- The transmission lines should be modeled with a lumped PI model considering phase couplings.
- The transformers should be modeled using three ideal single-phase transformers including phase couplings. Different connection matrices of transformers should also be implemented using *Dommel*'s method for representing various windings of transformers.
- The active and reactive components of static loads should be modeled using equivalent inductance and resistance.

The interested reader is referred to [25] for further details and more valuable information about lumped three-phase models.

The models presented so far are only suitable for short lines with lumped models. As described earlier, for transient simulation, short lines are represented by the PI model, which does not include the distributed nature of the elements. For this reason, the frequency dependency of line parameters is not considered and the models are just valid for short lines. On the contrary, it has to be kept in mind that if the distribution network includes long lines, it is necessary to consider distributed models of line parameters too.

In long lines, the traveling time of the electrical waves passing through the line is more than the standard time step of the simulation. Hence, one criterion for considering long line models is to take into account the longer traveling time of transient waves than the time step of the TSE algorithm. In such a situation, frequency-dependent models or *Bergeron* models [33,34] will be good alternatives. The frequency-dependent models can be implemented if it is necessary to consider line geometry. Otherwise, the Bergeron model is the usual choice for short circuit analysis in transient simulations.

With regard to the TSE simulation time step, it is possible to determine a length limit for lines, which should be determined using distributed models. Toward this aim, by assuming $\Delta t = 50\,\mu s$, the minimum length for long line considerations can be easily calculated as Eq. (58).

$$\frac{\text{length}}{C^{\text{light}}} > t^{\text{travelling}}(\Delta t = 50\,\mu s) \Rightarrow \text{length} > 15\,\text{km} \tag{58}$$

In Eq. (58), C is the traveling wave speed within the line, which translates into the light speed. Hence, for lines longer than 15 km with $\Delta t = 50\,\mu s$, it is necessary to use the long line model in the TSE algorithm. More often than not, the time step of $50\,\mu s$ is considered a standard criterion in such transient studies. The reader is referred to [33–35] for further details and additional discussion on how to insert the complete distributed line models into the TSE problems.

5 Conclusion

In this chapter, a set of state estimation techniques for studying and analyzing the power quality of distribution systems (i.e., HSE, VsSE, and TSE) was introduced. All three subproblems of HSE, VsSE, and TSE as the main categories of the power quality state estimation problem were mathematically modeled and their solution methods were then explained. Due to the importance of power quality issues in future distribution systems and also the lack of complete measurements for the entire distribution network, state estimation techniques seem to be essential for determining the power quality indices of the network.

To put it in a nutshell, in HSE, the steady-state behavior of the distribution network against harmonic distortions is analyzed. Among its advantages, determining harmonic content in the network and the identification and localization of harmonic loads are the main applications of HSE algorithms. In VsSE, short-term voltage issues such as voltage sags and swells are identified. In this vein, the number of occurred voltage sags at unmonitored areas can be determined using the recorded voltage sags of the meters. In TSE, the transient behavior resulting from various disturbances such as system faults is described. This analysis allows describing transient behaviors at unmonitored areas. In the case of full implementation, the TSE model includes all other state estimation algorithms, that is, HSE and VsSE. However, due to the required high sampling frequencies, the TSE models are still a little far from fully fledged implementation.

Acknowledgments

This work was supported by the *Niroo Research Institute* (NRI) under contract CPBDA01/T05. The authors would like to thank *Niki Moslemi* (NRI), *Dr. Azin Shojaei* (TAVANIR Company), and *Vahid Mottaghi* (Esfahan Electricity Power Distribution Company) for their valuable comments and feedback on this research project, and also the editor for his valuable comments and suggestions. The authors hope the readers enjoy this chapter and become inspired by the many research opportunities and challenges lying ahead.

References

[1] G. Heydt, Identification of harmonic sources by a state estimation technique, IEEE Trans. Power Delivery 4 (1) (1989) 569–576.
[2] F. Zhang, S. Zelingher, Power system harmonic state estimation, IEEE Trans. Power Delivery 9 (3) (1994) 1701–1709.
[3] Z.-P. Du, J. Arrillaga, N. Watson, S. Chen, Identification of harmonic sources of power systems using state estimation, IEE Proc. Gener. Transm. Distrib. 146 (1) (1999) 7–12. IET.
[4] G. D'Antona, C. Muscas, S. Sulis, State estimation for the localization of harmonic sources in electric distribution systems, IEEE Trans. Instrum. Meas. 58 (5) (2009) 1462–1470.
[5] W. Press, S. Teukolsky, W. Vetterling, B. Flannery, Numerical Recipes in Fortran 77: The Art of Scientific Computing, Cambridge Univ Press, Cambridge, 1992.
[6] C. Madtharad, S. Premrudeepreechacharn, N.R. Watson, Power system state estimation using singular value decomposition, Electr. Pow. Syst. Res. 67 (2) (2003) 99–107.
[7] N. Watson, J. Arrillaga, Z. Du, Modified symbolic observability for harmonic state estimation, IEE Proc. Gener. Transm. Distrib. 147 (2) (2000) 105–111.
[8] H. Abdi, Singular value decomposition (SVD) and generalized singular value decomposition, in: Neil J. Salkind (Ed.), Encyclopedia of Measurement and Statistics, SAGE Publications, 2007, pp. 907–912.
[9] J. Blanco-Solano, J.F. Petit-Suárez, G. Ordóñez-Plata, N. Kagan, Voltage sag state estimation using compressive sensing in power systems, in: 2019 IEEE Milan PowerTech, IEEE, 2019, pp. 1–6.

[10] B. Wang, W. Xu, Z. Pan, Voltage sag state estimation for power distribution systems, IEEE Trans. Power Syst. 20 (2) (2005) 806–812.
[11] G. Olguin, F. Vuinovich, M.H. Bollen, An optimal monitoring program for obtaining voltage sag system indexes, IEEE Trans. Power Syst. 21 (1) (2006) 378–384.
[12] J. Lucio, E. Espinosa-Juarez, A. Hernandez, Voltage sag state estimation in power systems by applying genetic algorithms, IET Gener. Transm. Distrib. 5 (2) (2011) 223–230.
[13] Y. Liao, Fault location observability analysis and optimal meter placement based on voltage measurements, Electr. Pow. Syst. Res. 79 (7) (2009) 1062–1068.
[14] M. Avendaño-Mora, J.V. Milanovic, Monitor placement for reliable estimation of voltage sags in power networks, IEEE Trans. Power Delivery 27 (2) (2012) 936–944.
[15] M.H. Bollen, Understanding power quality problems, in: Voltage Sags and Interruptions, IEEE press, 2000.
[16] G. Olguin, Voltage Dip (sag) Estimation in Power Systems Based on Stochastic Assessment and Optimal Monitoring, Chalmers Tekniska Hogskola (Sweden), 2005.
[17] L. Conrad, K. Little, C. Grigg, Predicting and preventing problems associated with remote fault-clearing voltage dips, IEEE Trans. Ind. Appl. 27 (1) (1991) 167–172.
[18] C. Becker, et al., Proposed chapter 9 for predicting voltage sags (dips) in revision to IEEE Std 493, the gold book, IEEE Trans. Ind. Appl. 30 (3) (1994) 805–821.
[19] M.H. Bollen, Fast assessment methods for voltage sags in distribution systems, IEEE Trans. Ind. Appl. 32 (6) (1996) 1414–1423.
[20] I.B. Cruz, A.P. Lavega, J.R. Orillaza, Overview of methods for voltage sag performance estimation, in: 2016 17th International Conference on Harmonics and Quality of Power (ICHQP), IEEE, 2016, pp. 508–512.
[21] I.B. Cruz, A.P. Lavega, J.R. Orillaza, Overview of voltage sag profile estimation, in: TENCON 2015–2015 IEEE Region 10 Conference, IEEE, 2015, pp. 1–6.
[22] E. Espinosa-Juarez, A. HernÁndez, A method for voltage sag state estimation in power systems, IEEE Trans. Power Delivery 22 (4) (2007) 2517–2526.
[23] M.H. Bollen, Algorithms for characterizing measured three-phase unbalanced voltage dips, IEEE Trans. Power Delivery 18 (3) (2003) 937–944.
[24] K. Yu, N. Watson, Identification of fault locations using transient state estimation, in: Proc. Int. Power System Transient (IPST2005), 2005.
[25] A. Farzanehrafat, N.R. Watson, Power quality state estimator for smart distribution grids, IEEE Trans. Power Syst. 28 (3) (2013) 2183–2191.
[26] A. Farzanehrafat, N.R. Watson, Review of power quality state estimation, in: 2010 20th Australasian Universities Power Engineering Conference, IEEE, 2010, pp. 1–5.
[27] K. Kent, N.R. Watson, An approximate method for transient state estimation, IEEE Trans. Power Delivery 22 (3) (2007) 1680–1687.
[28] N. Watson, K. Yu, Transient state estimation, in: 2008 13th International Conference on Harmonics and Quality of Power, IEEE, 2008, pp. 1–6.
[29] H.W. Dommel, Electromagnetic Transients Program (EMTP) Theory Book, Bonneville Power Administration, 1987.
[30] N. Watson, A. Farzanehrafat, Identification of the sources of transient disturbances, in: International Conference on Power Systems Transients, IPST2013, 2013, pp. 21–25.
[31] N.R. Watson, A. Farzanehrafat, Three-phase transient state estimation algorithm for distribution systems, IET Gener. Transm. Distrib. 8 (10) (2014) 1656–1666.
[32] N. Suyaroj, S. Premrudeepreechacharn, N.R. Watson, Transients state estimation with measurement noise, in: 2015 12th International Conference on Electrical Engineering/Electronics, Computer, Telecommunications and Information Technology (ECTI-CON), IEEE, 2015, pp. 1–6.

[33] A. Castellanos-Escamilla, N. Watson, R. Langella, Transient state estimation for transmission systems, in: 2020 19th International Conference on Harmonics and Quality of Power (ICHQP), IEEE, 2020, pp. 1–7.
[34] N. Suyaroj, S. Premrudeepreechacharn, N.R. Watson, Transient state estimation with the Bergeron transmission line model, Turk. J. Electr. Eng. Comput. Sci. 25 (2) (2017) 806–819.
[35] A.C. Escamilla, Modelling Nonlinear and Distributed Elements in Transient State Estimation, University of Canterbury, 2015.

SECTION 9

Reactive power compensation in active distribution network

CHAPTER 9

Reactive power management in active distribution network

Baseem Khan
Department of Electrical and Computer Engineering, Hawassa University, Hawassa, Ethiopia

Contents

1 Introduction

The regulation of reactive power to enhance the productivity of the AC supply network is referred to as reactive power compensation (VAR). The amount of idle power generated by reactive elements such as inductors or capacitors is known as the reactive component of power. The real part of

Active Electrical Distribution Network
https://doi.org/10.1016/B978-0-323-85169-5.00005-8

the power must transit via the utility grid to reach the user, which necessitates the use of a reactive component. One of the finest expenditures for lowering energy expenses with a quick payback is power factor correction [1].

In certain scenarios, the reactive component of power is critical to the proper working of the utility grid. The active part cannot be delivered if the network voltage is low. The potential levels required for active power to conduct appropriate work are provided by reactive power. In utility grid management and voltage profile management, the reactive component is required to transport the active part via the grid network to users using various means. For the system administrator, the compensation of reactive power gives a chance to reduce loss and enhance the voltage level. Because capacitors and other electrical devices seem to be more economically justifiable than series VAR compensating devices, they are commonly used for reactive power correction in the utility grid. The requested energy and the appropriate placement of deployed capacitors in the event of distinct distribution network operation modes are two essential elements in the compensation of reactive power. An electricity network with optimal reactive power control can have less active power loss, a better voltage profile, and lower reactive power generating costs.

The distributed generation resources and microgrid system are low-potential sources. The distribution grid consists of small-capacity generators, power storage, controlled and unregulated demands, localized and central controllers, and a communication infrastructure. Microgrids are miniature counterparts of traditional networks along with distributed energy sources. They freely distribute locally generated power to a community of customers. These are usually connected to a conventional system, but they can also run in isolated mode when necessary. As a result, they are thought of as active adaptable cells in a utility grid. Various renewable energy (RE) sources are being linked to electricity grids in increasing numbers across the globe. Electrical networks are presently facing several security and operational dependability concerns as a consequence of the variable working of these resources, in conjunction with the rapidly growing total demand. Such sources of power may now be used as reactive power generators thanks to new improved grid-interfacing techniques. Load frequency control (LFC) is a network that controls tie-line exchange plans and maintains a nearly fixed frequency of demand among generation sources. Because the power quality is so important, an LFC is essential for a utility network.

In an AC grid, there are two types of power: reactive and active. The average output provided is zero when the reactive component is transported from the generator to the demand and subsequently returned from the demand to the generator. To maintain the network potential within an acceptable range, reactive power correction is necessary. Dynamic and static reactive power may be categorized based on the speed and ability to absorb or create reactive power. Competitive processes have a lower impact on reactive power. Reactive power correction is a type of auxiliary service that system operators must get to keep grid stability inside the voltage limitations specified in [2].

The utility grid is an AC network, and the majority of the demands we encounter in our daily lives require reactive power. As a result, reactive power compensation is defined as the control of the reactive component to improve the AC system's performance. Its adjustment may be seen from two perspectives: demand and voltage support. From a load perspective, the goal is to achieve a higher-quality power factor and a true power balance. The reactive power flowing via the grid-connected microgrid must be adequately managed and balanced in both circumstances. The compensation strategy of reactive power is available in the reactive power planning of a utility grid, allowing the active power loss to be decreased and the network voltage level and power factor to be enhanced.

2 Related work

In a utility grid, reactive power production is commonly utilized to improve voltage profiles. The economic benefit of reactive power production, on the other hand, must be calculated because it has an impact on the network's frequency control capacity in [3]. For every form of energy demand to operate, real power is required. Reactive power is required to modify the voltage profile of the network, which is necessary to supply the active power from the generator to the demands. The required real power can't be delivered if the network's potential is insufficient. As a result, a reactive power component is utilized to keep the potential high for active power to accomplish productive work in [4].

A utility grid is an AC network, and the majority of the demands in everyday lives are inductive, requiring the reactive component of power. At the demand side, extra reactive power is required to manage the potential level or increase the power factor. Generally, there are two aspects to the issue of reactive power compensation, that is, compensation of demand and voltage support. The objectives of compensation of demand are to raise the network power factor and equalize the real power consumed. To reduce voltage variation at a line's termination, voltage support is typically necessary. The reactive component that flows through the network can be controlled effectively in both cases. In a utility grid, the compensation of reactive power enhances grid stability by raising the highest active power that could be delivered in [5].

The most economically attractive and successful technological method for both classic and novel issues at various potential levels in a grid has emerged through the compensation of reactive power using appropriate mechanisms. In recent years, the requirement of VAR compensation at the demand has grown [6].

The terms voltage and reactive power have been strongly linked to maintaining the voltage profile of the overall network. The system should have enough reactive power to meet the grid demand requirements and minimize system losses; further, voltage control of the utility grid is primarily accomplished by keeping a constant reactive power output, circulation, and usage.

In the operation and control of power systems, reactive power regulation has become one of the most accomplished works. The power quality can be harmed by several characteristics of the utility grid and its demands.

Reactive power should be kept constant in the utility grid; otherwise, it would degrade the network voltage profile and potentially collapse the voltage. As a result, optimizing and compensating the grid is critical for enhancing the voltage profile of the network, the power factor, and the efficacy of the electricity. It is good for lowering losses in the network. In [7], it is the most important energy-saving technological solution with a low initial cost and fast recovery.

As a result, static VAR compensators could be utilized to correct it, minimizing its flow among the demands and the generator and therefore enhancing the grid's voltage stability. The regulation of the reactive component of power to improve the power quality of the supply is known as compensation of demand. VAR producers coupled in series or in parallel could be utilized for reactive power adjustment in [8].

Throughout the operation, the reactive power generated by the alternator is stored in a reactor or capacitor. The reactive power varies between the producer and the demand at double the rated charge of the frequency. The majority of the demands, however, are inductive loads that consume reactive power, causing a low lagging power factor in [9].

Reactive power is absorbed by transformers, induction motors, and other inductive components from the grid during operating hours, reducing the network power flow problem to some degree. The requirement of reactive power would be met by a compensating device, which will reduce the flow of reactive power within the network, reduce system loss due to reactive power flow, and enhance the network's operational state. In [10], this phenomenon is referred to as compensation of reactive power.

The fundamental rule of compensation of reactive power is divided into three areas: voltage stratification, facility grid division, and indigenous balance. Reactive power should be regulated within the system; otherwise, it may exacerbate the voltage of the system and potentially collapse the network's voltage. Furthermore, if the potential is lower, instrumentation use will be reduced, resulting in an increase in actual loss. As a result, optimization of the grid and correction are quite important for improving the quality of supply, energy issues, and thus the strength of the service offering. It's also beneficial for reducing network losses.

To overcome the problem of voltage stability, the management of dynamic reactive power reserves dependent on optimal load flow and the Bender's decomposition method can be utilized. The effect of fault locations and fault duration on the induction motor, generator response on the unbalanced system, reaction of various demand models under different network scenarios affects the reactive power compensation.

Compensation for both basic wave reactive power and reactive power at a disturbance will be included in the reactive power compensation. Because the provider is responsible for this, and if it surpasses the mentioned amount, various customers/companies compel consumers to pay for it and minimize electricity usage. If the customers improve the power factor, they must spend energy prices for the total load in kVA plus the units utilized. In [11], an increase in the plant's capability is achieved by improving the power factor.

When it comes to the management and planning of a utility grid with higher photovoltaic and wind energy integration, the compensation of reactive power is a critical challenge. The required operating and management measures to be performed in terms of improving the voltage level and stability in a utility grid are referred to as the management of reactive power. At the design phase, the effective management of reactive power could be obtained by coordinating an optimal regulation of the voltage value at the source side and at the VAR settings. During the reactive power delivery phase, effective reactive power allocation might be accomplished by coordinating optimal regulatory oversights of the voltage reference value at the source point and at the VAR configurations in [12].

The most vital point is to limit the flow of reactive power and, as a consequence, power prices. In comparison to using a filter and compensation device, the most suitable approach for correcting the flow of reactive power is to utilize a power capacitor that is both affordable and effective. The individual, collective, and mass power factor correction approaches depend on the amount of consumers; however, capacitor banks have poor efficacy and less dependability [13].

One of the most efficient strategies to minimize electricity usage and enhance power quality is to use reactive power support. The advantages are as follows:

- Minimize consumer expenses and boost income.
- Reduce system losses.
- Enhance network capability and minimize the installation cost of new infrastructure.
- Enhance power factor.
- Increase the availability of power.

Power electronic equipment is utilized to enhance and regulate the mini/microgrid level. Any equipment linked in parallel or series with demand and is able to deliver the load of reactive power is referred to as a reactive power compensation mechanism in [14].

The production of reactive power is a separate service that is directly tied to ensure an appropriate potential level and power quality. Reactive power generation and transmission are intertwined with actual power generation and transmission. Reactive power can originate from alternators, but it can also be produced from other components/sources such as capacitors, with transmission network operators often deciding. The reactive power supply could be fast or slow depending on the production source. Reactive power could be utilized for two things: maintaining voltages in a safe range under typical operating levels, and stopping voltage instability under a severe grid situation.

3 Reactive power management techniques

These techniques can be broadly categorized under two categories.

3.1 Decentralized method (DM)

Independent reactive power managing devices are put directly in the system in decentralized compensation reactive power. It is possible to compute the ideal capacity of managing equipment and implement wherever needed for various time periods.

3.2 Centralized method (CM)

The whole system's unique managing devices are all centralized. This will be utilized to keep node potentials and power factors stable. For centralized compensation, reactive power control units are utilized. For compensation, the present market standard is to adopt a decentralized technique. Independent managing devices are put directly in the system for the distributed compensation of reactive power. Purchasing the managing equipment and installing it wherever needed for various time periods in [15] is quite expensive and unprofitable.

The decentralized reactive power compensation techniques are preferable due to economic and technical restrictions. Further, reliability and flexibility are the major concerns in decentralized reactive power compensation.

Because of these issues, a centralized solution will become more appropriate for distribution grid integrated microgrid systems. In [16], centralized compensation can supply reactive power to various demands to enhance the quality of service for grid integrated distributed energy resources and microgrids.

4 Reactive power compensation methods in the grid-connected system

Planners indicate that a voltage control technique will also be installed to agitate the reality of active power and dynamical energy ways. According to power system experts, only a small change in voltage over a particular system will result in voltage collapse and other problems. Components of the utility grid, such as transformers, supply a good amount of reactive power in phases to perform the reactive power compensation [17].

Reactive power compensation can be performed in three different ways:

1. Series
2. Shunt
3. Dynamic

These classes principally aided in the development of ways to connect the capacitance bank to the system. Voltage-ampere compensation techniques, both series and shunt, are used to alter the intrinsic electrical properties of an electrical facility. Shunt compensation adjusts the equal load electric characteristic while series compensation adjusts the electrical phenomenon attribute of the distribution network.

4.1 Series compensation

On the other hand, series capacitors have no control over the current flow. The load current is always sent via the series capacitor bank because it is linked serially to the demand. The system's voltage control has been enhanced. A series capacitance bank, on the other hand, has a distinct drawback. The voltage across the capacitor is additionally elevated up to 15 times its nominal value during fault situations. As a result, the protective instrumentation for series capacitance must be developed and made sophisticated. As a result, the use of series capacitor banks is limited to the higher-voltage circuit [2].

4.2 Shunt compensation

A shunt capacitance is introduced, attracting current and causing the voltage level to rise. It draws a nearly constant amount of leading current that is superimposed on the demand current, reducing reactive sections of the demand and thus improving the system's capability concern. Furthermore, capacitance banks installed on the load side with low-required power are unable to exchange their extra capacity with demand sites that have a high-required power. As a consequence, capacitance banks installed in active distribution network do not appear to be utilized quickly for reactive power adjustment [18].

4.3 Dynamic compensation

The goal of dynamic reactive compensation is to combine the shunt and series compensation systems' compensating characteristics. Dynamic reactive power compensators are an instrument capable of measuring the reactive power utilized inside the innovation that it's installed in and giving the reactive power required in a very progressive manner, from zero to maximum electrical phenomena and maximum inductive. They'll be used in single-phase and three-phase systems, substations, distribution lines, and at a specific client's door.

5 Requirement of reactive power management

There are three reasons that managing reactive power and controlling voltage are critical [19].

(a) Both user and equipment are assumed to function inside a voltage range, often within 5% of the nominal voltage. Different types of technology perform poorly at low-voltage levels; induction motor drives could be overheated and destroyed, and several electronic devices will not work.

(b) The transmission line and generating sources absorb reactive power. Reactive power flows must be decreased to optimize the value of actual power that can be moved through an overloaded transmission link. The real-power capability of a generator can be limited by a reactive power generating plant.

(c) Real-power losses are incurred when reactive power is moved via the power grid. To compensate for these losses, both capacity and power must be provided.

6 Basic principle of reactive power compensation

The reactive power regulation and compensation method has three aspects at its core [20]:

- Voltage stratification.
- Power grid partitioning.
- Local equilibrium.

7 Need of reactive power compensation

There are three types of reasons why flexible reactive power compensation is required [20]:

(i) The importance of ensuring the stability of the alternator. Because the rotor angles of synchronous machines fluctuate fast, voltage management using reactive power compensation can have a good stabilizing impact on the system during disruptions. The transient and dynamic instability of a network could both be improved. It's also feasible to use regulated compensators to intentionally push voltages beyond their typical steady-state boundaries for many seconds after a failure or other large disruption to boost the stabilizing effect even further.

(ii) To provide excellent service to demands, voltage must be controlled between reasonable boundaries around the intended steady-state value. This might be essential to achieve a potential enhancement in as little as a few cycles of the power frequency followed by certain rapid variations in the demand or network structure as a consequence of changeover activities. A few seconds of enhancement will be sufficient for other voltage problems. Voltage abnormalities that have not been addressed, even if they are just temporary, are also improved.

(iii) The requirement to alter system voltage levels to avoid unneeded reactive power flows on the grid. Reactive power adjustment could be utilized to keep network losses to a bare minimum in this case. While reactive compensation should be updated or altered on a regular basis to ensure minimum losses, the modifications may be performed only sometimes and take multiple minutes to finish.

8 Reactive power compensation techniques

The centralized approach will be vital for the regulating procedure in the active distribution network where dispersed power resources and a microgrid are coupled. By absorbing reactive power at the point of common coupling, the effect of employing reactive power compensation to regulate the potential increase impact and therefore enabling an increase in renewable generation is accomplished. Only the reactive energy absorbed is insufficient to keep the voltage within acceptable boundaries in this instance, hence active electricity production is reduced [21].

8.1 Regulation of reactive power of synchronous generator

The alternator's reactive power is controlled by its field current, which is constantly changing. Let's assume that a synchronous generator is linked to the system's ability to supply reactive power.

8.2 Compensation of reactive power by capacitors

To compensate for capacitive loss, shunt or series capacitors will be connected with regard to loading. For an industrial utility grid, a common reactive power compensation scheme is used. The location is essentially decided by the justification for compensation. The justification for compensation type is the primary determinant of placement.

- Direct.
- Cluster.
- Central compensation at low voltage (LV).
- Central compensation at high voltage (HV).

8.3 Static VAR compensators (SVCs)

Static VAR compensators (SVCs) are designed to compensate reactive power for regulating the voltage at a specific node. SVCs have been widely employed in comparison to other shunt flexible alternating current transmission system (FACTS) devices because of its advantages of simplicity, low losses, low harmonic output, and inexpensive cost. SVC is a shunt-linked

FACTS controller with an outcome that is frequently modified to transmit capacitive or inductive currents with the associated network. This current is managed to handle particular utility grid characteristics at a particular bus. It's an adjustable device that controls the flow of electricity through a reactor and is regulated by back-to-back thyristor valves. The thyristor has played a crucial role in the realization of the SVC and the modification of its reactive power flow control. It could be utilized as a switch or a continuously regulated valve by controlling the firing angle. It's worth noting that the SVC current can have some harmonic content, which is something to keep in mind when using the style approach in [17].

8.3.1 Component of SVC

Fig. 1 shows the various components of the SVC.

- Thyristor-controlled reactor (TCR).
- Thyristor-switched capacitor (TSC).
- Harmonic filter (FC).
- Mechanically switched capacitor bank (MSC) or reactor bank (MSR).

8.3.2 Operation of SVC

The SVC concept is based on controlling shunt susceptance by changing the thyristor's firing angle. The SVC's primary goal is to maintain a specified voltage profile. The SVC in [22] provides some steady-state potential control inside the steady state to maintain a specified value at a voltage bus. Fig. 2 shows the operation of SVC.

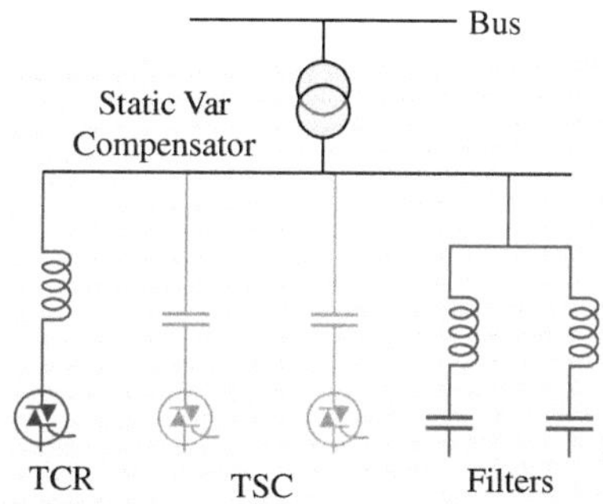

Fig. 1 Component diagram of SVC.

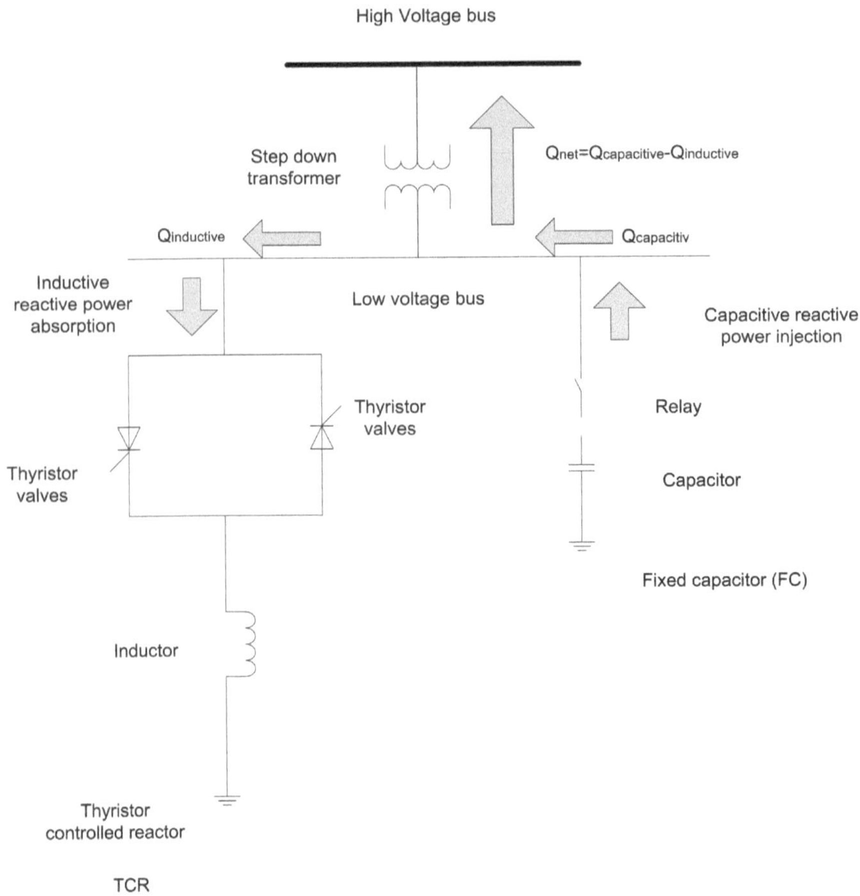

Fig. 2 Operation of SVC.

9 Artificial intelligence techniques for reactive power/voltage control

Different artificial intelligence (AI) techniques are utilized to control the reactive power flow in the utility grid system.

9.1 Expert system

This is a category of AI approaches that is well established. Expert systems (ES) are primarily composed of information and thinking mechanisms that deliver expert-level responses via arguments dependent on certain fields of expert information in a topic to replicate human expert judgment.

This technique is commonly utilized in voltage and reactive power management to combine existing knowledge or experience of voltage and reactive power control with regulations, producing knowledge and understanding of expert systems [23].

9.2 The fuzzy theory

The original fuzzy set theory, with the addition of linguistic variables and fuzzy logic approximation reasoning, combined with an intelligent reasoning technological network is known as fuzzy theory (FT). Fuzzy control is a set of real ways for simulating the natural fuzzy reasoning and decision-making procedure and regulating it as per established principles.

Network potential and reactive power regulation by power system degeneration, operational circumstances, and system factors frequently modify the parameters of reactive power and demand under a variety of scenarios, making it hard to precisely estimate the impact [24].

The following are the main issues with fuzzy intelligence for controlling the voltage and reactive power: (1). Basic fuzzy data processing reduces the network's management efficiency while increasing the precision of convolution layers, leading the regulation to broaden the scope of the search and slow down decision-making (2). It is hard to determine exact control objectives in the absence of a systematic design methodology. (3). Control the rules of selection.

9.3 Artificial neural network

An artificial neural network (ANN) is a huge amount of simple artificial neurons joined collectively in a specific manner that simulates human transmission and processing of essential information properties.

In comparison to fuzzy, ES and ANN are defined by neurons, and they both should cope with weight issues related to tacit knowledge. They have the following benefits:

- Metadata of distributed storage.
- Robust fault tolerance.
- Capacity to learn, understanding of self-organization may be obtained.
- To fulfill the data processing needs.

This technology may be used in a generation station to obtain better outcomes in terms of voltage and power factor management as well as reducing capacitor and main transformer tap adjusting times [25]. ES, ANN, and FT show the AI application in a utility grid for the voltage and reactive power regulation.

10 Conclusion

As reactive power management is the critical issue in present power systems, this chapter provided a detailed overview of the basic concept of reactive power requirements with generation and control. Different methods for reactive power compensation and voltage control were discussed in detail. Further, different artificial intelligence techniques such as expert systems, fuzzy theory, and artificial neural networks were discussed for reactive power and voltage control in a utility grid system.

References

[1] V.M. Awasth, V.A. Huchche, Reactive power compensation using D-STATCOM, in: 2016 International Conference on Energy Efficient Technologies for Sustainability ICEETS, 2016, pp. 583–585, https://doi.org/10.1109/ICEETS.2016.7583821.

[2] N.K. Roy, H.R. Pota, M.J. Hossain, Reactive power management of distribution networks with wind generation for improving voltage stability, Renew. Energy 58 (2013) 85–94, https://doi.org/10.1016/j.renene.2013.02.030.

[3] N.M. Tabatabaei, A.J. Aghbolaghi, N. Bizon, F. Blaabjerg, Reactive Power Control in AC Power Systems, Springer, 2017.

[4] S. Supsomboon, K. Hongthanapach, A simulation model for machine efficiency improvement using reliability centered maintenance: case study of semiconductor factory, Model. Simul. Eng. 2014 (2014), https://doi.org/10.1155/2014/956182.

[5] A. Águila Téllez, G. López, I. Isaac, J.W. González, Optimal reactive power compensation in electrical distribution systems with distributed resources. Review, Heliyon 4 (8) (2018), https://doi.org/10.1016/j.heliyon.2018.e00746.

[6] T. Baldwin, T. Hogans, S. Henry, F. Renovich, P. Latkovic, Reactive power compensation for voltage control at resistance welders, in: IEEE Systems Technical Conference on Industrial and Commercial Power 2005, 2005, pp. 1–9, https://doi.org/10.1109/ICPS.2005.1436351.

[7] S. Aminzadeh, M. Tarafdar Hagh, H. Seyedi, Reactive power management for microgrid frequency control, Int. J. Electr. Power Energy Syst. 120 (2020) 105959, https://doi.org/10.1016/j.ijepes.2020.105959. March 2019.

[8] H. Hu, W. Wu, S. Xiao, M. Tan, C. Han, Summary on Reactive Power Compensation Technology and Application, no. ICICCI, 2015, pp. 225–229, https://doi.org/10.2991/icicci-15.2015.48.

[9] F. Ayalew, S. Hussen, G.K. Pasam, Reactive power compensation: a review, Int. J. Eng. Appl. Sci. Technol. 03 (11) (2019) 1–7, https://doi.org/10.33564/ijeast.2019.v03i11.001.

[10] G. Chicco, G. Gross, Current issues in reactive power management: a critical overview, in: IEEE Power Energy Society 2008 Genernal Meeting Conversion and Delivery Electrical Energy 21st Century, PES, no. August 2008, 2008, https://doi.org/10.1109/PES.2008.4596731.

[11] H. Amaris, M. Alonso, C.A. Ortega, Reactive Power Management of Power Networks with Wind Generation, Lect. Notes Energy, vol. 5, 2013, https://doi.org/10.1007/978-1-4471-4667-4.

[12] H. Hu, W. Wu, S. Xiao, M. Tan, C. Han, Reactive power compensation technology and application, in: 2nd International Conference on Intelligent Computing and Cognitive Informatics, 2015, pp. 1–27.

[13] F. Muhammad, Reactive power compensation by power capacitor method, Eng. Technol. Open Access J. 1 (3) (2018) 92–95, https://doi.org/10.19080/etoaj.2018.01.555565.
[14] L. Liudvinavičius, The methods of reactive power compensation in the 25 kV, 50 Hz contact network, Transp. Probl. 13 (1) (2018) 59–68, https://doi.org/10.21307/tp.2018.13.1.6.
[15] E. Rokrok, M. Shafie-khah, J.P.S. Catalão, Review of primary voltage and frequency control methods for inverter-based islanded microgrids with distributed generation, Renew. Sustain. Energy Rev. 82 (2018) 3225–3235, https://doi.org/10.1016/j.rser.2017.10.022.
[16] M.S. Mahmoud, S. Azher Hussain, M.A. Abido, Modeling and control of microgrid: an overview, J. Franklin Inst. 351 (5) (2014) 2822–2859, https://doi.org/10.1016/j.jfranklin.2014.01.016.
[17] V. Huchche, Reactive power compensation of the grid connected PV system, Helix 9 (6) (2019) 5830–5840, https://doi.org/10.29042/2019-5836-5840.
[18] S.S. Kanojia, S. Rajyaguru, Reactive power compensation for LV distribution network, Int. J. Innov. Technol. Explor. Eng. 8 (6) (2019) 1734–1741.
[19] M.E. Elkhatib, R. El Shatshat, M.M.A. Salama, Decentralized reactive power control for advanced distribution automation systems, IEEE Trans. Smart Grid 3 (3) (2012) 1482–1490, https://doi.org/10.1109/TSG.2012.2197833.
[20] U. Holmberg, T. Sjögren, J. Hellström, Comparing Centralized and Decentralized Banking: A Study of the Risk-Return Profiles of Banks, 2012, pp. 1–33.
[21] S.X. Chen, Y.S.F. Eddy, H.B. Gooi, M.Q. Wang, S.F. Lu, A centralized reactive power compensation system for LV distribution networks, IEEE Trans. Power Syst. 30 (1) (2015) 274–284, https://doi.org/10.1109/TPWRS.2014.2326520.
[22] H. Boudjella, F. Gherbi, Modelling and Simulation of Static Var Compensator (SVC) in Power System Studies By Matlab, vol. 53, 2013, pp. 1689–1699. no. 9.
[23] Y. Saratkar, A. Pachori, Reactive power compensation in transmission line using fuzzy control method : a survey, Int. J. Adv. Comput. Res. 3 (1) (2013) 1–4.
[24] G. Sheng, G. Tu, Y. Luo, Application of artificial intelligence techniques in reactive power/voltage control of power system, Electron. Sci. Technol. Appl. 1 (1) (2014) 5, https://doi.org/10.18686/esta.v1i1.2.
[25] R. Hooshmand, M. Joorabian, Application of artificial neural networks in controlling voltage and reactive power, Sci. Iran. 12 (1) (2005) 99–108.

SECTION 10

Active distribution network reliability assessment and enhancement

CHAPTER 10

Reliability improvement of distribution using distributed generation in smart grid: A case study

Takele Ferede Agajie[a,b] **and Mesfin Fanuel**[a,b]
[a]Debre Markos University, Debre Markos, Ethiopia
[b]Hawassa University, Hawassa, Ethiopia

Contents

1 Introduction

The electric power system is significantly important for a country because it determines the level of economical, technological, and social advancement of a nation [1–4]. A power distribution network supplies power to customers but aging infrastructure, poor design and operation practices, radial operating status, and high exposure to environmental conditions are the main factors responsible for reliability problems. To improve customer reliability, different methods can be used. A reliability assessment of the distribution

Active Electrical Distribution Network
https://doi.org/10.1016/B978-0-323-85169-5.00022-8

network is performed to find a suitable enhancement method to get better reliability and to reduce the costs of energy due to outages.

To evaluate the distribution network reliability, reliability indices are utilized. The dependability of a distribution system is a measure of the consistency and quality of the power supply to customers, which is primarily determined by the interruption profile and is based on system topology and component reliability data. In order to measure the system reliability indices, different reliability parameters are employed in the distribution system.

There are a variety of indicators used to assess the dependability of the power system as a whole. IEEE standard 762 addresses generating reliability indices, IEEE standard 859 addresses transmission facility reliability indices, and IEEE standard 1366 addresses distribution reliability indices [5]. Simple measures such as tree trimming on a regular basis, the installation of lightning arresters, the use of animal guards, the replacement of overhead bare conductors with covered conductors or underground cables, protection scheme modification, and so on have traditionally improved distribution system reliability. In addition, there are various traditional techniques for improving reliability as well as sophisticated reliability enhancement measures that are now classified as smart grid technology. The existing Ethiopian grid has various problems that affect the reliability of the distribution system, including poor protection, poor infrastructure, line loss, poor efficiency, voltage profile, high interruption rate, etc. Because of these, power blackouts or power shading have increased in recent times. Many consumers complain that the frequency as well as the duration of power cuts has also risen, with outages lasting many hours or even days. The problem is particularly worse in industrial areas more so than residential ones. Because of this, customer satisfaction is decreasing. This book chapter utilized the practical data of the medium voltage feeder's interruption of the Injibara Ethiopian Electric Utility District. For the Injibara Radial Distribution Network, the monthly average number of interruptions (frequency) and duration of interruptions for the years 2013, 2014, and 2015 were 23.84 interruptions and 20.74 h, respectively.

The main causes of interruptions are distribution permanent earth fault, distribution permanent short circuit, distribution temporary earth fault, distribution temporary short circuit, and operational issues, such as when medium voltage feeders are interrupted voluntarily for maintenance. Furthermore, the other operational issues are load transfer and new transformer installation. The power system is composed of many elements that are exposed to failures and require removal from operation for maintenance. It is not possible to prevent all component failures in the electric system, but it is possible to minimize them. To find relevant alternative improvement techniques for this problem, a historical reliability assessment is an important task.

2 Traditional grid power distribution networks

The traditional power grid is basically the interconnection of various power system elements such as synchronous machines, power transformers, transmission lines, transmission substations, distribution lines, distribution substations, and different types of loads. They are located far from the power consumption area and electric power is transmitted through long transmission lines [6].

Presently, the grid is facing a multitude of challenges that can be outlined in three categories. The first is that there are infrastructural problems because the system is out of date and unprepared to deal with increasing demand. As a result, network congestion is occurring much more frequently because it does not have the ability to react to such issues in a timely fashion. Ultimately, such imbalances can lead to blackouts, which are extremely costly for utilities because they spread rapidly due to the lack of communication between the grid and its control centers. The second issue is the need for more information and transparency for customers to make optimal decisions relative to the market to reduce their consumption during the most expensive peak hours. Finally, the third issue is the present grid's inflexibility, which prevents the development of renewable energy or other technologies that might make it more sustainable. Renewable energy sources such as wind and solar in particular are intermittent, posing a serious challenge for a system that does not transmit data to control centers quickly. The smart grid solves all these issues by enhancing communications technology, which has significant advantages for both the supply and demand sides of the power market [7]. The current utility grid is a centralized system in which electricity travels in one direction from generation resources to the consumer via the transmission-distribution system, as depicted in Fig. 1.

3 Smart grid power distribution networks

The smart electrical distribution network distributes the electrical energy in reliable manner. It utilized different smart grid technologies to distribute the electrical energy in more adequate way. Information and communication

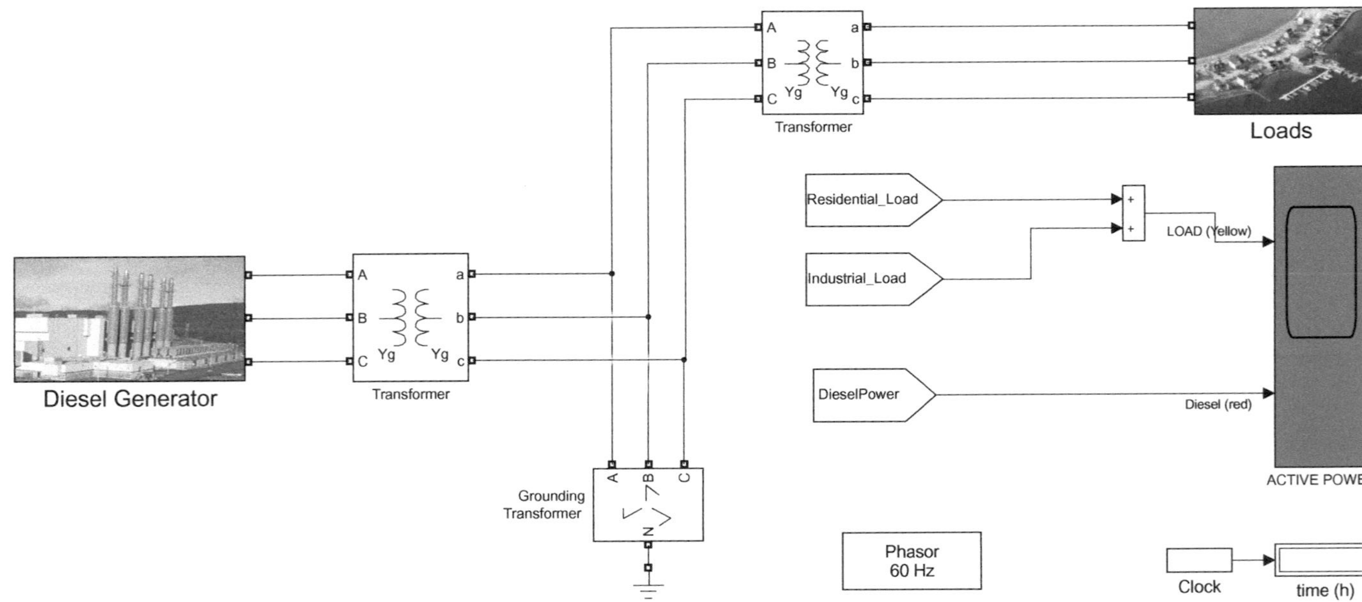

Fig. 1 Layout of a traditional grid power distribution network.

technologies are the backbone of smart distribution network. To put it another way, it's an electrical grid that combines electric infrastructure with information technology [8].

The phrase "smart grid" refers to a traditional electric power system that has been upgraded with modern technology for reasons such as increased dependability, ease of control and management, and the integration of distributed energy resources and electricity market activities [5]. Further, a smart distribution grid is an electricity distribution network that can intelligently integrate all available energy resources with the actions of all users connected to it, in order to efficiently deliver sustainable, economic, and secure electricity supplies [9,10]. Fig. 2 presents the layout of smart grid power distribution networks.

4 Comparisons of traditional and smart grid power distribution networks

The smart grid has many distinguishing benefits over the traditional grid, including new levels of power transmission and distribution. The aspects of handling uncertainties, failure prediction, and automatic system control make power restoration easier. Data acquisition and information transfer attain a level of stability with digital bi-directional communication and system interconnectivity while the existing distribution operates at the level of electromechanical devices and analog communication [11]. The smart grid answers the question of reliability, which is the concern of this book chapter, with high priority to customers that enables rapid power restoration to as many customers as possible [12]. A comparison between traditional distribution and smart grid power distribution networks shows:

Traditional grid	Smart grid
Electromechanical	Digital
One-way communication	Two-way communication
Centralized generation	Distribution generation
Manual monitoring	Self-monitoring
Manual restoration	Self-healing
Slow response to power quality	Power quality a priority
Limited customers	Many customers given priority
Vulnerable to attacks	More or less secured from attacks

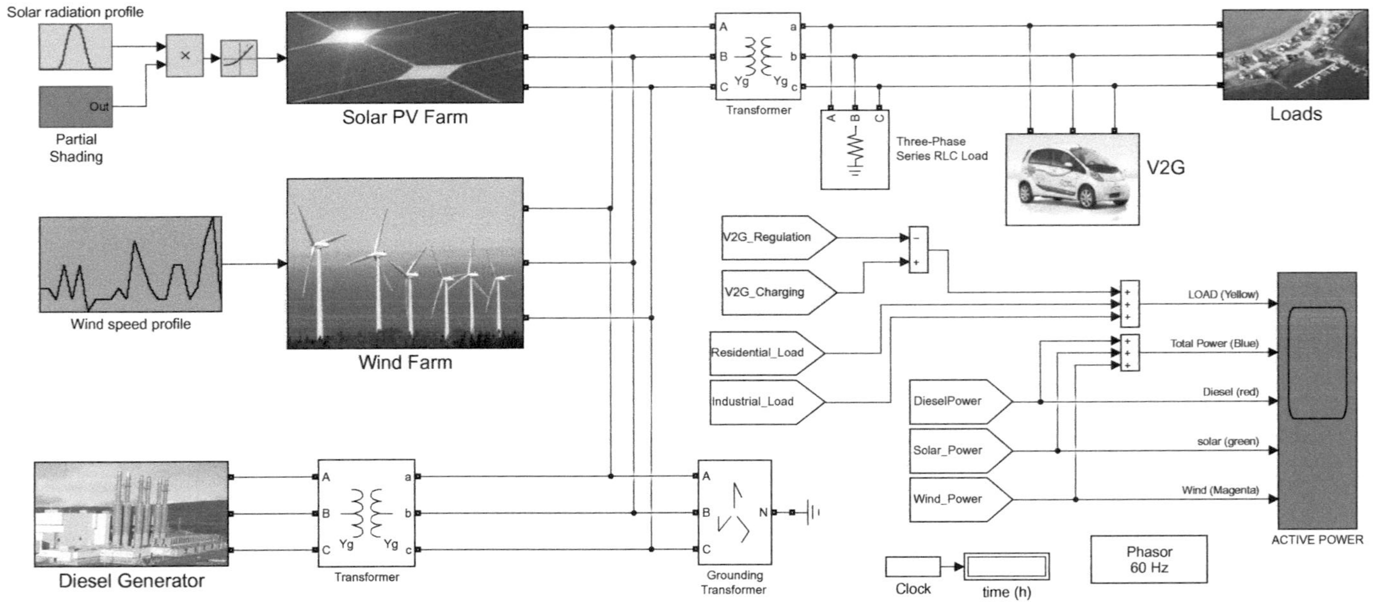

Fig. 2 Layout of smart grid power distribution networks.

5 Reliability indices in distribution system

In the distribution system, two types of reliability indices are employed to evaluate system performance: load point reliability indices and system reliability indices [13]. Reliability indices used for assessing a distribution system's load point are average failure rate λ_i (failure rate/year), average outage duration r_i (h), and annual outage time Ui (h/year).

These indices are expressed in Eqs. (1)–(3).

$$\lambda i = \sum_{i=1}^{n} \lambda \tag{1}$$

$$Ui = \sum_{i=1}^{n} \lambda_i r_i \tag{2}$$

$$r_i = \frac{Ui}{\lambda_i} = \frac{\sum_{i=1} \lambda_i r_i}{\sum_{i=1}^{n} \lambda} \tag{3}$$

5.1 Customer-oriented reliability indices

These reliability indices comprise the system average interruption frequency index (SAIFI, interruptions/customer/year), the system average interruption duration index (SAIDI, hours/customer/year), and the energy not supplied (ENS, kWh/year). In addition, these indices can be used for reliability prediction analysis and for the identification of weak PCC links or chronological changes within the power system network.

Several studies have been done on the reliability assessment of power systems using a minimal cut set, which is a step-by-step approach in evaluating the reliability of a composite power system. This was tested with an IEEE 6 bus, a 14 bus and a single area IEEE RTS 96 system. It was validated using the classical node elimination method, conditional probability, and the Monte Carlo simulation method [14]. In an independent data acquisition center (IDAC) for reliability evaluation of power distribution companies in Nigeria using the System Average Interruption Duration Index (SAIDI) and System Average Interruption Frequency Index (SAIFI) [15]. The analytical method is utilized for investigating the effects of a hybrid system consisting of photovoltaic (PV), wind turbine generator (WTG), electric storage system (ESS), and diesel generator on distribution system reliability. This method was validated using the RBTS distribution network [16].

5.2 Analytical reliability assessment of distribution network

Reliability assessment helps customers and investors make informed decisions about the performance of the power system utility [17,18]. A standard IEEE set of equations is used for the analytical evaluation of system reliability indices, as defined in [13].

- **System average interruption duration index (SAIDI) [h/year, cust]**: This is the average duration of interruption per customer served.

$$\mathrm{SAIDI} = \frac{\text{Sum of customers interrupt durations}}{\text{Total number of customers served}} = \frac{\sum UiNi}{\sum Ni} \tag{4}$$

where, Ui = annual expected outage time of load point i,
Ni = total number of customers interrupted.
SAIDI can also be expressed as

$$\mathrm{SAIDI} = \mathrm{ASUI}*8760 \tag{5}$$

- **Customer average interruption duration index (CAIDI) [h/int]**: This is the average duration of interruption for customers during a year.

$$\mathrm{CAIDI} = \frac{\text{Sum of customers interrupt durations}}{\text{Total number of customer interrupt}} = \frac{\sum UiNi}{\sum \lambda iNi} \tag{6}$$

where, Ni = total number of customers interrupted,
λi = failure rate of each load point.

- **System average interruption frequency index (SAIFI) [int/year, cust]**: This reliability index is used to measure the average frequency and duration of supply interruptions.

$$\mathrm{SAIFI} = \frac{\text{Total number of customer interrupt}}{\text{Total number of customers served}} = \frac{\sum \lambda iNi}{\sum Ni} \tag{7}$$

where, λi = the failure rate of the load point i,
Ni = the number of customers at load point i.

- **Customer average interruption frequency index (CAIFI) [int/year, cust]**: This can be expressed as defined in Eq. (8).

$$\mathrm{CAIFI} = \frac{\text{Total number of customer interrupt}}{\text{Total number of customer interrupt}} = \frac{\sum \lambda iNi}{\sum Nai} \tag{8}$$

Nai = Number of customers interrupted at load point i.

CAIFI is mostly useful in observing chronological trends in the reliability of a particular distribution system:

$$\text{Failurerate}(\lambda) = \frac{1}{MTTR} \tag{9}$$

- **Expected repair rate (μ) occurrence per year**: This is the frequency of repair in a calendar year and is denoted by (μ). It can be expressed mathematically as in Eq. (10).

$$\mu = \frac{\text{MTTR}}{8760} \tag{10}$$

- **Mean time to failure (MTTF)**: Mean time to failure is the probable time (in years) the component will remain in a failed condition.

$$\text{MTTR} = \text{MTTF} = \frac{1}{\lambda} \tag{11}$$

- **Mean time between failures (MTBF)**: This is the expected time in years a component fails.

  It can be expressed mathematically as defined in Eq. (12).

$$\text{MTBF} = \frac{\text{MTTF} + \text{MTTR}}{8760} \tag{12}$$

- **Average service availability index (ASAI)**: This can be expressed mathematically as in Eq. (13).

$$\text{ASAI} = \frac{\text{customer hours of available service}}{\text{customers hours demanded}} = \frac{8760 - \sum UiNi}{8760} \tag{13}$$

- **Average energy not supplied (AENS)**: This can be expressed mathematically as in Eq. (14).

$$\text{AENS} = \frac{\text{Total Energy not supplied}}{\text{Total number of customers served}} = \frac{\sum LiUI}{\sum Ni} \tag{14}$$

Where Li is the average load connected to load point i.

➢ **Average service unavailability index (ASUI)**: This can be expressed mathematically as defined in Eq. (15).

$$\text{ASUI} = 1 - ASAI = \frac{\text{Customer hours of un} - \text{available service}}{\text{Customers hours demanded}} = \frac{\sum UiNi}{8760} \quad (15)$$

$$\text{Total monthly hours(TH)} = \text{No of days} \times 24\text{h} \quad (16)$$

$$\text{Total outage in a month(TM)} = \text{No of outage} \times \text{No of outage hour} \quad (17)$$

$$\text{Total service hour in a month} = \text{TH} - \text{TM} \quad (18)$$

$$\text{Mean time between failure (MTBF)} = \text{MTTF} + \text{MTTR} \quad (19)$$

➢ **Power system availability:** This can be expressed as Eq. (17).

$$\text{Availability} = \frac{\text{MTBF}}{\text{MTBF} + \text{MTTR}} \quad (20)$$

$$\text{Unavailability} = 1 - \frac{\mu}{\lambda + \mu} = 1 - \frac{m}{m + r} \quad (21)$$

$$\text{Loss of load expectation (LOLE)} \sum_{i=1}^{n} P_i(C_i - L_i)\text{day/period} \quad (22)$$

where,
μ = expected failure rate,
r = mean time to repair = MTTR = $\frac{1}{\mu}$,
Available capacity on day i,
L_i = Forecast peak load on day i,

$$P_i(C_i - L_i) = \text{Probability of loss of load on loadi, Probability of average}$$

$$\text{power availability} = \left(\sum P \times \left(\frac{\mu}{\lambda + \mu}\right)\right) \quad (23)$$

Mathematical model for power security function $S\ (t)$ is given as defined in Eq. (21).

$$S(t) = \sum_i P_i(t) Q_i(t) \quad (24)$$

where, $P_i(t)$ = Probability that the system is operating in its normal operating state,

$Q_i(t)$ = Probability that there is a breach of system security at time t in the future.

6 Existing Injibara distribution system reliability indices evaluation

Reliability indices are statistical aggregations of reliability data for a well-defined set of loads, components, or customers. Most reliability indices are average values of a particular reliability characteristic for an entire system, operating region, substation service territory, or feeder. The distribution system reliability indices indicate the annual average performance of the network in terms of interruption frequency and duration. The basic function of a distribution system is to supply electrical energy from a substation to the individual customer load points. Service continuity is, therefore, an important criterion in a distribution system. Service continuity can be described by three basic load point indices and a series of system indices [19]. The three basic load point reliability indices usually used are the average failure rate λ, the average outage time r, and the average annual unavailability or average annual outage time U. It should be noted that these indices are not deterministic values but are expected values of an underlying probability distribution and hence are long-run average values. The three primary load point indices are fundamentally important parameters. They can be aggregated to provide an appreciation of the system performance using a series of system indices. Fig. 3 shows the existing Injibara power distribution network.

The additional indices that are most commonly used are defined in the following sections. Table 1 presents the base case Injibara power distribution network system indices.

From Table 1, it can be concluded that with the base case system indices (no DG), SAIDI is 621.789 h/customer year, suggesting that the system's average interruption duration for each customer is 621.789 a year; SAIFI is 630.128125 inter./customer year, marking the system's average interruption frequency for each customer during a year; CAIDI is 0.982 h/customer interruption, suggesting that the system's average interruption duration for the customers that experience interruption is 0.982 h a year while system availability and unavailability are 92.90% and 7.10%, respectively. Also, the expected energy not supplied (EENS) of a system due to the failures is 3268.927 MWh/year and the energy not supplied per customer is 306 kWh/year. Finally, the expected interruption cost of the system is $2.86881344 million/year. Therefore, the system reliability indices show that the Injibara distribution substation is an unreliable distribution system.

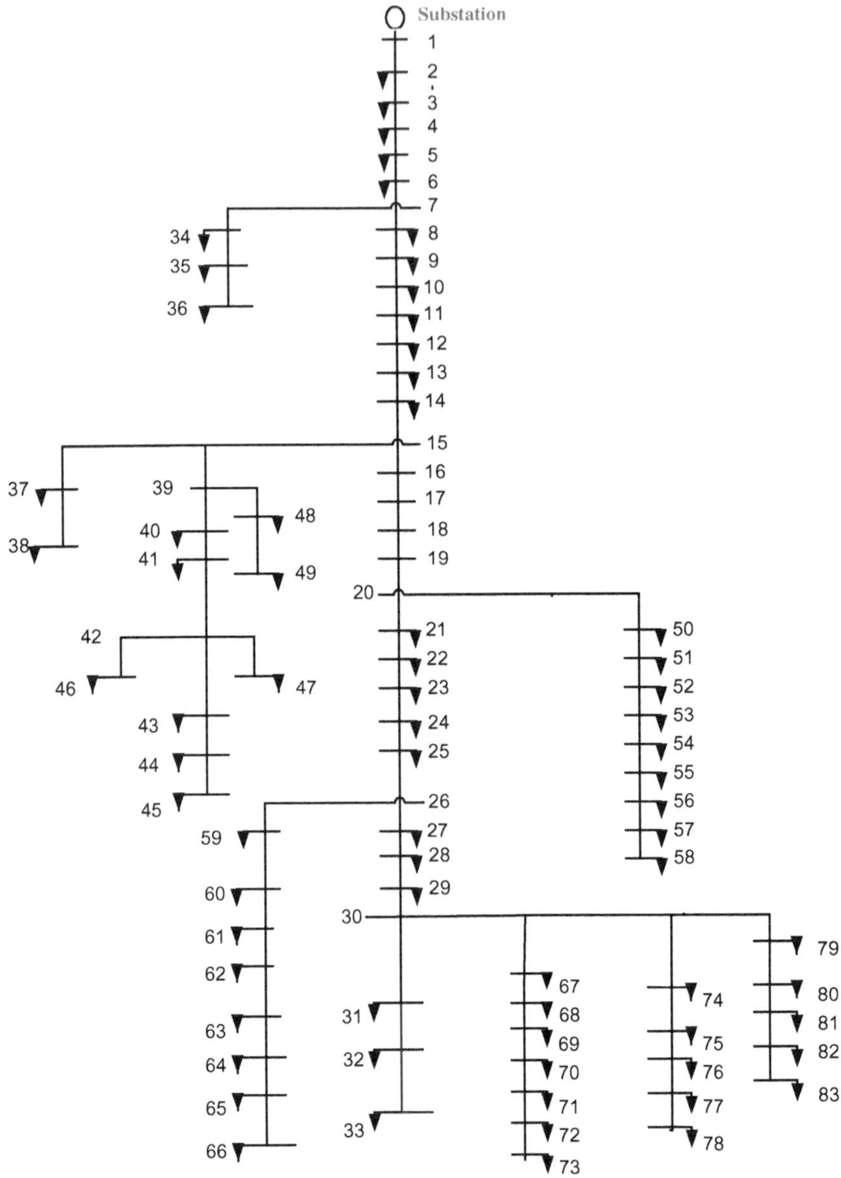

Fig. 3 Existing Injibara power distribution network.

Table 1 Base case Injibara power distribution network system indices.

Indices	Units	Value
SAIFI	Inter./customer Year	630.209793
SAIDI	h/customer Year	621.789
CAIDI	h/customer interruption	0.982
ASAI	%	92.90195582
ASUI	%	7.09804418
EENS	MWh/year	3268.927
EIC	m$/year	2.86881344
AENS	kWh/ca. year	306

7 Distribution network reliability improvement methods

The main purpose of reliability data quantification and information extraction is to take reliability improvement measures. The reliability of the distribution system can be improved by increasing distribution system protection, decreasing equipment failure, system automation, installment of reclosing and switching devices, and system reconfiguration. The restoration time of a momentary outage will be short. Hence, the duration that an outage will last will be diminished. Similarly, system configuration produces effective improvement in reliability. During faults that last a long time, the distribution system can be configured into a set of network topologies. The distribution network will have an alternative supplying network after reconfiguration. These topologies provide an alternative path for power flow by sectioning the faulted part. Configuration can be important during maintenance and a permanent fault event. During the disconnection of a portion of the distribution system for maintenance, the existence of an alternative way of supplying customers enables reconfiguration, as shown in Fig. 4. Hence, customers experience little outage time. Additionally, reclosing and switching devices provide patterns to help localize fault points and disconnect faulted lines. This leads to just a few customers being affected. Sectionalizing devices also enable choosing a supply path during a contingency. By changing the status (ON/OFF) of the sectionalizing switches, network reconfiguration is established. Such capability improves reliability by reducing the duration and frequency of interruption. The study considers automation, the installment of reclosing and switching devices, and system reconfiguration for reliability improvement. One of the methods of reliability enhancement is to increase the level of protection of a distribution system. A increase in protection gives the option of a selective protection system; thus, any failure in some part of the distribution network may not affect other parts of distribution [20]. Selective automation of the distribution system can also improve its reliability by more than half by placing an automatically controlled switch midway in the distribution system; any downstream fault away from the switch cannot affect customers in the upstream [21]. Hence, this improves the reliability by 50% for faults that occur in the downstream portion. This also is related with designing alternative power supplying laterals for emergency cases. Any fault occurring between normally opened and normally closed can be cleared by opening the middle normally closed switch. Similarly, any fault between the normally closed switches can be isolated by opening the remotely controlled middle switch. As explained

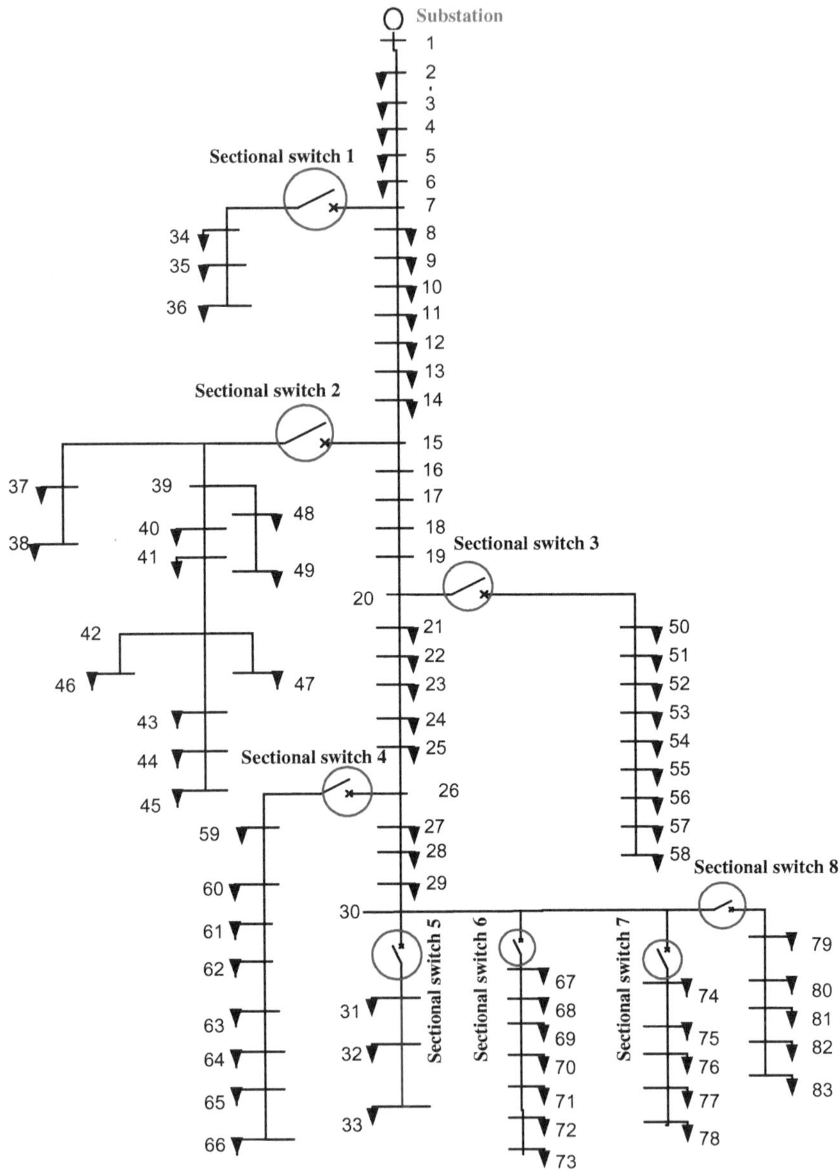

Fig. 4 Existing Injibara power distribution network with sectional switches.

above, the network reconfiguration capability of a system enables customer power access and a reduction in the unserved energy of the system. Sectionalizing and a tie switch system option create a power path searching alternative with the existing laterals. This improves the supply reliability and power availability. It's also related to a fault searching mechanism with remote control capability to identify which line to connect and disconnect. The resulting reconfigured network can have multiobjective functions such as achieving maximized reliability and reduced power loss [22,23]. In this study, the distribution system's reliability was improved by incorporating smart grid technologies such as optimal installation of sectional switches and incorporating distributed generation with optimal sizing and location into existing distribution networks.

7.1 Optimal placement of sectional switches in the distribution network

The main consideration of this study is to introduce distribution sectionalizing switch optimal placement that explicitly considers the customer outage costs as well as sectionalizing switch investment, operation, and maintenance costs. Furthermore, the importance of the customer types and load density in determining the sectionalizing switch placement strategy in distribution systems is investigated through several sensitivity analyses and by step-by-step inserting each sectional switch in each branch. Distribution sectionalizing switch installation brings major benefits to distribution utilities. However, its implementation requires economic justification because it is usually site-specific and should be targeted to areas where improved performance provides measurable benefits [24]. The benefits of sectionalizing switches can be quantified in terms of reduced outage duration and reduced number of customers affected during permanent faults by the fast restoration of power to customers. This can be done by isolating the faulty part of the network by switching sectionalizing devices and supplying the unfaulty part from other routes. Hence, the distribution network configuration, the number and type of customers upstream and downstream of the sectionalizing switch, etc., can dramatically affect the benefits of sectionalizing device placement. Here, the objective is to find the optimal number and location of sectionalizing switches to minimize customer outage costs in conjunction with sectionalizing switch capital investment, installation, and annual operation and maintenance. In this study, the optimal placement of the sectional switch is determined by considering each branch and taking the optimal values of indices according to the standards. Fig. 4 presents the existing Injibara power distribution network with sectional switches.

7.2 Optimal sizing and siting of distributed generation on existing distribution network

A number of approaches and algorithms have been proposed for siting and sizing of DGs. The majority of the previous works in this area aim to reduce active power losses and improve the voltage profile [25,26]. The solution methods applied to the problems can be broadly classified as mathematical, heuristic and metaheuristic, and hybrid [27]. Mathematical techniques including mixed integer linear programming [28,29], mixed integer second-order cone programming [30], and multiperiod optimal power flow [31] have been used in the literature to resolve DG sizing and placement planning. In [32,33], the power flow algorithm was used to determine the optimum DG size at all load buses, assuming every load bus is connected to the DG source. However, such a method is inefficient due to a number of load flow computations. The genetic algorithm (GA) is used to determine the size and location of DG, as is presented in [34–36]. GA is suitable for multiobjective problems such as DG allocation and can give near-optimal results, but it is computationally demanding and slow in convergence. In Ref. [37], an analytical method to place DG in radial as well as meshed systems to minimize the power loss of the system was presented. In this method, separate expressions for radial and meshed network systems are derived and a complex procedure based on phasor current was proposed to solve the location problem. However, this method only optimizes the location and considers the size of the DG as fixed.

In this paper, the proposed technique for optimal sizing and siting of DG units in an active distribution network system is performed by the weighted sum harmony search algorithm.

7.2.1 Weighted sum harmony search algorithm

A harmony search algorithm (HSA) is a metaheuristic optimization algorithm stimulated from the managing process of music in search of the perfect harmony sound [38]. This paper adopts and implements a grid strategy as a secondary criterion in one of the most efficient population-based metaheuristic algorithms for a multiobjective problem for optimal siting and sizing of DGs. It is simple for programming, faster in convergence, and efficient in searching for the optimal solution among other heuristic algorithms. The procedural steps of the proposed HSA for optimal DG sizing and placement result in the lowest active and reactive power loss and voltage deviation, as shown in the flowchart in Fig. 5.

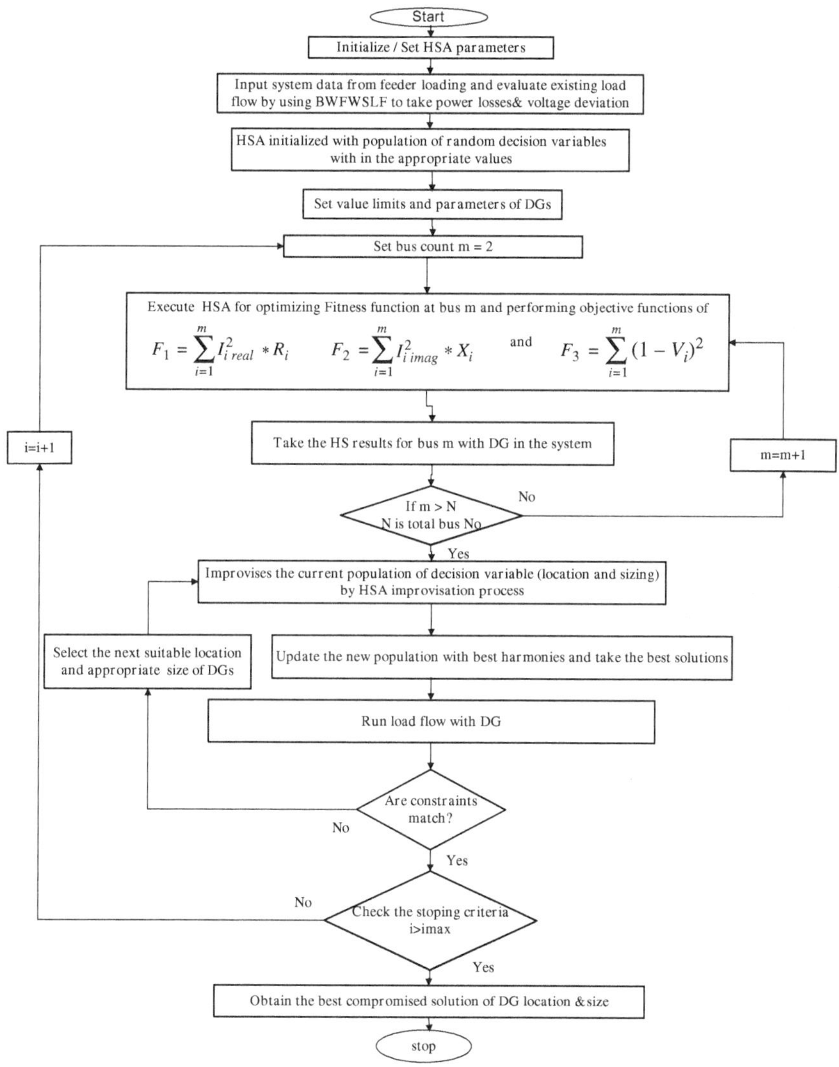

Fig. 5 Proposed weighted sum HSA for optimal DG sizing and placement.

The main aim is to determine the optimal DG locating and sizing by minimizing the planning objectives instantaneously for the existing Injibara distribution networks. The objective functions are active power loss, reactive power loss, and voltage deviation, and the constraints are bus voltage, DG capacity, and branch thermal limit (rated current).

The optimal placement and sizing of DGs to solve the problem of distribution network reliability improvement is highly dependent on their optimal placement and sizing. After sizing and placing as shown in Fig. 6, the DG in the network, the total voltage deviation, and the active and reactive power losses are reduced from 15.6075 pu, 519.3944 kW and 518.3686 kvar to 1.027 pu, 95.398 kW, and 90.978 kvar, respectively. With the incorporation

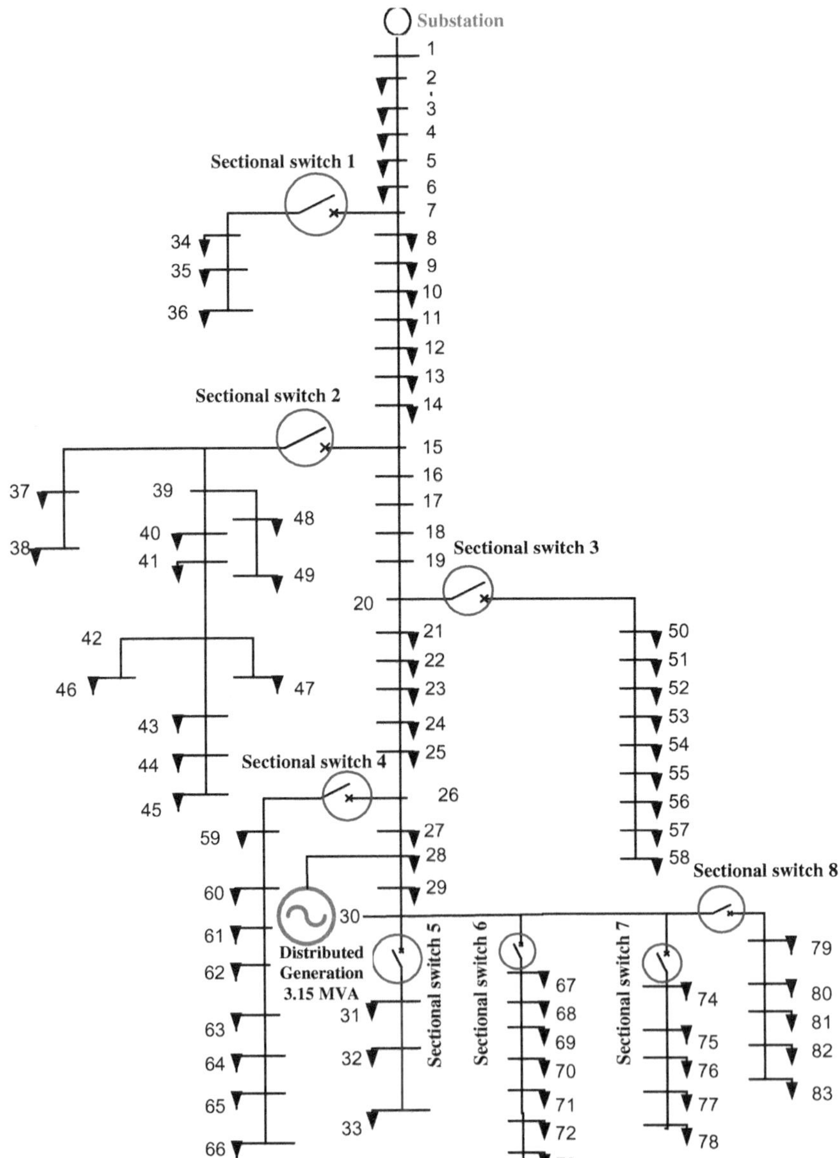

Fig. 6 Proposed Injibara power distribution networks with DG and sectional switches.

of the proposed work, reliable power is delivered with an improved bus voltage profile.

After applying the aforementioned smart grid technologies on existing distribution networks, the reliability indices will become optimized.

As shown in Table 2, the line losses should be reduced for stable and enhanced voltage profiles, which is only possible by distributing active and reactive power support to the system. Economically, it is not possible to deliver support on an individual bus. Therefore, the most suitable site and size of DG should be selected by using optimization techniques from where maximum benefits could be achieved. The quantity of active power from DG changes from 0% to 100% of the total active load, with generation and voltage constraint. The main objective in selecting DG size and location is to minimize the total system power losses by injecting active power. The relationship between DG size and losses follows the parabolic curve. Thus, in this study, DG is selected for its size and location optimally by employing a weighted sum harmony search algorithm. Once DG is installed at 15 kV bus, the expected energy not supplied (EENS) and (expected interruption cost (EIC) are improved by 97.98% and 97.87% , respectively. This is because the circuit will be supplied by the DG unit. However, the DG would be used when there is a failure of the grid, which is why reliability indices are slightly better. There are clearly significantly improvements when the DG is placed at any other point distant from the supply point. SAIFI and SAIDI are reduced while DG is further from the supply point. On the other hand, CAIDI tends to be slightly higher when the distance from the supply point increases. The EIC index presents the cost due to outages.

Table 2 System indices after applying DG and sectional switches on the distribution network.

Indices	Base case (No DG)	(3.15 MW) DG at 15 kV bus number 28
SAIFI	630.128125	11.759062
SAIDI	618.477	12.680
CAIDI	0.982	1.078
ASAI (%)	92.93976058	99.85524675
ASUI (%)	7.06023942	0.14475325
EENS (MWh/year)	3268.927	65.945
AENS (KWh/ year ca.)	306	6
EIC (million $/year)	2.86881344	0.0578733927

8 Conclusions

The distribution system is the core part of a power system network that delivers power to the end user. Reliable power delivery plays a key role in profitability and customer satisfaction, but achieving reliable power delivery has been a major challenge. The problem facing electric power utilities is that the power demand is increasing, but power problems in the distribution system occur several times due to overloading, short circuits, improper allocated distribution transformers, poor maintenance, and limited protective devices. An unreliable power supply not only slows down or damages production or shuts down the plant, but it also leads to equipment damage, additional maintenance, and damaged reputations. This study investigates the proper placement of sectional switches as well as DG location and size to solve these problems for reliability assessments of the Injibara power distribution system, based on reliability improvement, voltage profile improvement, and power loss reduction.

References

[1] Y. Jibril, K.R. Ekundayo, Reliability assessment of 33kV kaduna electricity distribution feeders, Northern Region, Nigeria, in: Proceedings of the World Congress on Engineering and Computer Science, vol. 1, 2013.
[2] P.I. Obi, K.J. Offor, G.C. Chidolue, Reliable and efficient power supply in Nigeria through national integrated power projects and independent power projects: a case study of Onitsha metropolis, Int. J. Adv. Res. Technol. 2 (2013) 5.
[3] S. Mwale, I.E. Davidson, SADC power grid reliability-a steady-state contingency analysis and strategic HVDC interconnections using the N-1 criterion, in: Proceedings of the 2nd International Symposium on Energy Challenges and Mechanics, Aberdeen, UK, 2014.
[4] I. Oluwafemi, T. Laseinde, A.O. Salau, Design and construction of a 0.5 kW solar tree for powering farm settlements, Int. J. Mech. Eng. Technol. 10 (2019) 6.
[5] S. Kazemi, Reliability Evaluation of Smart Distribution Grids (Doctoral dissertations), Aalto University, 2011.
[6] R.A. Walling, et al., Summary of distributed resources impact on power delivery systems, IEEE Trans. Power Delivery 23 (3) (2008) 1636–1644.
[7] Z. Xue-song, C. Li-qiang, Y.-j. Ma, Research on smart grid technology, in: 2010 International Conference on Computer Application and System Modeling (ICCASM 2010), vol. 3, IEEE, 2010.
[8] S. Manandhar, Reliability Analysis of Smart Distribution System and Optimization Of Automatic Line Switches (Master thesis), The University of Tennessee at Chattanooga, May 2013.
[9] V. Giordano, G. Fulli, A business case for smart grid technologies: a systemic perspective, Energy Policy 40 (2012) 252–259.
[10] X.I.A.O. Shijie, Consideration of technology for constructing Chinese smart grid, Autom Electr. Power Syst. 9 (33) (2009) 1–4.
[11] T. Gebreegziabher, Study on Smart Grid System for Improvement of Power Distribution System Reliability Case Study: Addis Ababa District (Diss), Addis Ababa University, 2014.

[12] N. Sabpayakom, S. Sirisumrannukul, Power losses reduction and reliability improvement in distribution system with very small power producers, Energy Procedia 100 (100) (2016) 388–395.
[13] T.B. Kumar, O.C. Sekhar, M. Ramamoorty, Composite power system reliability evaluation using modified minimal cut set approach, Alex. Eng. J. 57 (4) (2018) 2521–2528.
[14] J.O. Dada, Towards standardization of deregulated electricity market communications in Nigeria, Int. J. Comput. Appl. 975 (2015) 8887.
[15] T. Adefarati, R.C. Bansal, The impacts of PV-wind-diesel-electric storage hybrid system on the reliability of a power system, Energy Procedia 105 (2017) 616–621.
[16] G. Dumakude, A.G. Swanson, R. Stephen, I.E. Davidson, Power distribution system reliability improvement through the application of smart technology, in: 24th South African Universities Power Engineering Conference, Vereeniging, South Africa, vol. 24, 2016, pp. 1–8.
[17] T. Adefarati, R.C. Bansal, J.J. Justo, Reliability and economic evaluation of a microgrid power system, Energy Procedia 142 (2017) 43–48.
[18] R.E. Brown, Electric power distribution reliability, second ed., CRC press, Taylor & Francis Group, Boca Raton, FL, 2017. 33487–2742.
[19] L. Ardito, et al., Smart grid technologies in Europe: an overview, Energies 6 (1) (2013) 251–281.
[20] A.S. Bouhouras, G.T. Andreou, D.P. Labridis, A.G. Bakirtzis, Selective automation upgrade in distribution networks towards a smarter grid, IEEE Trans. Smart Grid 1 (3) (2010) 278–285, https://doi.org/10.1109/TSG.2010.2080294.
[21] B. Amanulla, S. Chakrabarti, S.N. Singh, Reconfiguration of power distribution systems considering reliability and power loss, IEEE Trans. Power Delivery 27 (2) (2012) 918–926.
[22] M.R. Farooqi, P.K. Agarwal, K.R. Niazi, Multi-objective distribution feeder reconfiguration, in: Proceedings of the 7th WSEAS/IASME International Conference on Electric Power Systems, High Voltage, Electric Machines, Venice, Italy, 2007.
[23] A. Bagheri, H. Monsef, H. Lesani, Electrical power and energy systems integrated distribution network expansion planning incorporating distributed generation considering uncertainties, reliability, and operational conditions, Int. J. Electr. Power Energy Syst. 73 (2015) 56–70.
[24] J. Northcote-Green, R.G. Wilson, Control and Automation of Electrical Power Distribution Systems, CRC Press, 2017.
[25] I.-K. Song, et al., Operation schemes of smart distribution networks with distributed energy resources for loss reduction and service restoration, IEEE Trans. Smart Grid 4 (1) (2012) 367–374.
[26] C. Lueken, P.M.S. Carvalho, J. Apt, Distribution grid reconfiguration reduces power losses and helps integrate renewables, Energy Policy 48 (2012) 260–273.
[27] D.K. Singh, S. Sharma, D. Balana, Multiple benefits analysis of restructuring and distributed generators placement in a distribution network, in: 2015 International Conference on Electrical, Electronics, Signals, Communication, and Optimization (EESCO), 2015.
[28] H. Haghighat, B. Zeng, Distribution system reconfiguration under uncertain load and renewable generation, IEEE Trans. Power Syst. 31 (4) (2015) 2666–2675.
[29] G. Munoz-Delgado, J. Contreras, J.M. Arroyo, Joint expansion planning of distributed generation and distribution networks, IEEE Trans. Power Syst. 30 (5) (2015) 2579–2590.
[30] Dorostkar-Ghamsari, M. Reza, et al., Value of distribution network reconfiguration in the presence of renewable energy resources, IEEE Trans. Power Syst. 31 (3) (2015) 1879–1888.
[31] F. Capitanescu, et al., Assessing the potential of network reconfiguration to improve distributed generation hosting capacity in active distribution systems, IEEE Trans. Power Syst. 30 (1) (2014) 346–356.

[32] N.S. Rau, Y.-h. Wan, Optimum location of resources in distributed planning, IEEE Trans. Power Syst. 9 (4) (1994) 2014–2020.
[33] J.-O. Kim, et al., Dispersed generation planning using improved Hereford ranch algorithm, Electr. Pow. Syst. Res. 47 (1) (1998) 47–55.
[34] K.-H. Kim, et al., Dispersed generator placement using fuzzy-GA in distribution systems, in: IEEE Power Engineering Society Summer Meeting, vol. 3, IEEE, 2002.
[35] A. Silvestri, A. Berizzi, S. Buonanno, Distributed generation planning using genetic algorithms, in: PowerTech Budapest 99. Abstract Records. (Cat. No. 99EX376), IEEE, 1999.
[36] G. Carpinelli, et al., Distributed generation siting and sizing under uncertainty, in: 2001 IEEE Porto Power Tech Proceedings (Cat. No. 01EX502), vol. 4, IEEE, 2001.
[37] C. Wang, M. Hashem Nehrir, Analytical approaches for optimal placement of distributed generation sources in power systems, IEEE Trans. Power Syst. 19 (4) (2004) 2068–2076.
[38] E.X. Yang, Harmony Search as a Metaheuristic Algorithm. In the Music-inspired Harmony Search Algorithm, Springer, 2009, pp. 1–14.

SECTION 11

Reconfiguration of active distribution network

CHAPTER 11

Optimal network reconfiguration of active distribution network

Baseem Khan[a] **and Ziad M. Ali**[b]
[a]Department of Electrical and Computer Engineering, Hawassa University, Hawassa, Ethiopia
[b]College of Engineering at Wadi Addawasir, Prince Sattam Bin Abdulaziz University, Wadi Addawasir, Saudi Arabia

Contents

1 Introduction

Production, transmission, and distribution are all typical components of an electricity system. An electricity industry prioritizes network demand and also customer needs by guaranteeing minimal operational cost under conditions that are appropriate to ensure the smooth running and reliability of electric power [1]. An electricity distribution infrastructure is made up of multiple substations linked by feeders, which are electric conductors that transmit electricity from a substation to satisfy load needs along their course. This is a component of the power distribution system, which delivers electricity to end consumers. As a consequence, it is the largest component of the power network, and it is mostly essential for power losses. According to

Active Electrical Distribution Network
https://doi.org/10.1016/B978-0-323-85169-5.00003-4

some research, as much as 13% of the overall electricity produced is wasted as a result of losses in the distribution grid [2].

Power distribution system long-term planning includes a number of studies with the goal of determining when and where a device should be installed. The typical planning process is 10 years or longer. It is known that the addition of the problem's multimodal character and a large range of instances, including both intermediary decades and the perspective year, increases the complexity of the selected problem.

The growth of the system, which is not only connected to the reinforcing of the current network but also involves substantial changes to the topology of the network, raises several key technical challenges. Furthermore, when it comes to optimal planning, the remedy for unconnected systems must be addressed from the outset of the procedure [3].

Distribution system (DS) expansion planning aims to maximize reinforcing techniques with reducing total energy losses and expense, all while adhering to a number of performance and reliability restrictions. It has been one of the most complicated issues, with a long history of continual attempts and advancements toward better answers. In emerging nations throughout the world, where energy consumption has been rising in recent days, distribution expansion planning is a major issue. Substantial contributions in the field of power management, on the other hand, have significantly slowed the pace of growth in power consumption. Nevertheless, rapid expansion appears to be unavoidable in the foreseeable future. Constructing and strengthening distribution substation units, the deployment of distributed generator (DG) units, and the installation or replacement of distribution lines are all part of the DS expansion planning [4].

The electric distribution utility is a critical component of the grid system, which suffers from significant energy losses, voltage variation, and reliability concerns. The voltage profile must be maintained, losses must be reduced, and reliability must be improved using various approaches. One such approach is network reconfiguration of the electric distribution grid. Proper network reconfiguration using tie/sectional switches or reclosers can improve the voltage profile as well as the system's dependability, resulting in acceptable loss reduction.

Furthermore, the widespread use of DG resources creates additional obstacles in the management and planning of a distribution grid, such as increased energy loss, aggregation of harmonic distortion, overloading of the equipment, and quality issues of supply. As a result, there is a lot of space for development. To meet these difficulties, different innovations are necessary to deal with a changing system, create a robust distribution grid with higher renewable integration, and provide dependable and economical network operation. Operators of electric grids fail to define the sources of system losses; fail to apply proper methods to reduce losses, operational expenses, and greenhouse gases; and fail to maintain losses at a minimum via appropriate distribution grid planning using low emission techniques [5,6]. Considering their clean, green, and environmental friendliness, varying green energy sources such as wind and solar are considered potential choices. Nevertheless, integrating renewable energy sources into advanced distribution networks is heavily reliant on advancements in power storage techniques as well as enhanced restrictions to encourage the utilization of storage devices with green energy sources [7].

The optimum functioning of these resources in stable and unstable active distribution systems was explored recently [8,9]. Numerous methodologies, including minimizing loss [10] or yearly expenditure [11] in a network, balancing demand [12], improving the voltage profile [12], and enhancing the hosting capacity of renewables [13] in the utility grid, have been adjusted for the optimum working of the system. Such optimum issues were solved using a variety of single and multiobjective approaches. By utilizing Pareto optimality, an optimization problem was constructed in Ref. [14] for minimizing power losses, load balancing, and enhancing distributed generation (DG) integration. Using the various objective functions independently, four DGs and network reconfiguration (NR) were appropriately scaled to achieve this goal. Furthermore, because reversals of powers are permitted, the given goals were not adequately managed concurrently utilizing NR alone, resulting in consecutive distributed generator integration and increased energy losses. Rather than a specific tie-line, a lossless soft open point was ideally assigned after selecting the optimum arrangement among Pareto options. The soft open point's installation was more successful than the NR deployment in terms of decreasing energy losses and demand balancing. It also improved the developed soft open point capacity to move DG-supplied energies from lighter to heavier demanded lines. To reduce both the operating expenses of the utility grid and the storage system capital cost, a single objective optimal problem is written in Ref. [15]. Comparative research was presented to explore the benefits of using specific techniques when developing storage devices. Hourly network reconfiguration, soft open points, and distributed generator deployment are among the approaches. In this study, two types of distributed generators were used: DG depending on inverters and DG running at the unity power factor. In terms of overall cost, DG depending on inverters outperformed the unity power factor distributed generators. Also, this work included a short-term periodic network reconfiguration technique for optimizing the load flow equations and illustrated its advantages in battery storage planning. The appropriate size and location of soft open points as well as green distributed generators were strongly suggested in this research for enhanced battery storage development [16–18].

2 Related work in network reconfiguration

Due to the enormous number of discrete switching parts, the challenge of distribution grid reconfiguration is a very complicated, multimodal, nondifferentiable optimization problem. Furthermore, in a massive distribution grid, the radial limitation often adds to the complexity of the reconfiguration issue [5]. The challenge of distribution grid reconfiguration is a mixed-integer nonlinear programming problem that falls within the domain of nondeterministic combinatorial optimization techniques [6]. Traditional approaches for tackling reconfiguration issues in a massive distribution grid, such as mixed-integer linear programming (MILP), have indeed been utilized. However, these techniques are likely to settle to a local optima rather than a global solution.

2.1 Classical optimization methods

In Ref. [19], network reconfiguration was accomplished by utilizing a customized linear programming approach for a lower load flow issue, in which the simplex technique was adapted to resolve a distribution grid layout that minimized losses while remaining within line capacity restrictions. The program begins with each node's power balancing calculations, ignoring system losses and voltage limitations.

With all these presumptions, a fast and simple linear programming method could be used to find a realistic option to the radial distribution system reconfiguration issue with the least amount of loss. However, while this method can help with loss minimization, it can't help with other goals such as reducing the amount of variable speed events. Furthermore, this strategy only produces a suboptimal result. In Ref. [20], the same technique was used with a DG integrated distribution grid, with the goal defined as the weighted total of absolute power flows across all network branches, energy production from each controlled DG system, and, if required, the demand being limited by DR operations. As a result, only the active energy integration and line impedances were taken into account throughout the optimization process.

In Ref. [21], the authors described an MILP strategy for reducing losses and the amount of switching frequency. The aim was minimized to losses, which included line overloading and the amount of control activities (including appropriate line-switching procedures). The linear programming with mixed variables was developed. For problems with convex constraints, the deterministic "branch-and-bound" decomposition approach can yield optimum outputs. Franco et al. [22] proposed an MILP technique to solve network reconfiguration for the distribution grid with dispersed generating. The load flow was modeled by utilizing linear assumptions in relation to the active and reactive components of the voltage, bearing in mind the network's usual operating circumstances. This MILP technique was successfully deployed to sample the network and two practical systems. The demands were modeled as constant power, current, and impedance, respectively.

2.2 Heuristic algorithms

Numerous studies have used heuristic methods to solve the issue of NR to minimize losses. In Ref. [23], the aim of NR was to decrease losses and rebalance the demands, and several searching methods were utilized to find answers. To ensure the network's radial layout, a branch interchange mechanism was implemented. The sensitivity assessment performed to determine whether a line must be opened or closed at each stage determines the variations amid the search techniques. As a result, the network's radial architecture restriction is enforced indirectly by metaheuristics rather than directly in the model. Metaheuristic strategies use knowledge from a performance measure to identify answers to optimization issues through trial and error. The techniques utilized for NR include the evolutionary technique [24], the genetic algorithm [25,26], simulated annealing [27,28], and ant colony optimization.

3 Network reconfiguration problem development

This section discusses the distribution network reconfiguration problem in detail.

3.1 Power loss of distribution system

The impedance of the wires is a source of system losses. Copper losses [29], which are functions of the square of current flowing through the line, account for the majority of active power losses in the utility grid.

$$P_{loss} = I^2 * R \tag{1}$$

where I represents the current and R represents the resistance of the conductor.

3.2 Power flow analysis

The current at node n is given by [29],

$$I_n = \frac{P_n + Q_n}{V_n} \tag{2}$$

The complex load flow equations are as follows,

$$P_n = \sum_{n=1}^{N_L} |Z_{n,n+1} V_n V_{n+1}| \cos(\theta_{n,n+1} + \delta_n - \delta_{n,n+1}) \tag{3}$$

$$Q_n = -\sum_{n=1}^{N_L} |Z_{n,n+1} V_n V_{n+1}| \sin(\theta_{n,n+1} + \delta_n - \delta_{n,n+1}) \tag{4}$$

Power loss between buses n and $n+1$ is as follows,

$$P_{loss\,(n,\,n+1)} = R_{n,n+1}\left(\frac{P^2{}_{n,n+1} + Q^2{}_{n,n+1}}{|V^2{}_{n+1}|}\right) \tag{5}$$

$$P_{Tloss} = \sum_{n=1}^{N_L} P_{loss(n,\,n+1)} \tag{6}$$

Power loss after reconfiguration is,

$$P'_{loss\,(n,\,n+1)} = R_{n,n+1}\left(\frac{P'^2{}_{n,n+1} + Q'^2{}_{n,n+1}}{|V'^2{}_{n+1}|}\right) \tag{7}$$

$$P'_{Tloss} = \sum_{n=1}^{N_L} P'_{loss(n,\,n+1)} \tag{8}$$

Reduction in the net power loss is presented as,

$$\Delta P_{Tloss} = \sum_{n=1}^{N_L} P_{loss(n,\,n+1)} - \sum_{n=1}^{N_L} P'_{loss(n,\,n+1)} \tag{9}$$

3.3 Problem formulation

The goal of the distribution grid reconfiguration problem is to modify the network's design by changing the position of present links and sectionalizing switches with the goal of reducing or enhancing specific operational attributes such as reliability, power losses, and voltage profile [30,31].

3.3.1 Objective function

$$OF_1 = \min\,(SAIFI, SAIDI, EENS) \tag{10}$$

where,

$$SAIFI = \frac{Total\ number\ of\ interruption}{Total\ number\ of\ customer\ served}\ \mathrm{int}/cus/yr \tag{11}$$

$$SAIDI = \frac{Total\ duration\ of\ all\ interrupions\ in\ customers}{total\ nomber\ of\ customer\ served}\ h/\mathrm{int}/yr \tag{12}$$

$$EENS = \sum L_i * U_i \tag{13}$$

$$OF_2 = \min\,(P_{Tloss}) \tag{14}$$

$$OF_3 = \sum_{n=1}^{N_L}\left(K_n * R_n\left(\frac{P^2{}_n * Q^2{}_n}{V^2{}_n}\right)\right) \tag{15}$$

where SAIFI represents system average interruption frequency index, SAIDI represents system average interruption duration index, EENS represents expected energy not supplied, L_i is the average connected load at load point i, and U_i is the average annual outage time at load point i, $n \in N_L$, subject to the following constraints.

3.3.2 Constraints

Voltage limits,

$$V_{m,\min} \leq V_m \leq V_{m,\max} \tag{16}$$

Feeder's capability limits in the nth branch;

$$K_n/P_n/ \leq P_{nmax} \quad n \in N_L \tag{17}$$

$$K_n/Q_n/ \leq Q_{nmax} \quad n \in N_L \tag{18}$$

$$K_n/I_n/ \leq I_{nmax} \quad n \in N_L \tag{19}$$

$$T_{switch} = (N_L - N_{bus}) + 1 \tag{20}$$

where, Q_n represents reactive power in the nth branch, P_n is active power in the nth branch, V_m is the magnitude of voltage at node m, K_n is the status topology of the branches (if branch n is closed $K_n=1$ and if it is open $K_n=0$), N_L is the set of branches in the network, N_{bus} is the total number of buses, and T_{switch} is the total number of switches.

4 Different techniques utilized for distribution network reconfiguration

There are several techniques used by different researchers for optimal network reconfiguration. A few of these techniques are:

4.1 Improved genetic algorithm [32]

Because of its stability and the necessity to pick just a few parameters, the method used in this study is dependent on selected improved genetic algorithm optimization. Some rules have been specified in the developed program to construct only viable persons during start up and to rectify infeasible individuals when they surface across the initial search. Fig. 1 presents a flowchart of the improved genetic algorithm for optimal network reconfiguration.

4.2 Multiobjective shark's smell (MSSO) [3]

The SSO is based on the shark's habit of attacking its prey once it detects blood in a search region. This motivates the simulation of a search process with the goal of finding the optimal output in a particular search region. Fig. 2 presents a flowchart of the multiobjective shark's smell optimization for optimal network reconfiguration.

4.3 Improved particle swarm optimization (PSO) [33]

Kennedy and Eberhart published the first particle swarm optimization (PSO) approach in 1995 [34]. It is a population-based metaheuristic optimization approach. Each swarm in a PSO's search region tracks a defined inertia and velocity via related iterations. The direction and speed of the velocity are in tune based on the swarm's most recent local best knowledge and previous best knowledge in its neighborhood. Fig. 3 presents a flowchart of the improved PSO method for optimal network reconfiguration.

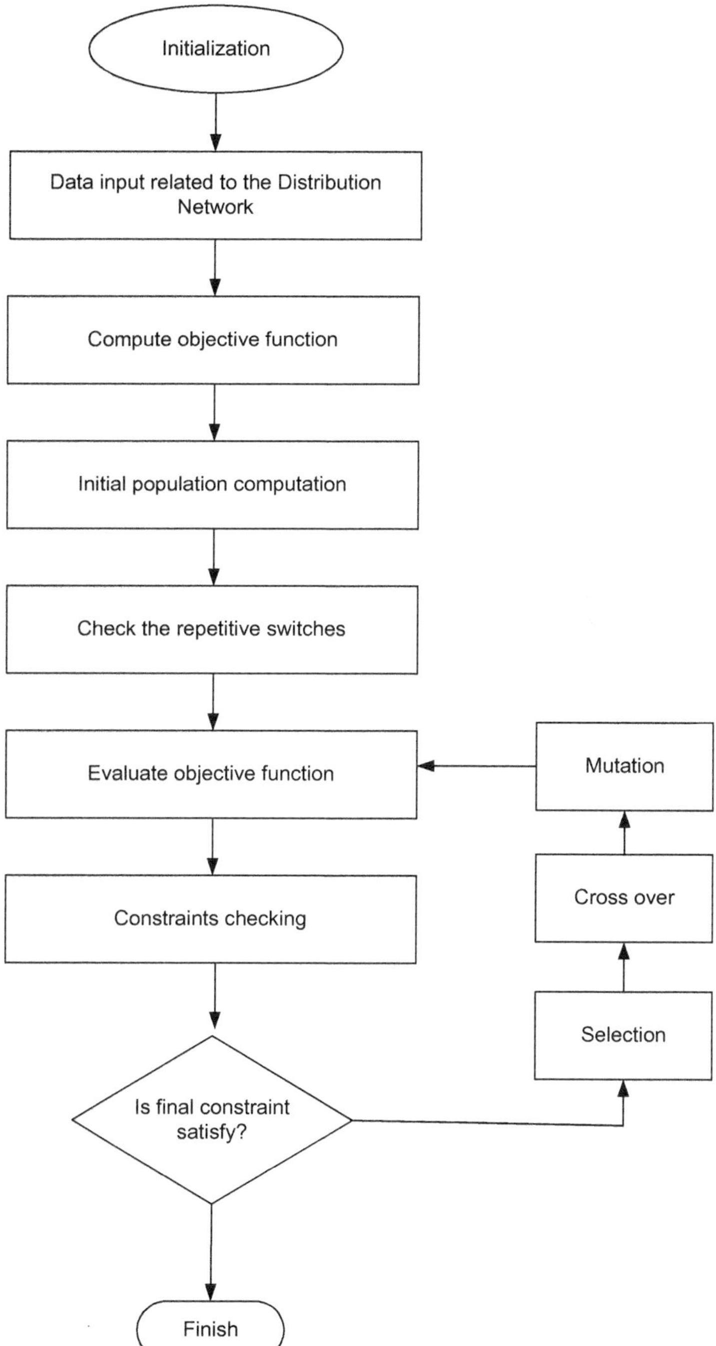

Fig. 1 Flowchart of improved genetic algorithm.

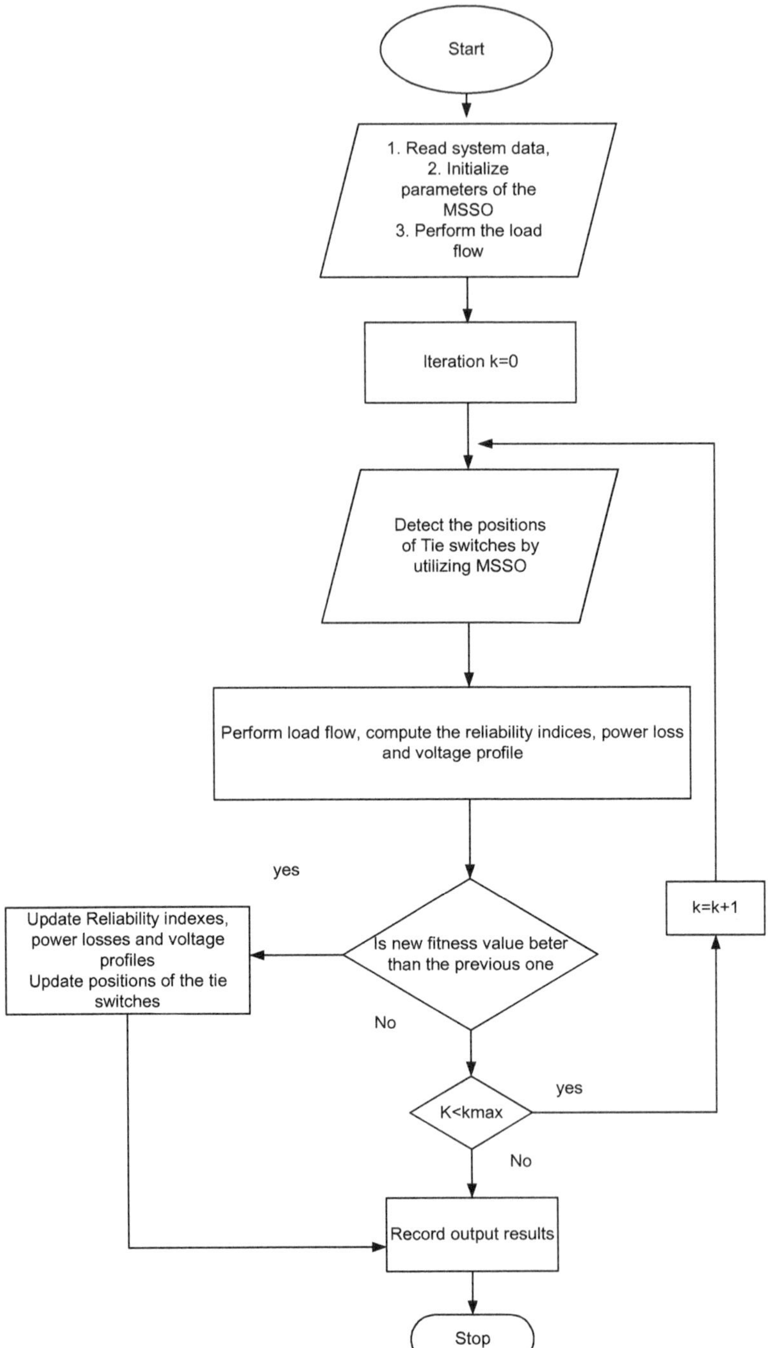

Fig. 2 Flow chart of multiobjective shark's smell optimization.

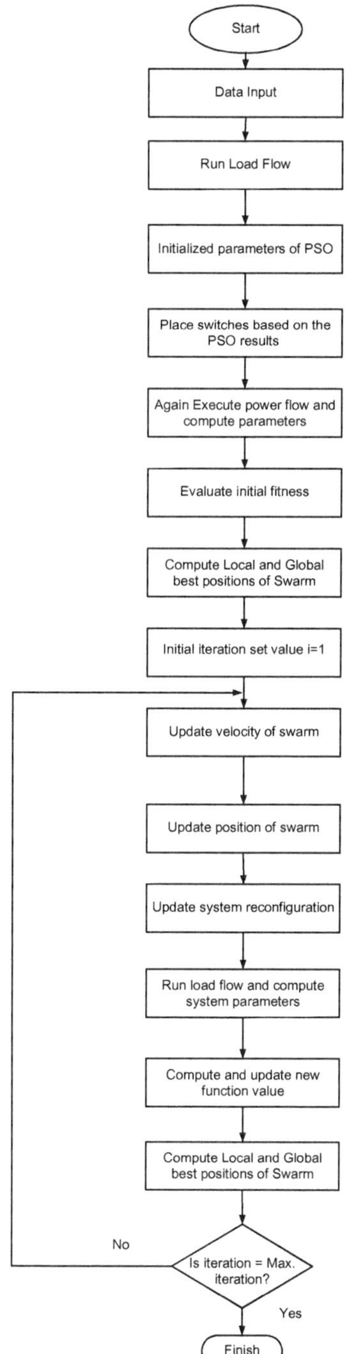

Fig. 3 Improved PSO algorithm for network reconfiguration.

5 Conclusion

This chapter provided a high-level overview of the distribution grid's optimal network reconfiguration problem. The optimal network reconfiguration problem is explained in depth. Different optimization strategies such as GA, PSO, and SSO are also explored to tackle the optimal network reconfiguration problem.

References

[1] T.F. Agajie, B. Khan, J.M. Guerrero, O.P. Mahela, Reliability enhancement and voltage profile improvement of distribution network using optimal capacity allocation and placement of distributed energy resources, Comput. Electr. Eng. 93 (2021) 107295.
[2] T.F. Agajie, B. Khan, H.H. Alhelou, O.P. Mahela, Optimal expansion planning of distribution system using grid-based multi-objective harmony search algorithm, Comput. Electr. Eng. 87 (2020) 106823.
[3] D. Anteneh, B. Khan, O.P. Mahela, H.H. Alhelou, J.M. Guerrero, Distribution network reliability enhancement and power loss reduction by optimal network reconfiguration, Comput. Electr. Eng. 96 (part A) (2021) 107518.
[4] M.H. Shariatkhah, M.R. Haghifam, Using feeder reconfiguration for congestion management of smart distribution network with high DG penetration, in: CIRED 2012 Workshop: Integration of Renewables into the Distribution Grid, 2012, pp. 1–4, https://doi.org/10.1049/cp.2012.0863.
[5] S. Ismael, S. Abdel Aleem, A. Abdelaziz, A. Zobaa, Probabilistic hosting capacity enhancement in non-sinusoidal power distribution systems using a hybrid PSOGSA optimization algorithm, Energies 12 (2019) 1018.
[6] J.M. Home-Ortiz, O.D. Melgar-Dominguez, M. Pourakbari-Kasmaei, J.R.S. Mantovani, A stochastic mixed-integer convex programming model for long-term distribution system expansion planning considering greenhouse gas emission mitigation, Int. J. Electr. Power Energy Syst. 108 (2019) 86–95.
[7] H. Zsiborács, N.H. Baranyai, A. Vincze, L. Zentkó, Z. Birkner, K. Máté, G. Pintér, Intermittent renewable energy sources: the role of energy storage in the European power system of 2040, Electronics 8 (2019) 729.
[8] C. Wang, G. Song, P. Li, H. Ji, J. Zhao, J. Wu, Optimal siting and sizing of soft open points in active electrical distribution networks, Appl. Energy 189 (2017) 301–309.
[9] W. Cao, J. Wu, N. Jenkins, C. Wang, T. Green, Benefits analysis of soft open points for electrical distribution network operation, Appl. Energy 165 (2016) 36–47.
[10] L. Zhang, C. Shen, Y. Chen, S. Huang, W. Tang, Coordinated optimal allocation of DGs, capacitor banks and SOPs in active distribution network considering dispatching results through bi-level programming, Energy Procedia 142 (2017) 2065–2071.
[11] C. Long, J. Wu, L. Thomas, N. Jenkins, Optimal operation of soft open points in medium voltage electrical distribution networks with distributed generation, Appl. Energy 184 (2016) 427–437.
[12] Q. Qi, J. Wu, C. Long, Multi-objective operation optimization of an electrical distribution network with soft open point, Appl. Energy 208 (2017) 734–744.
[13] H. Ji, P. Li, C. Wang, G. Song, J. Zhao, H. Su, J. Wu, A strengthened SOCP-based approach for evaluating the distributed generation hosting capacity with soft open points, Energy Procedia 142 (2017) 1947–1952.

[14] Q. Qi, J. Wu, L. Zhang, M. Cheng, Multi-objective optimization of electrical distribution network operation considering reconfiguration and soft open points, Energy Procedia 103 (2016) 141–146.
[15] L. Bai, T. Jiang, F. Li, H. Chen, X. Li, Distributed energy storage planning in soft open point based active distribution networks incorporating network reconfiguration and DG reactive power capability, Appl. Energy 210 (2018) 1082–1091.
[16] H. Ji, C. Wang, P. Li, J. Zhao, G. Song, F. Ding, J. Wu, An enhanced SOCP-based method for feeder load balancing using the multi-terminal soft open point in active distribution networks, Appl. Energy 208 (2017) 986–995.
[17] H. Ji, C. Wang, P. Li, G. Song, J. Wu, SOP-based islanding partition method of active distribution networks considering the characteristics of DG, energy storage system and load, Energy 155 (2018) 312–325.
[18] P. Li, H. Ji, C. Wang, J. Zhao, G. Song, F. Ding, J. Wu, Coordinated control method of voltage and reactive power for active distribution networks based on soft open point, IEEE Trans. Sustain. Energy 8 (2017) 1430–1442.
[19] A. Abur, A modified linear programming method for distribution system reconfiguration, Electr. Power Energy Syst. 18 (1996) 469–474.
[20] G. Celli, M. Loddo, F. Pilo, A. Abur, On-line network reconfiguration for loss reduction in distribution networks with distributed generation, in: 18th International Conference and Exhibition on Electricity Distribution, CIRED 2005. (Turin, Italy) Jun. 2005, 2005, pp. 6–9.
[21] B. Moradzadeh, K. Tomsovic, Optimal distribution system reconfiguration and restoration without heuristics, in: 17th Power System Computation Conference, (Stockholm), July 2011, pp. 1–7.
[22] J.F. Franco, M.J. Rider, M. Lavorato, R. Romero, A mixed integer LP model for the reconfiguration of radial electric distribution systems considering distributed generation, Electr. Power Syst. Res. 97 (4) (2013) 51–60.
[23] M.E. Baran, F.F. Wu, Network reconfiguration in distribution systems for loss reduction and load balancing, IEEE Trans. Power Delivery 9 (1989) 1401–1407.
[24] A. Delbem, A. de Carvalho, N.G. Bretas, Main chain representation for evolutionary algorithms applied to distribution system reconfiguration, IEEE Trans. Power Syst. 20 (2005) 425–436.
[25] J. Mendoza, R. Lopez, D. Morales, E. Lopez, P. Dessante, R. Moraga, Minimal loss reconfiguration using genetic algorithms with restricted population and addressed operators: real application, IEEE Trans. Power Syst. 21 (2006) 948–954.
[26] E.M. Carreo, R. Romero, A. Padilha-Feltrin, An efficient codification to solve distribution network reconfiguration for loss reduction problem, IEEE Trans. Power Syst. 23 (2008) 1542–1551.
[27] A. Vargas, M.E. Samper, Real-time monitoring and economic dispatch of smart distribution grids: high performance algorithms for DMS applications, IEEE Trans. Smart Grid 3 (2) (2012) 866–877, https://doi.org/10.1109/TSG.2012.2187078.
[28] S. Bruno, S. Lamonaca, M. La Scala, U. Stecchi, Integration of optimal reconfiguration tools in advanced distribution management system, in: 2012 3rd IEEE PES Innovative Smart Grid Technologies Europe (ISGT Europe), 2012, pp. 1–8, https://doi.org/10.1109/ISGTEurope.2012.6465873.
[29] F. Miranda Rodrigues, L. Ramos de Araujo, D.R.R. Penido, Optimization of reliability through switch reconfiguration in distribution systems, IEEE Lat. Am. Trans. 17 (06) (2019) 972–982.
[30] R. Pegado, Z. Ñaupari, Y. Molina, C. Castillo, Radial distribution network reconfiguration for power losses reduction based on improved selective BPSO, Electr. Power Syst. Res. 169 (2019) 206–213.

[31] A.A. Abou El-Ela, R.A. El-Sehiemy, N.K. El-Ayaat, Multi-objective binary particle swarm optimization algorithm for optimal distribution system reconfiguration, in: 2019 21st International Middle East Power Systems Conference (MEPCON), Cairo, Egypt, 2019, pp. 435–440.
[32] A.S. Abubakar, K.R. Ekundayo, A.A. Olaniyan, Optimal reconfiguration of radial distribution networks using improved genetic algorithm, Niger. J. Technol. Dev. 16 (1) (2019).
[33] B. Kumar, B.K. Saw, A.K. Bohre, Optimal distribution network reconfiguration to improve the system performances using PSO with multiple-objectives, in: 2020 International Conference on Computational Intelligence for Smart Power System and Sustainable Energy (CISPSSE), 2020, pp. 1–6.
[34] R. Eberhart, J. Kennedy, A new optimizer using particle swarm theory, in: MHS'95. Proceedings of the Sixth International Symposium on Micro Machine and Human Science, IEEE, 1995.

SECTION 12

Demand side response techniques

CHAPTER 12

Demand response aggregation

Baseem Khan and Issaias Gidey Hagos
Institute of Technology, Hawassa University, Hawassa, Ethiopia

Contents

1 Introduction

Demand response (DR) aggregators contribute significantly in the user market by providing personalized, automated controls for demands and devices that allow remote access while bringing choices and social behaviors into consideration. Such aggregators can bridge the data communication technology gaps that exist in today's power networks. DR aggregators can offer administrators a cost-effective way to reduce grid infrastructure needs while also allowing them to incorporate renewable energy (RE) technology [1]. These are entrants into the electricity industry that assist in incorporating

Active Electrical Distribution Network
https://doi.org/10.1016/B978-0-323-85169-5.00008-3

load-side solutions by using improvements in information and communications technology (ICT) and smart metering technology to bring new products that involve and motivate consumers to engage in the electricity industry [2]. Power storage devices and electric vehicles (EVs) could be employed for variable demand or power production in a DR scheme in this way. Some DR schemes integrate distributed energy sources with a storage device, such as rooftop solar photovoltaics (PVs), with the objective of recharging the battery bank at times of cheap energy costs or substantial excessive energy output from distributed energy sources [3]. While benefitting from a greater power tariff, an aggregator analyzes the discharge of Battery Energy Storage System (BESS) throughout a congested or maximum price of electricity to fulfill its legal responsibility of lowest cost assurance. Users may obtain greater value from existing solar PV systems with the aid of an aggregator, according to the research. A DR or electric vehicle aggregator can also collaborate with operators of electric car systems to optimize their accessible energy and storage [4]. Despite the fact that EV aggregators have always been in the pilot phase of development and have yet to be mechanically implemented in the industry, their variable energy production and storage capacities make them a valuable resource to the electrical network, and they are anticipated to become key players in the long term [5].

DR benefits the energy industry in a unique manner. Moving demand away from peak times–that is, shifting from production availability to network sufficiency–increases network efficacy and reduces the requirement for peak production investment. It may also be used to balance wind and solar electricity at a cheap cost [6]. It increases stable operation and lowers loads for coal and gas-fired spinning reserves, as generators must burn fuel continuously to be able to introduce more energy on an urgent basis. In regions where system capability is restricted, DR can also decrease the need for local network expenses by moving usage away from peak times. The much more important aspect of DR is that it rewards customers for the worth of their demand-side flexibility. Lastly, by enabling third-party network operators to join and encouraging forward-thinking, service-oriented businesses, DR improves competitiveness in the market [7].

Demand response achieves these goals by giving control signals and/or financial incentives to customers—residential, commercial, or industrial—to reduce or alter their consumption at opportune periods. As a result, demand response allows end users to directly profit from market competition as well as the smart grid. The term "commercial" refers to all structures and enterprises that are not directly industrial or residential in nature. To put it another way, municipal structures, Small and mid-size enterprises (SMEs), and companies such as hotels and office spaces are all included [8].

2 Type of DR

DR techniques can be classified as follows [9]:
- Incentive-based or explicit DR techniques.
- Price-based or implicit demand response.

2.1 Incentive-based or explicit DR techniques

These techniques are basically incentive-based. The total cumulative demands are provided to energy markets, supplying similar amenities to resources at the supply end and getting the same rates of prices for those servicers. This is most commonly seen in the proper balance, capability, or bulk electricity industry. Customers get direct compensation when they wish to modify their usage, which is usually prompted by the implementation of balancing measures, pricing variations in energy, or a network restriction. Individually or via a contract with an aggregator, users can profit from their flexibility in power use. A third-party aggregator or the customer's retailer can be the latter. Direct DR necessitates the development of established protocols between balancing responsible parties and independent aggregators as well as suitable programmed participation rules, communication, and measurement technology [10].

2.2 Price-based or implicit demand response

Customers can choose to be connected to variable energy rates or variable system tariffs (or both). These systems present a portion of the energy price and/or transmission over multiple time periods. Customers can respond to those rate differences depending on their own capabilities and limitations. It is clear that neither DR type can be substituted by the other DR. Numerous consumers take part in incentive-based DR via an aggregator while also getting involved in price-based DR via somewhat dynamic tariffs [11].

Each necessity and advantage is distinct and builds on the previous one. Within the markets, both are utilized for a variety of purposes. Further, both DR schemes are worth varying amounts of money. Although engaging in a variable costing method will generally result in a cheaper bill, engaging in a direct DR program will result in direct revenue. Direct DR, maybe most crucially, offers network administrators a useful and dependable strategy to ensure adjusting demand to alleviate operational difficulties.

Price-based DR is not really a tactical instrument; instead, it enables customers to profit on bulk electricity industry rate changes toward the degree that they really are capable and willing. As a result, a variable tariff scheme prohibits a user from participating in a balanced market. These are now the most lucrative income streams for customers.

Further, price-based DR necessitates sufficient customer communication and measuring instruments, revised operator transaction and payment processes, and customer feedback/communication methods. It also need proper information, including through online sites, such that users are not punished by variable rates that are not tailored to actual utilization behavior; it also needs appropriate customer protection mechanisms for vulnerable individuals.

3 Advantages of aggregation

Several customers lack the ability to participate actively inside the electricity market because they are too small, for instance, to deal with the complexities. To engage, organizations will need the assistance of an aggregator. Aggregators pool a variety of demands with varied characteristics and, as part of the pooling process, offer backup for single demands, boosting the quality and reliability and lowering the danger for specific participants. Aggregation providers are important in fostering active demand-side engagement and DR [12].

Customer flexibility is "aggregated" by the aggregator in order to "construct" dependable DR solutions. Companies work out deals between corporate, business, and domestic power users to pool their capacity to cut power use and/or transfer demands at a moment's notice. They combine several specific customer demands into a unique "pool" of aggregated controlled demand and offer it as a specific asset. Ventilation, heating systems, water boilers, crushers, industrial facilities, pumping stations, refrigerators, and other demands fall into this category.

It's critical to understand that accumulating customer demands necessitates a set of extremely particular skills that are unique to this position. To illustrate, in order to accommodate unique needs in a certain market, the aggregator needs substantial industrial expertise to determine the additional flexibility in diverse sectors, technological assets, and procedures, and the constraints of such flexibilities. Customers are frequently unaware of their own versatility capability, necessitating professional assistance. Aggregation can attain quality standards similar to or greater than production and meet market criteria for dependability. Because the grid operators may combine specific consumers, the aggregated demand-side capability can be used as a specific asset. This gives the network administrator a wide range of options. Among the most important advantages of aggregation is the variety of the aggregated portfolio and assurances that the aggregator will provide the pledged capability although certain specific customers are unable to do so [13].

By developing personalized, automatically generated control mechanisms for customer demands and equipment that enable remote access even with trying to take priorities and social behavior tendencies into account, DR aggregators play a critical role in trying to tap into the end user market. Such aggregators can close the technology and information mismatch that electricity systems are now experiencing. DR aggregators can concurrently supply operators with a cost-effective method for lowering network requirements and a mechanism of incorporating green technologies. DR aggregators seem to be newcomers to the electricity industry that help to integrate DS techniques by leveraging developments in communication and information techniques as well as smart meter facilities to design new ones that interact and inspire end users to engage in the electricity industry. In this sense, primary batteries and electric cars can be used as a flexible demand or a form of electricity in a DR scheme. Several DR schemes, such as a rooftop solar PV with a battery energy storage system, aim to recharge storage during periods of low electricity tariffs or large excess power output from distributed energy resources (DERs) [13].

An aggregator considers battery energy storage system discharge throughout a congested or peak price of power to fulfill its legal requirement of cheapest price assurance despite profiting from the greater energy value. The study also shows that with the help of an aggregator, end users may get more value out of their solar PV installations. Additionally, aggregators may work with proprietors of electric vehicles (EVs) and vehicle-to-grid (V2G) devices to maximize their accessible energy and storage that could be classified as a DR or V2G aggregator. Because V2G aggregators are in the prototype phase and about to be practically deployed in the industry, their flexible energy output and storage capabilities make them a valuable resource in the energy market, and they have been expected to be important participants in the near future. Aggregators of electric vehicles must take into account the characteristics of their functioning that essentially "ramp up" and "ramp down" at high levels with short reaction times. EVs may also link to and disengage from the power network for charging [13].

4 DR schemes

Administrators in electrical systems deploy DR schemes to keep the network economical and secure at periods of peak load, peak distributed energy resource production, or peak energy tariffs. Users respond to controller instructions by reducing or transferring specified demands or production in return for an inducement or compensation in such systems. The advantages of managing these demands or productions include savings on electricity bills and incentives for users; secured volatility in the market; savings on power system infrastructure; efficiency; enhanced power system stability and reliability while minimizing variable costs throughout peak timings; and offering operational flexibility that could be utilized to incorporate RES [14].

The classification of DR schemes into "implicit" or "explicit" techniques is a common practice. Users choose whether to participate for implicit initiatives, which convey high power tariffs to them. The users receive a defined compensation for a certain demand or production curtailment or transfer in explicit DR schemes. Implicit DR schemes could only be engaged in the electricity industry because network administrators have little influence over the result. However, explicit DR schemes are predefined agreements that provide some command and may thus engage in electricity, capability, and reactive power management service markets as needed. Bidding at the demand side, which allows big users or aggregators of DR to actively bid significant amounts of controllable demand into electricity tenders as a substitute for existing generating supplies, allows DR to engage in the bulk electricity sector in some areas.

If indeed the DR offer is acceptable, the central controller will deploy it based on the need to reduce the electricity price of power generation. Because the expenses of allowing DR are far less than the price of operating a power-producing facility, bidding on the demand side through DR has the potential to successfully replace conventional producers from the bulk electricity industry. Network administrators utilize reactive power management facilities in real time to preserve system stability and dependability in the event of unforeseen disruptions or supply-demand fluctuations that create resilience concerns. A capacity system enables DR initiatives to also be managed as a different source of power purchasing from the bulk power electricity sector, and people can be used to reserve the capacity of future load forecasts or in response to a backup. Such marketplaces improve power system efficiency while lowering electricity prices, as DR schemes utilized for power reserves can offer cheaper tariff models compared to conventional power generation. Hardly larger industrial users have generally engaged in DR schemes because they are more financially sustainable for facilitating DR and have sophisticated systems of controllable demands [14].

Bulk users could reduce demands such as illumination, heaters, air flow, and air conditioners, but their demand shedding possibilities are also determined by their specific manufacturing applications, which might allow for the closure of certain processes. In accordance with the needs of DR instructions, commercial entrants can modify illumination, washers, conditioners, air conditioning, refrigeration, heaters, and pool pumping systems. Demand response regulation over heating and cooling as well as water heaters has been used by some domestic customers to engage in DR schemes.

Residential user engagement has indeed been restricted because the allowable price of DR is greater for such customers. Home facilities are locations of individual connectedness, making it difficult to encourage engagement of residential users if it implies interrupting their style of living, particularly if the bill money saved is considered insignificant by the user and thus not worth the hassle. Moreover, residential consumers were previously unable to perceive a flexible power tariff and had flat fee meters fitted on their venue. With the emergence of smart meters, this really is evolving [14] (Fig. 1).

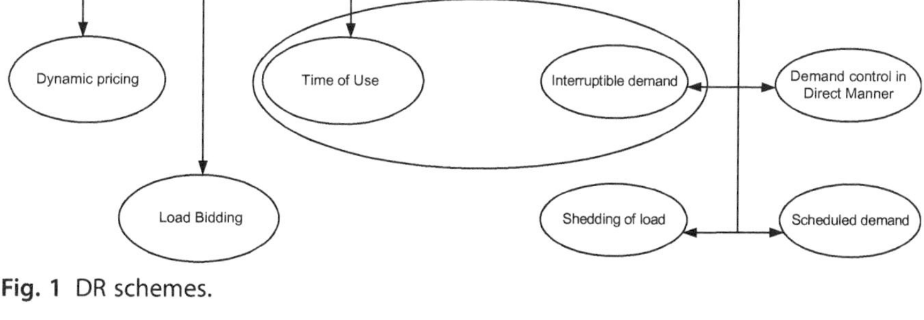

Fig. 1 DR schemes.

4.1 Energy storage and electric vehicle technology

In the context of energy storage and EVs, aggregators would like to establish a bilateral agreement only with EVs and/or battery holders in which the aggregator would monitor and control the EV in exchange for a reward. As an added incentive for interacting shareholders while alleviating societal issues about rechargeable battery damage, aggregators frequently substitute and retain the battery. The majority of the existing literature concludes that EV and/or battery aggregators are very suitable to supply reactive power management facilities to network administrators, with an emphasis on the reserve capacity for providing regulation in frequency, and that will be the primary business concentration for this category of aggregator. Even though conventional large-capacity power stations can provide lower-priced sources of energy than DR aggregators focused on electric vehicles/batteries, the reserve capacity tariff requires payouts for "backup" capability as well as payouts for the "real" power dispatched, allowing electric vehicles/battery aggregators to compete effectively [15].

4.2 Smart meters

These are instruments that help end users of electricity get information such as device power utilization statistics, load curves, time varying tariffs, disruption occurrences, overvoltages, phase loss, and asymmetry. Customers may now react to energy signals and make wiser power utilization choices as a result of their intellect, allowing them to become regular players in the electricity industry. Because conventional energy markets are not designed to observe distribution systems, smart meters include a supervision system that is searchable and viewable, which is critical for DER connection success.

Despite the fact that the deployment of such a technique is a critical enabler for residential participation in DR, these consumers are difficult to motivate for the reasons stated above. As a result, maximizing resident potential necessitates the development of personalized capabilities that help customers take part without trying to make demand shedding too complex. A network administrator might not even be able to bear the extra burden of creating such personalized profile information for occupants, which might necessitate evaluating their specific utilization patterns and choices to successfully enable their involvement [16].

4.3 Demand response aggregator

In the power industry, DR aggregators operate as third-party intermediaries among energy industry players and electricity users in the industrial, business, and domestic sectors. Such intermediaries recognize that end users can supply system capability by maintaining demand/productions during critical times such as peak demand or transferring specific processes to a lower priced time of each day. The aggregator makes use of this capability by interacting with plenty of end users to build up sufficient aggregated capability to meet the demands for joining the bulk power industry.

Network administrators as well as other key stakeholders purchase DR aggregator capability as supplementary facilities, buffers of capacity, or balancing measures. As a result, DR aggregators give the electricity market a way to capture potential capabilities that might not have been recognized or considered important enough to join the network as individual pieces. This is especially beneficial for administrators who just want to acquire additional capability utilization as the percentage of RE on the grid grows [17].

4.4 Virtual power plant (VPP)

The notion of a virtual power plant (VPP) is presented to combine the benefits of demand response schemes and distributed energy resources in a group of consumers who may also participate in the energy sector. Corporate VPP (CVPP) and technological VPP (TVPP) are two types of VPPs. Power network constraints are included in the TVPP power management approach, but not in CVPP. Such VPPs employ a bidding technique that takes into account their resources and electrical industry conditions. Furthermore, battery storage-incorporated VPPs could add to the sufficiency and dependability of a production source in a low-voltage system, where the charging and discharging of battery storage must be determined by an optimization technique. Fig. 2 presents the implementation of DR Schemes [17].

4.5 Communication modules

These are also required for data transfer between the HEMS/BEMS system, equipment/demand controllers, sensing, graphical user interface, and relevant players to be effective. Conventional load management solutions send and receive information using wired network modules and standards such as power line carriers, fiberoptics, and Internet standards. These modes of communication, though, have greater implementation expenses connected with their hardware needs than their wireless equivalents. New progress in wireless connectivity technologies and protocols such as Zigbee, 6LoWPAN, WiFi, Bluetooth, and Z-Wave have enabled a much more cost-effective and flexible type of communication that matches the evolving power system's dispersed topology [18].

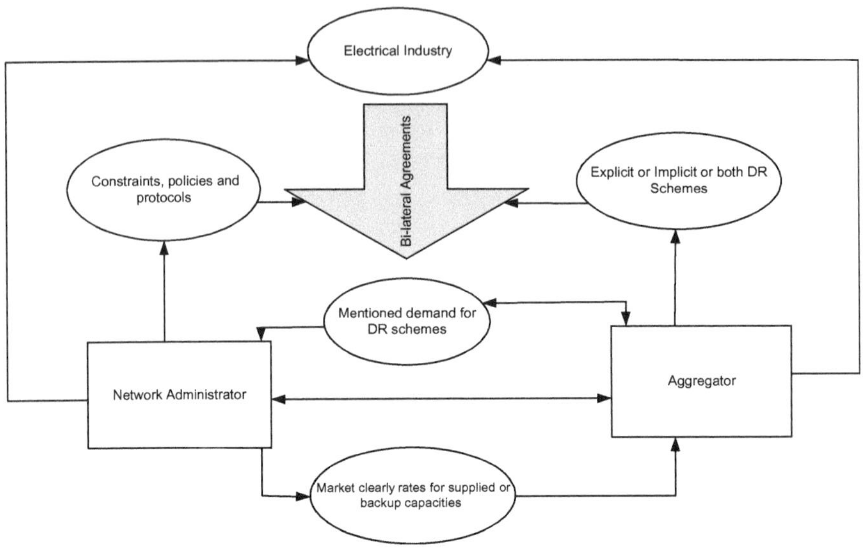

Fig. 2 DR scheme implementation.

5 Different considerations regarding implementation of DR schemes

5.1 Technical considerations

DR aggregators should be able to interact successfully with end users. Aggregators ought to be able to independently connect equipment or predefined demands that the end user has specified as well as execute demand management to recover a certain DR capability. Aggregators also need to provide users with a rational graphical user interface (GUI) to interact with DR messages and enable a certain tier of user customization.

Home energy management systems (HEMS) and building energy management systems (BEMS) have been developed as a result of advancements in information and communication technologies and smart metering, which provide dynamic conditions that permit the effective control of sensitive demands and offer strong communication abilities. Smart metering is described as a system that detects, analyzes, gathers, and evaluates information on electricity characteristics and the quality statistics that go with them. The technology can interact with meters and central/distributed controllers on a predetermined timetable [17].

5.2 Economic considerations

The economical features of DR aggregators in relation to the capability to produce income, investment and trading activity as well as implementation, service, and operating expenses are influenced by a variety of factors, which can be classified as follows:

1. The type of economy as well as the circumstances and community in which it exists.
2. Market access constraints.
3. Does the location have any updates or infrastructural facilities?
4. The density of the population of the geographic region.

The administrator site has signed up to engage in disaster recovery initiatives that get launched in both the auxiliary company and capacity markets.

Due to a shortage of information and the expense of establishing enhanced monitoring, domestic users typically have greater incremental costs connected to DR schemes. Based on knowledge from past experience with DR schemes and their inclination to have some preexisting equipment, aggregators may find that the extra cost of permitting industrial and commercial users to engage in a DR scheme is significantly lower [19].

Aggregators could also help network administrators save money by deferring network development by using DR aggregators to modify the demand characteristics to save network assets from overheating through peak times. For instance, a company can use a DR aggregator to entice consumers to engage in a DR scheme by raising their demands amid significant RE sources and/or lowering their demands amid peak usage.

5.3 Social and political considerations

In several ways, the political and social consequences of DR aggregators are interwoven. As a result, sociopolitical issues and their interrelationships, are discussed in detail. The study demonstrates how state rollouts of DR key enablers, such as smart meters, promote market involvement by facilitating a user's active engagement via scale economies. Some users may consider it an infringement of privacy to have their device energy usage rates observed and modified by different energy market players.

As a result, it is the aggregator's responsibility to guarantee that the appropriate security solutions are implemented to alleviate this societal issue. Difficult problems in the Internet of Things and privacy security are cybersecurity and intrusion/theft/fraud detection in smart meters.

Retailers can use the aggregator's facilities to mitigate their exposures toward economic uncertainty by enabling the aggregator to manage their customers' peak load. DR aggregators also include a form of peak usage mitigation that may compete with peak generators because they have lower marginal costs and produce less carbon dioxide than conventional generating techniques. Another benefit is that DR aggregators usually reply quicker, whereas traditional generators almost always have operational constraints that limit their capacity to rapidly alter their energy generation [20].

6 Strengths and weaknesses of DR schemes

6.1 Strengths

The following are some of a DR aggregator's major strengths [21]:

- The incremental cost of allowing DR for urban and business users is modest. Furthermore, the expenses of permitting bidirectional power in battery-only cars are quite cheap.
- Technological advancements have lowered the price of equipment and broadened the variety of products and devices that can be utilized for disaster recovery. In addition, battery storage technology production prices are falling, giving V2G aggregators additional opportunities to take advantage of the existing technology.
- Enabling a high number of domestic DRs helps diversify the investment and reduce risk.
- DR aggregators have reduced peak load operation and maintenance expenses than conventional generation stations, allowing them to provide fair rates.
- By giving access to the common market, DR aggregators can increase capital productivity.
- DR aggregators offer a capacity resource with a low carbon impact.
- EVs offer quick reaction times and high power capacity, giving V2G/battery aggregators strategic benefits over conventional producers and peak generators.

6.2 Weaknesses

The following are a few of a DR aggregator's potential flaws [21].

- The incremental costs of enabling each existing residential end users are high.

- Qualified technical personnel are expected, but they come at a considerable cost to the company.
- Users may be irritated by having to modify their usage habits.
- Consumer preference conduct has a significant impact on obtainable profits.
- Starting and ongoing costs of marketplace participation and beginning fees are considerable.
- Because domestic users are generally unaware of dynamic energy rates, appropriate advertising involvement is required.

7 Conclusions

This chapter examined DR aggregators from a variety of viewpoints, including legal, financial, societal, and technical. As a result, a thorough examination of a standard DR aggregator's capabilities, shortcomings, possibilities, and dangers has been undertaken. This research confirms that the DR aggregator has the ability to play a vital role in generating value for a power network and consumers as a whole. This article discussed the role of DR aggregators in the energy sector as well as the benefits that these aggregators bring. It also explored the idea of who obtains such benefits and what possible difficulties this concept may bring.

References

[1] B. Shen, G. Ghatikar, Z. Lei, J. Li, G. Wikler, P. Martin, The role of regulatory reforms, market changes, and technology development to make demand response a viable resource in meeting energy challenges, Appl. Energy 130 (2014) 814–823.

[2] A.M. Carreiro, H.M. Jorge, C.H. Antunes, Energy management systems aggregators: a literature survey, Renew. Sustain. Energy Rev. 73 (2017) 1160–1172.

[3] A. Papalexopoulos, C. Hansen, R. Frowd, A. Tuohy, E. Lannoye, Impact of the transmission grid on the operational system flexibility, in: Proceedings of the 2016 Power Systems Computation Conference (PSCC), Genoa, Italy, 20–24 June, 2016, pp. 1–10.

[4] S.M. Nosratabadi, R.-A. Hooshmand, E. Gholipour, A comprehensive review on microgrid and virtual power plant concepts employed for distributed energy resources scheduling in power systems, Renew. Sustain. Energy Rev. 67 (2017) 341–363.

[5] Z. Ma, A. Badi, B.N. Jørgensen, Market opportunities and barriers for smart buildings, in: Proceedings of the 2016 IEEE Green Energy and Systems Conference (IGSEC), Long Beach, CA, USA, 6–7 November, 2016, pp. 1–6.

[6] Z. Ma, J.D. Billanes, M.B. Kjærgaard, B.N. Jørgensen, Energy flexibility in retail buildings: From a business ecosystem perspective, in: Proceedings of the 2017 14th International Conference on the European Energy Market (EEM), Dresden, Germany, 6–9 June, 2017, p. 6.

[7] B. Li, J. Shen, X. Wang, C. Jiang, From controllable loads to generalized demand-side resources: a review on developments of demand-side resources, Renew. Sustain. Energy Rev. 53 (2016) 936–944.

[8] S. Annala, S. Honkapuro, Demand response in Australian and European electricity markets, in: Proceedings of the 2016 13th International Conference on the European Energy Market (EEM), Porto, Portugal, 6–9 June, 2016, pp. 1–5.
[9] N.G. Paterakis, O. Erdinç, J.P.S. Catalão, An overview of Demand Response: key-elements and international experience, Renew. Sustain. Energy Rev. 69 (2017) 871–891.
[10] H. Du, S. Liu, Q. Kong, W. Zhao, D. Zhao, M.G. Yao, A microgrid energy management system with demand response, in: Proceedings of the 2014 China International Conference on Electricity Distribution (CICED), Shenzhen, China, 23–26 September, 2014, pp. 551–554.
[11] N. O'Connell, P. Pinson, H. Madsen, M. O'Malley, Benefits and challenges of electrical demand response: a critical review, Renew. Sustain. Energy Rev. 39 (2014) 686–699.
[12] M. Minou, G.D. Stamoulis, G. Thanos, V. Chandan, Incentives and targeting policies for automated demand response contracts, in: Proceedings of the 2015 IEEE International Conference on Smart Grid Communications (SmartGridComm), Miami, FL, USA, 2–5 November, 2015, pp. 557–562.
[13] R. Walawalkar, S. Fernands, N. Thakur, K.R. Chevva, Evolution and current status of demand response (DR) in electricity markets: insights from PJM and NYISO, Energy 35 (2010) 1553–1560.
[14] Z. Liu, X.J. Zeng, Q. Ma, Integrating demand response into electricity market, in: Proceedings of the 2016 IEEE Congress on Evolutionary Computation (CEC), Vancouver, BC, Canada, 24–29 July, 2016, pp. 2021–2027.
[15] S. Sorrell, Reducing energy demand: a review of issues, challenges and approaches, Renew. Sustain. Energy Rev. 47 (2015) 74–82.
[16] J. Son, R. Hara, H. Kita, E. Tanaka, Energy management considering demand response resource in commercial building with chiller system and energy storage systems, in: Proceedings of the 2014 International Conference on Power Engineering and Renewable Energy (ICPERE), Bali, Indonesia, 9–11 December, 2014, pp. 96–101.
[17] V. Fabi, G. Spigliantini, S.P. Corgnati, Insights on smart home concept and occupants' interaction with building controls, Energy Procedia 111 (2017) 759–769.
[18] L.J. De Vries, R. Verzijlbergh, Organizing flexibility: How to adapt market design to the growing demand for flexibility, in: Proceedings of the 2015 12th International Conference on the European Energy Market (EEM), Lisbon, Portugal, 19–22 May, 2015, pp. 1–5.
[19] Z. Ma, H.T.A. Friis, C.G. Mostrup, B.N. Jørgensen, Energy flexibility potential of industrial processes in the regulating power market, in: Proceedings of the 6th International Conference on Smart Cities and Green ICT Systems, Porto, Portugal, 22–24 April, 2017, pp. 109–115.
[20] A. Picciariello, C. Vergara, J. Reneses, P. Frías, L. Söder, Electricity distribution tariffs and distributed generation: quantifying cross-subsidies from consumers to prosumers, Util. Policy 37 (2015) 23–33.
[21] I. Lamprinos, N.D. Hatziargyriou, I. Kokos, A.D. Dimeas, Making demand response a reality in Europe: policy, regulations, and deployment status, IEEE Commun. Mag. 54 (2016) 108–113.

Further reading

J. Li, Z. Wu, S. Zhou, H. Fu, X.P. Zhang, Aggregator service for PV and battery energy storage systems of residential building, CSEE J. Power Energy Syst. 1 (2015) 3–11.

SECTION 13

Smart metering

CHAPTER 13

Hybrid signal processing technique to design smart meter for detection of voltage amplitude disturbances

Surendra Singh[a], Avdhesh Sharma[a], and Om Prakash Mahela[b]
[a]Department of Electrical Engineering, Faculty of Engineering and Architecture, Jai Narain Vyas University, Jodhpur, Rajasthan, India
[b]Power System Planning Division, Rajasthan Rajya Vidyut Prasaran Nigam Ltd., Jaipur, Rajasthan, India

Contents

Abbreviations

AHOM absolute Hilbert output matrix
APSO adaptive particle swarm optimization
ASOM absolute values Stockwell output matrix
CVT curvelet transform
DWT discrete wavelet transform
EL extreme learning
FCM fuzzy c-means clustering
FES fuzzy expert system
FFNN feed-forward neural network
GT Gabor transform

Active Electrical Distribution Network
https://doi.org/10.1016/B978-0-323-85169-5.00009-5

GWC	Gabor wavelet coefficients
HPQDLI	hybrid power quality detection and location index
HSPT	hybrid signal processing technique
HT	Hilbert transform
IEEE	Institute of Electrical and Electronics Engineers
MATLAB	matrix laboratory
MI	momentary interruption
ML	machine learning
MNN	modular neural network
NN	neural network
PNN	probabilistic neural network
PQ	power quality
PQI	power quality detection index
RBDT	rule-based decision tree
RBFNN	radial basis function neural network
RE	renewable energy
RT	ridgelet transform
RTDS	real-time digital simulator
RV	reconstructed voltage signal
SRE	signal reconstruction error
ST	Stockwell transform
STFT	short time Fourier transform
WFD	PQ disturbance detection weight factor
WFDL	PQ disturbance detection and location weight factor
WT	wavelet transform

1 Introduction

The power quality (PQ) indicates the degree to which the voltage waveform is approaching the standard specifications. Major power quality events with a distribution network include harmonic distortions, frequency variations, and voltage magnitude-related disturbances such sags, swells, momentary interruptions (MI), spikes, and notches [1]. In recent decades, smart meters for monitoring PQ disturbances have become essential tools for utilities. Further, continuously increasing the integration of nonlinear loads and generators to power grids using power electronics technology initiated the requirement of smart PQ monitoring devices [2]. The voltage-related PQ disturbances are commonly observed with a power supply system. These can be monitored using the signal processing-based hybrid methods [3]. Commonly applied signal processing approaches for the detection of disturbances include wavelet transform (WT), discrete wavelet transform (DWT), Gabor transform (GT), Short time Fourier transform (STFT), Hilbert transform (HT), Stockwell transform (ST), curvelet transform (CVT), and ridgelet transform (RT) [4]. The commonly used techniques for the categorization of PQ disturbances include the probabilistic neural network (PNN), modular neural network (MNN), neural network (NN), feed-forward neural network (FFNN), fuzzy c-means (FCM) clustering, adaptive particle swarm optimization (APSO), fuzzy expert system (FES), rule-based decision tree (RBDT), image processing, extreme learning (EL), and machine learning (ML) [5]. A hybrid technique using the ST and machine learning method to detect different PQ events is designed in Ref. [6]. Due to the good time-frequency resolution features of ST, a proposed algorithm has high efficiency and it requires fewer features compared to the WT-based PQ identification algorithm. In Ref. [7], the authors designed an approach to identify and classify PQ events with a power distribution network using the radial basis function neural network (RBFNN) and WT. PQ classification is achieved using fewer features in comparison to the conventional techniques. Weights are so selected that high classification accuracy is achieved using RBFNN. A PQ recognition technique using ST in the presence of wind and solar energy in the utility grid under various operational scenarios is reported in Refs. [8,9]. A hybrid method using the ST and FCM to detect various operational events and PQ disturbances associated with these events in a utility grid with high wind energy penetration is designed in Ref. [10]. Operational events are rated in terms of associated PQ disturbances using a PQ index. The effectiveness of the algorithm is analyzed on an IEEE-13 bus test system and real-time validation is performed using a real-time digital simulator (RTDS). In Ref. [11], the authors introduced a method to identify PQ issues pertaining to a utility grid with high penetration of solar energy in various loads. Harmonic analysis is also performed using the hardware set-up.

After a detailed study of the above-discussed literature, it is pointed out that voltage magnitude-related PQ disturbances commonly occur on the distribution network due to the nonlinear nature of loads, high RE penetration, operational events, and faulty events. Further, the hybrid use of signal processing methods helps to design an efficient online PQ monitoring smart meter. The following are the main contributions of this paper:

- A hybrid signal processing technique using GT, ST, and HT is designed for a smart PQ monitoring meter to detect voltage sags, swells, and MIs.
- A PQI and HPQDLI are used to detect and locate the voltage magnitude-related PQ disturbances accurately.
- Features computed from PQI and HPQDLI are used for the categorization of the disturbances using decision rules.

- The proposed hybrid algorithm is more efficient compared with the DWT method, specifically in noisy conditions.

This paper is organized into seven sections. General introduction, literature review, and research contribution are discussed in the Section 1. Section 2 details the mathematical formulation used for the generation of investigated PQ disturbances. The designed hybrid PQ identification algorithm is detailed in Section 3. A discussion of simulation results is included in Section 4 and the categorization of PQ events is discussed in Section 5. A performance comparative study is included in Section 6 and the conclusion is made in Section 7.

2 Generation of PQ disturbances

Voltage with a pure sinusoidal nature, voltage sags, swells, and MIs are simulated using the mathematical formulation reported in Refs. [3,12] which is described below.

- Voltage signal without any disturbance

$$V(t) = A\sin(\omega t)$$

- Voltage signal with sag disturbance

$$V(t) = [1 - \alpha(u(t - t_1) - u(t - t_2))]\sin(\omega t)$$

- Voltage signal with swell disturbance

$$V(t) = [1 + \alpha(u(t - t_1) - u(t - t_2))]\sin(\omega t)$$

- Voltage with MI disturbance

$$V(t) = [1 - \alpha(u(t - t_1) - u(t - t_2))]\sin(\omega t).$$

In the above equations, the nomenclature used is as follows: A—amplitude of signal, f—frequency of signal, V—voltage signal, T—time period ($T = 1/f$), ω—angular frequency, $u(t)$—unit step function, and t—time instant. Standard parameters and simulated parameters used for this study are available in Ref. [12].

3 Hybrid signal processing technique to design smart PQ monitoring meter

The hybrid signal processing technique (HSPT) for smart PQ monitoring meters to detect voltage amplitude-related disturbances is designed by the hybrid use of GT, ST, and HT. The detailed mathematical expression of GT, ST, and HT used for the study is available in Refs. [13–16].

The recognition of voltage magnitude-related PQ disturbances is completed in two steps. The first step involves the detection of PQ disturbances and the second step involves the categorization of these PQ disturbances.

3.1 Detection of voltage magnitude-related PQ disturbances

The following are the steps to implement HSPT to detect voltage amplitude-related PQ disturbances:

- Generate voltage with PQ disturbance (V) using the mathematical formulation detailed in Section 2.
- Decompose voltage with the PQ disturbance using GT, compute the complex Gabor wavelet coefficients, compute the absolute values, and designate this as GWC.
- Reconstruct the signal with the PQ disturbance (RV) and compute the signal reconstruction error (SRE) by the difference between the original signal and the reconstructed signal.

$$SRE = V - RV$$

- Decompose the reconstructed signal (RV) using the Stockwell transform and compute the Stockwell output matrix (SOM). Compute the absolute values of the SOM to compute the absolute Stockwell output matrix (ASOM).

$$ASOM = abs(SOM)$$

- Compute the summation of values of every column of ASOM

$$SS = sum(ASOM)$$

- Decompose the reconstructed signal (RV) using the Hilbert transform and compute the absolute Hilbert output matrix (AHOM).

$$AHOM = hilbert(RV)$$

- Compute the PQ disturbance detection index (PQI) by the multiplication of AHOM with SS element to element. A PQ disturbance detection weight factor (WFD) is used to normalize the value of PQI. The value of WFD is considered equal to 0.555, which is selected after the testing method on different magnitudes of voltage sag, swell, and MI disturbances.

$$PQI = AHOM \times SS \times WFD$$

- Compute the hybrid power quality disturbance detection and location index (HPQDLI) by the multiplication of AHOM with SS and

GWC element to element. A PQ disturbance detection and location weight factor (WFDL) is used to normalize the value of the HPQDLI index. The value of WFD is considered equal to 166.67, which is selected after the testing method on different magnitudes of voltage sag, swell, and MI disturbances.

$$HPQDLI = AHOM \times SS \times \mathrm{GWC} \times WFDL$$

- An analysis of the plots of PQI and HPQDLI associated with a voltage signal having PQ disturbance and comparing these plots with the respective plots of pure sine wave, PQ disturbances related to voltage magnitude such as sag, swell, and MI can be identified effectively.

3.2 Categorization of voltage magnitude-related PQ disturbances

The voltage-related PQ disturbances are categorized using simple decision rules designed using the features computed from the PQI and HPQDLI. A detailed description of the decision rules is available in Ref. [17]. The computation of features is detailed below:

- F1: Variance of PQI

$$F1 = \mathrm{var}(PQI)$$

- F2: Variance of HPQDLI

$$F2 = \mathrm{var}(HPQDLI)$$

- F3: Standard deviations of ASOM

$$F3 = std(ASOM)$$

- F4: Standard deviations of AHOM

$$F4 = std(AHOM)$$

4 Simulation results and discussion

Simulation results for the detection of voltage magnitude-related PQ disturbances are discussed in this section. PQI, HPQDLI, and signal reconstruction plots for the investigated disturbances of voltage sag, voltage swell, and MI have been discussed. These plots for the sinusoidal waveform are considered as reference plots and are also discussed in this section.

4.1 Sinusoidal waveform

A voltage signal without a PQ disturbance is simulated and shown in Fig. 1A. This signal is decomposed using GT to compute the absolute values of GWC. The signal is reconstructed, and this is shown in Fig. 1B. SRE is computed by the difference between the original signal and the reconstructed signal, which is plotted in Fig. 2. The reconstructed signal (RV) is decomposed using the ST and the ASOM is calculated. The summation of the values of every column of ASOM (SS) is computed. Further, the RV signal is also processed using HT and the absolute values of the output matrix are computed (AHOM). The PQI index is computed by the multiplication of AHOM, SS, and WFD, which is detailed in Fig. 1C. HPQDLI is computed by multiplying the AHOM, SS, GWC, and WFDL, which is illustrated in Fig. 1D.

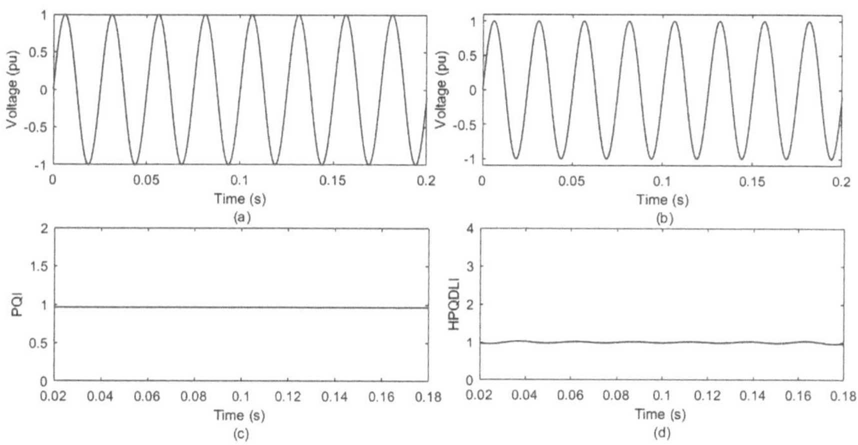

Fig. 1 Voltage without PQ disturbance: (A) Original voltage signal, (B) reconstructed voltage signal, (C) PQI plot, (D) HPQDLI plot.

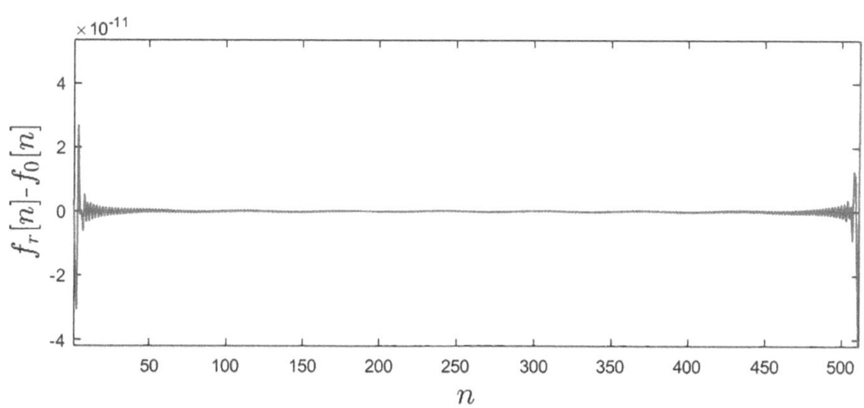

Fig. 2 Reconstruction error for voltage without a PQ disturbance.

It is inferred from Fig. 1A and B that the disturbance is not associated with the voltage signal. Fig. 1C shows that the magnitude of the PQI index is constant and equal to the unit over the entire time period of the signal, which indicates that signal is pure sinusoidal in nature and not associated with a PQ event. Fig. 1D shows that the magnitude of the HPQDLI index is constant and equal to the unit over the entire time period of the signal, which indicates that the signal is pure sinusoidal in nature and not associated with a PQ event. Fig. 2 shows that the reconstruction error is zero for a pure sinusoidal voltage signal.

4.2 Voltage sag

A voltage signal with a sag disturbance is simulated and shown in Fig. 3A. This signal is decomposed using GT to compute the absolute values of the GWC. The signal is reconstructed, which is detailed in Fig. 3B. SRE is computed by the difference between the original signal and the reconstructed signal, which is plotted in Fig. 4. The reconstructed signal (RV) is processed using ST and the ASOM is computed. The summation of values of every column of ASOM (SS) is computed. Further, the RV signal with sag event is also processed using the HT and the absolute values of output matrix are computed (AHOM). The PQI index is computed by the multiplication of AHOM, SS, and WFD, which is illustrated in Fig. 3C. HPQDLI is computed by the multiplication of AHOM, SS, GWC, and WFDL, which is illustrated in Fig. 3D.

It is inferred from Fig. 3A that the voltage magnitude has reduced between 0.05 s and 0.15 s, which indicates the availability of the sag disturbance. It is observed from Fig. 3B that the voltage magnitude of the reconstructed signal has been reduced between 0.05 s and 0.15 s, which detects the availability of the sag event. Fig. 3C shows that the magnitude of the PQI index is reduced between 0.05 s and 0.15 s, which indicates the availability of the sag event associated with voltage. Fig. 3D shows that the magnitude of the HPQDLI index is reduced between 0.05 s and 0.15 s, which indicates the availability of the sag event pertaining to voltage. Further, high peaks are seen with the plot HPQDLI at time instants of 0.05 s and 0.15 s, which indicates the start and end of the sag event. Hence, the proposed PQI and HPQDLI indices effectively identified and located the voltage sag event in the time domain. Fig. 4 shows that the reconstruction error is nonzero at the start and end of the voltage sag disturbance and zero at all other times. Hence, the plot for the reconstruction error effectively localizes the sag PQ disturbance in the time domain.

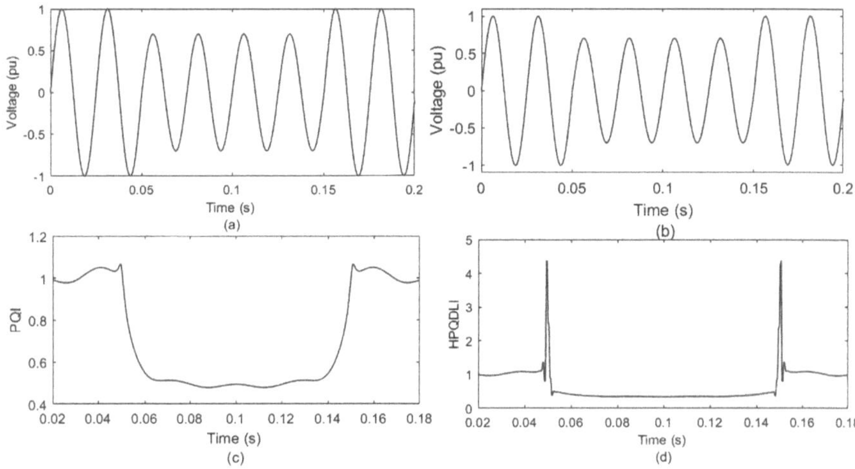

Fig. 3 Voltage with sag event: (A) Original voltage, (B) reconstructed voltage, (C) PQI plot, (D) HPQDLI plot.

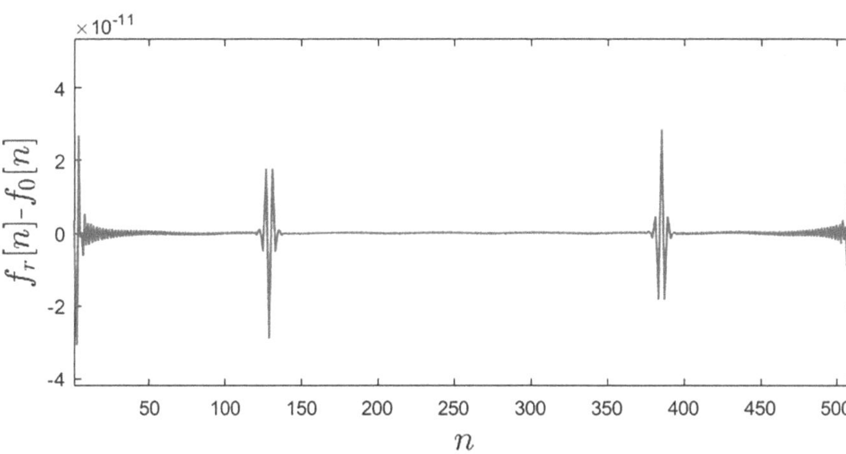

Fig. 4 Reconstruction error for voltage with sag disturbance.

4.3 Voltage swell

Voltage with a swell event is simulated and shown in Fig. 5A. This signal is decomposed using GT to compute the absolute values of GWC. The signal is reconstructed, which is illustrated in Fig. 5B. SRE is computed by the difference between the original signal and the reconstructed signal, which is plotted in Fig. 6. The reconstructed signal (RV) is processed using ST

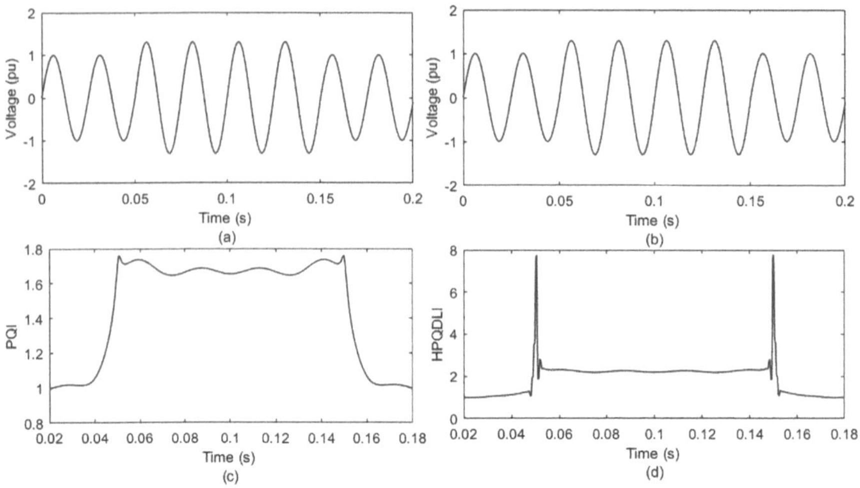

Fig. 5 Voltage with well event: (A) Original voltage, (B) reconstructed voltage, (C) PQI plot, (D) HPQDLI plot.

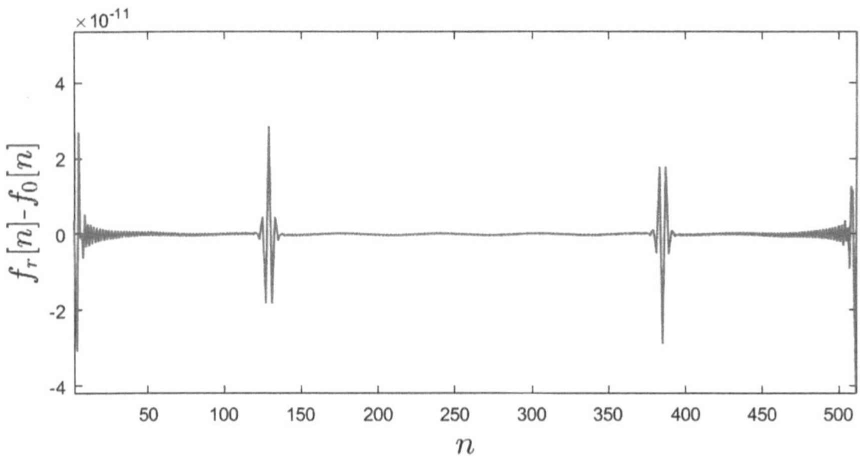

Fig. 6 Reconstruction error for voltage with well event.

and the ASOM is computed. The summation of values of every column of ASOM (SS) is computed. Further, the RV signal with swell disturbance is also processed using HT and the absolute magnitudes of output matrix are computed (AHOM). The PQI index is computed by the multiplication of AHOM, SS, and WFD, which is detailed in Fig. 5C. HPQDLI is computed by the multiplication of AHOM, SS, GWC, and WFDL, which is described in Fig. 5D.

It is inferred from Fig. 5A that the voltage has increased between 0.05 s and 0.15 s, which indicates the availability of the swell disturbance. It is observed from Fig. 5B that the voltage magnitude of the reconstructed signal has increased between 0.05 s and 0.15 s, which indicates the availability of the swell disturbance. Fig. 5C shows that the PQI index is increased between 0.05 s and 0.15 s, which indicates the availability of the swell event associated with voltage. Fig. 5D shows that the magnitude of the HPQDLI index is raised between 0.05 s and 0.15 s, which indicates the availability of the swell event associated with voltage. Further, high magnitude peaks are observed with the plot HPQDLI at the time instants of 0.05 s and 0.15 s, which indicates the start and end of the swell disturbance. Hence, the proposed PQI and HPQDLI indices effectively identified and located the voltage swell disturbance in the time domain. Fig. 6 shows that the reconstruction error is nonzero at the start and end of the voltage swell disturbance and zero at all other times. Hence, the plot for the reconstruction error effectively localizes the swell PQ disturbance in the time domain.

4.4 Momentary interruption

Voltage with the MI PQ event is simulated and shown in Fig. 7A. This signal is processed using GT to compute the absolute values of GWC. The signal is reconstructed, which is described in Fig. 7B. SRE is computed by the difference between the original signal and the reconstructed signal, which is plotted in Fig. 8. The reconstructed signal (RV) is processed using ST and the ASOM is computed. The summation of values of every column of ASOM (SS) is computed. Further, the RV signal with MI PQ disturbance is also processed using the HT and the absolute values of output matrix are computed (AHOM). The PQI index is computed by the multiplication of AHOM, SS, and WFD, which is detailed in Fig. 7C. HPQDLI is computed by the multiplication of AHOM, SS, GWC, and WFDL, which is described in Fig. 7D.

It is inferred from Fig. 7A that the voltage magnitude is reduced below 0.1 pu between 0.05 s and 0.15 s, which indicates the availability of the MI disturbance. It is observed from Fig. 7B that the voltage magnitude of the reconstructed signal has decreased below 0.1 pu between 0.05 s and 0.15 s, which indicates the availability of the MI event. Fig. 7C shows that the magnitude of the PQI index is reduced below 0.1 pu between 0.05 s and 0.15 s, which indicates the availability of the MI event associated with voltage. Fig. 7D shows that the magnitude of the HPQDLI index is reduced

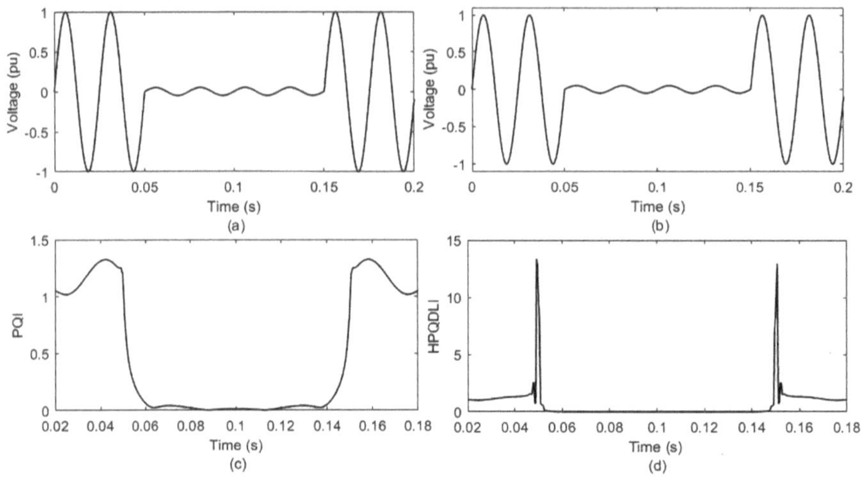

Fig. 7 Voltage with MI event: (A) Original voltage, (B) reconstructed voltage, (C) PQI plot, (D) HPQDLI plot.

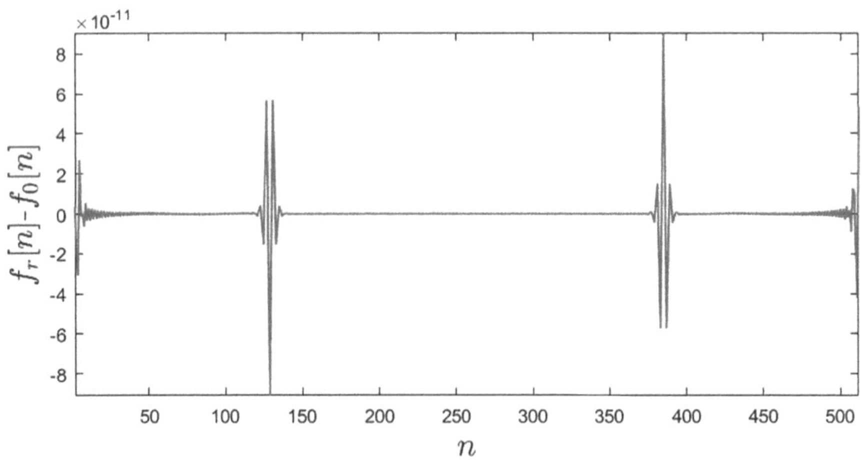

Fig. 8 Reconstruction error for voltage with MI disturbance.

between 0.05 s and 0.15 s, which indicates the availability of the MI disturbance associated with the voltage signal. Further, high magnitude peaks are observed with the plot HPQDLI at the time instants of 0.05 s and 0.15 s, which indicates the start and end of the MI disturbance. Hence, the proposed PQI and HPQDLI indices effectively identified and located the voltage MI disturbance in the time domain. Fig. 8 shows that the reconstruction error is nonzero at the start and end of the voltage MI disturbance and zero at all other times. Hence, the plot for the reconstruction error effectively localizes the MI PQ disturbance in the time domain.

5 Classification of voltage disturbances

Investigated voltage-related PQ disturbances have been categorized using the decision rules based on features F1, F2, F3, and F4. The numerical values of F1-F2 used for this study are included in Table 1. Decision rules used for the classification and classification tree are illustrated in Fig. 9. If F1 > 0.90, then a momentary interruption is detected. Further, for the sine wave features, F1, F2, and F3 are zero. Subsequently, if F1 > 0.2, then the voltage swell is observed; otherwise, a voltage sag is detected. Hence, all the investigated voltage magnitude-related disturbances have been recognized effectively using the proposed algorithm.

The classification accuracy of the algorithm is tested using 60 datasets of each disturbance computed by changing the parameters such as time of disturbance, change in voltage magnitude, signal time period, etc. Correctly detected signals, incorrectly detected signals, and the accuracy of classification are included in Table 2. It is inferred that the algorithm effectively identified the voltage magnitude-related PQ disturbances with an accuracy higher than 99%.

6 Performance comparative study

The performance of the proposed algorithm is compared with ST and the decision rules algorithm reported in Refs. [18,19]. The algorithm reported in Refs. [18,19] has an accuracy of 96%. However, the proposed method has

Table 1 Features for the classification of disturbances.

S. no.	Voltage disturbance	F1	F2	F3	F4
1	Voltage signal without PQ disturbance	0	3.0708	0	0
2	Voltage sag	0.1963	4.2687	0.2720	0.1487
3	Voltage swell	0.3552	5.5555	0.2709	0.1488
4	Momentary interruption	0.9651	11.4218	0.8171	0.4563

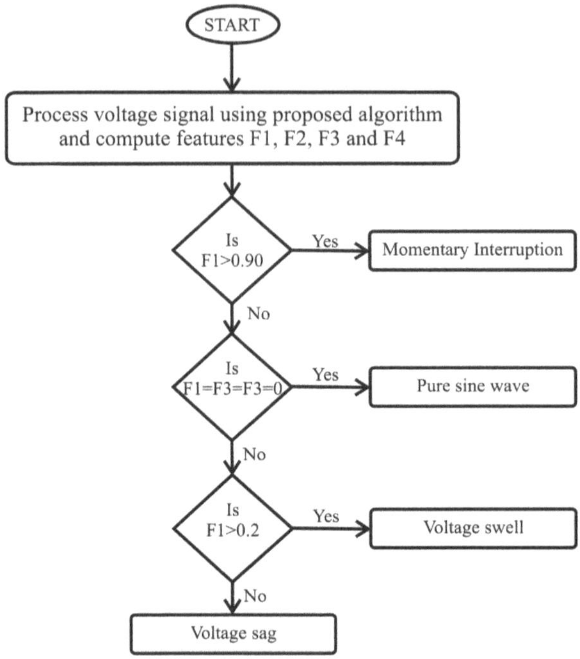

Fig. 9 Classification tree for disturbances.

Table 2 Features for the classification of disturbances.

S. no.	Voltage disturbance	Number of tested signals	Correctly classified	Wrongly classified	Classification accuracy (%)
1	Voltage signal without PQ disturbance	60	60	0	100
2	Voltage sag	60	59	1	98.33
3	Voltage swell	60	59	1	98.33
4	Momentary interruption	60	60	0	100
Overall Accuracy					99.165

an accuracy higher than 99%. Hence, the performance of the designed algorithm is better compared to the algorithms based on ST and the decision rules algorithm reported in Refs. [18,19]. As a future study, the performance of proposed algorithm may be compared with the techniques reported in Refs. [20–28].

7 Conclusion

This research designed a hybrid signal processing technique using GT, ST, HT, and decision rules for smart PQ monitoring meters to detect voltage magnitude-related PQ events such as sags, swells, and MIs. This was achieved using the proposed PQI, HPQDLI, and decision rules driven by various features. It is concluded that the designed hybrid signal processing technique effectively recognized and localized the voltage magnitude-related PQ disturbances. These disturbances have been categorized with an accuracy higher than 99%. The proposed hybrid algorithm performs better compared to the ST and decision rule-based algorithm reported in the literature.

References

[1] R. Machlev, A. Chachkes, J. Belikov, Y. Beck, Y. Levron, Open source dataset generator for power quality disturbances with deep-learning reference classifiers, Electr. Power Syst. Res. 195 (2021). paper no. 107152.

[2] R. Mahla, B. Khan, O.P. Mahela, A. Singh, Recognition of complex and multiple power quality disturbances using wavelet packet based fast kurtogram and ruled decision tree algorithm, Int. J. Model. Simul. Sci. Comput. (2021) 1–23, https://doi.org/10.1142/S179396232150032X. Paper No. 2150032.

[3] R. Kaushik, O.P. Mahela, P.K. Bhatt, B. Khan, S. Padmanabhan, F. Blaabjerg, A hybrid algorithm for recognition of power quality disturbances, IEEE Access 2 (2020) 229184–229200, https://doi.org/10.1109/ACCESS.2020.3046425.

[4] G.S. Chawda, A.G. Shaik, M. Shaik, P. Sanjeevikumar, J.B. Holm-Nielsen, O.P. Mahela, K. Palanisamy, Comprehensive review on detection and classification of power quality disturbances in utility grid with renewable energy penetration, IEEE Access 8 (2020) 146807–146830, https://doi.org/10.1109/ACCESS.2020.3014732.

[5] O.P. Mahela, A.G. Shaik, B. Khan, R. Mahla, H.H. Alhelou, Recognition of complex power quality disturbances using S-transform based ruled decision tree, IEEE Access 8 (2020) 173530–173547, https://doi.org/10.1109/ACCESS.2020.3025190.

[6] B.K. Panigrahi, P.K. Dash, J.B.V. Reddy, Hybrid signal processing and machine intelligence techniques for detection, quantification and classification of power quality disturbances, Eng. Appl. Artif. Intell. 22 (2009) 442–454.

[7] P. Kanirajan, V. Suresh Kumar, Power quality disturbance detection and classification using wavelet and RBFNN, Appl. Soft Comput. 35 (2015) 470–481.

[8] O.P. Mahela, B. Khan, H.H. Alhelou, S. Tanwar, Assessment of power quality in the utility grid integrated with wind energy generation, IET Power Electron. 13 (13) (2020) 2917–2925, https://doi.org/10.1049/iet-pel.2019.1351.

[9] O.P. Mahela, B. Khan, H.H. Alhelou, S. Tanwar, S.K. Padmanaban, Harmonic mitigation and power quality improvement in utility grid with solar energy penetration using distribution static compensator, IET Power Electron. 14 (2021) 912–922, https://doi.org/10.1049/pel2.12074.

[10] O.P. Mahela, B. Khan, H.H. Alhelou, P. Siano, Power quality assessment and event detection in distribution network with wind energy penetration using Stockwell transform and fuzzy clustering, IEEE Trans. Ind. Inf. 16 (11) (2020) 6922–6932, https://doi.org/10.1109/TII.2020.2971709.

[11] O.P. Mahela, A.G. Shaik, N. Gupta, M. Khosravy, B. Khan, H.H. Alhelou, S. Padmanaban, Recognition of the power quality issues associated with grid integrated solar photovoltaic plant in experimental frame work, IEEE Syst. J. (2020), https://doi.org/10.1109/JSYST.2020.3027203.
[12] O.P. Mahela, A.G. Shaik, Recognition of power quality disturbances using S-transform based ruled decision tree and fuzzy C-means clustering classifiers, Appl. Soft Comput. (Elsevier) 59 (2017) 243–257, https://doi.org/10.1016/j.asoc.2017.05.061.
[13] Q. Wang, C. Meng, C. Wang, Compressive sampling and reconstruction in shift-invariant spaces associated with the fractional Gabor transform, Def. Technol. (2021), https://doi.org/10.1016/j.dt.2021.04.003.
[14] O.P. Mahela, V.S. Bhati, G. Ahmad, B. Khan, P. Sanjeevikumar, A.R. Garg, R. Mahla, A protection scheme for distribution utility network in the presence of wind energy penetration, Comput. Electr. Eng. 94 (2021), https://doi.org/10.1016/j.compeleceng.2021.107324. Paper No. 107324.
[15] O.P. Mahela, A.G. Shaik, Power quality recognition in distribution system with solar energy penetration using S-transform and fuzzy C-means clustering, Renew. Energy 106 (2017) 37–51, https://doi.org/10.1016/j.renene.2016.12.098.
[16] G.S. Yogee, O.P. Mahela, K.D. Kansal, B. Khan, R. Mahla, H.H. Alhelou, P. Siano, An algorithm for recognition of fault conditions in the utility grid with renewable energy penetration, Energies 13 (9) (2020) 2383, https://doi.org/10.3390/en13092383.
[17] K. Khalili-Damghani, F. Abdi, S. Abolmakarem, Hybrid soft computing approach based on clustering, rule mining, and decision tree analysis for customer segmentation problem: real case of customer-centric industries, Appl. Soft Comput. 73 (2018) 816–828.
[18] M. Meena, O.P. Mahela, M. Kumar, N. Kumar, Detection and classification of complex power quality disturbances using Stockwell transform and rule based decision tree, in: IEEE PES International Conference on Smart Electric Drives and Power System (ICSEDPS-2018), G H Raisoni College of Engineering, Nagpur, India, 2018, https://doi.org/10.1109/ICSEDPS.2018.8536028. June 12–13.
[19] M. Meena, O.P. Mahela, M. Kumar, N. Kumar, Detection and classification of power quality disturbances using Stockwell transform and rule based decision tree, in: IEEE PES International Conference on Smart Electric Drives and Power System (ICSEDPS-2018), G H Raisoni College of Engineering, Nagpur, India, 2018, https://doi.org/10.1109/ICSEDPS.2018.8536064. June 12–13.
[20] O.P. Mahela, A.G. Shaik, Recognition of power quality disturbances using S-transform and fuzzy C-means clustering, in: IEEE International Conference and Utility Exhibition on Co-Generation, Small Power Plants and District Energy (ICUE 2016), BITEC, Bang Na, Bangkok, Thailand, 2016, https://doi.org/10.1109/COGEN.2016.7728955. September 14–16.
[21] R. Kaushik, O.P. Mahela, P.K. Bhatt, Events recognition and power quality estimation in distribution network in the presence of solar PV generation, in: 10th IEEE International Conference on Communication Systems and Network Technologies (CSNT2021), Apr 24–25, 2021, Bhopal, India, 2021.
[22] R. Kaushik, O.P. Mahela, P.K. Bhatt, Hybrid algorithm for detection of events and power quality disturbances associated with distribution network in the presence of wind energy, in: IEEE International Conference on Advance Computing and Innovative Technologies in Engineering (ICACITE 2021), Galgotias College of Engineering College and Technology, Greater Noida, India, 2021, https://doi.org/10.1109/ICACITE51222.2021.9404665. March 4–5.
[23] R.N. Shah, O.P. Mahela, S. Lohia, Recognition and mitigation of power quality disturbances in renewable energy interfaced hybrid power grid, in: 12th IEEE International Conference on Computational Intelligence and Communication Networks (CICN

2020), BIAS Bhimtal, Nainital, India, 2020, https://doi.org/10.1109/CICN49253.2020.9242623. September 24–25.

[24] V. Pandya, R.R. Choudhary, O.P. Mahela, S. Choudhary, Detection and classification of complex power quality disturbances using hybrid algorithm based on combined features of Stockwell transform and Hilbert transform, in: 2020 IEEE International Students' Conference on Electrical, Electronics and Computer Science (SCEECS 2020), MANIT Bhopal, India, 2020, https://doi.org/10.1109/SCEECS48394.2020.5. February 22–23.

[25] V. Pandya, S. Agarwal, O.P. Mahela, S. Choudhary, Recognition of power quality disturbances using hybrid algorithm based on combined features of Stockwell transform and Hilbert transform, in: 2020 IEEE International Students' Conference on Electrical, Electronics and Computer Science (SCEECS 2020), MANIT Bhopal, India, 2020, https://doi.org/10.1109/SCEECS48394.2020.4. February 22–23.

[26] O.P. Mahela, S. Agarwal, N.K. Saini, Power quality improvement in hybrid power system using synchronous reference frame theory based distribution static compensator with battery energy storage system, in: IEEE International Conference on Computing, Power and Communication Technologies (GUCON 2019), Galgotias College of Engineering & Technology (GCET), Greater Noida, India, 2019. September 27–28.

[27] R. Saini, O.P. Mahela, D. Sharma, Detection and classification of complex power quality disturbances using Hilbert transform and rule based decision tree, in: 2018 IEEE 8th Power India International Conference (PIICON 2018), NIT Kurukshetra, India, 2018, https://doi.org/10.1109/POWERI.2018.8704427. December 10–12.

[28] R. Saini, O.P. Mahela, D. Sharma, An algorithm based on Hilbert transform and rule based decision tree classification of power quality disturbances, in: 2018 IEEE 8th Power India International Conference (PIICON 2018), NIT Kurukshetra, India, 2018, https://doi.org/10.1109/POWERI.2018.8704465. December 10–12.

SECTION 14

Optimization of operational issues

CHAPTER 14.1

Optimal expansion planning of distribution system in smart grid environment

Takele Ferede Agajie[a], Yayehyirad Ayalew Awoke[a], and Dessalegn Bitew Aeggegn[b]
[a]Department of Electrical and Electronics Engineering, University of Buea, Buea, Cameroon
[b]School of Electrical and Computer Engineering, Debre Markos University, Debre Markos, Ethiopia

Contents

1 Introduction

An electric conductor carrying power from a substation to end users to meet load demands along its path that is the most extensive part of the electrical system in which it is primarily responsible for energy losses due to improper planning [1–3]; An electric conductor carries power from a substation to end users to meet the load along its path. It is an extensive part of the electrical system that is primarily responsible for the energy losses. Installing new distribution networks and strengthening current distribution networks are critical to meet the rapidly rising demand of consumers. Distribution networks, on the other hand, get the greatest share of strained circumstances due to their aim, planning, and operation margins [4]. The additional power

Active Electrical Distribution Network
https://doi.org/10.1016/B978-0-323-85169-5.00021-6

demand at all points of electric distribution systems necessitates the expansion of electricity supply chain components [5–7]. One of the steps taken by planners to deal with increased electric demands is distribution system expansion [8–10]. Distribution system (DS) expansion planning is chosen as this study's major topic of interest. The challenge of distribution system expansion planning involves the placement, size, and efficacy of distribution equipment installation.

As a result, the distribution system expansion planning challenge is a multiobjective problem with many dimensions, including investment and operating costs that must be minimized while meeting electrical and geographical restrictions. One of the fundamental and basic aspects of the power system expansion planning challenge is uncertainty. This problem is more difficult to solve in the context of optimum distribution network expansion planning in a smart grid setting with demand and generation forecasts. If the distribution is done correctly, planned growth will ensure that the power systems run smoothly, assisting in the development of the economy and society. Several approximation ways to solve the distribution-planning issue have been created over the last decades [11–17] as a result of various investigations on expansion problem methods. Initially, the study concentrated on typical planning problems such as substation feeder placement and distributed generation (if any) routing to reduce voltage variation (better voltage profile) at each node as well as distribution network power losses [18–20].

Increasing load needs, limited expansion choices, extremely competitive electricity market situations, and environmental problems all contribute to strained transmission and distribution network operations. Owing to their aim, planning, and operation margins, distribution networks get the greatest share of strained situations [21].

The major issue confronting Ethiopia's electric power providers today is that electricity consumption is quickly increasing while supply is restricted by economic, environmental, and other common issues. This has necessitated the development of more comprehensive explanations for additional system facilities as well as improvements in fabrication and power usage. The development and management of a distribution network system based on a consistency assessment methodology gives a chance to rationalize one of Ethiopia's most scrutinized and vulnerable economic sectors. This results in a high frequency of interruptions, a large numeric value of power losses, and voltage variations in Debre Markos distribution networks. The following are the main causes of the aforementioned issues: the power grid conductor size and transformer capacity are too small and outdated while networks were built without proper planning.

To address the aforementioned issues and provide reliable and uninterrupted electricity with a better power system profile, a good expansion plan for the Debre Markos DS is required, taking into account the optimal sizing and placement of distributed energy and sectional switches to meet current and future electricity demand.

The primary goal of DS planning is to provide for demand to increase revenue and reduce operating costs, which necessitates coping with typical issues such as power losses and the voltage profile. Optimal DS expansion planning has been a hot area of research. Due to the fast-growing electric power demand in Ethiopia, particularly in Debre Markos, additional circuits have been added to the existing DS under smart grid standards in recent years.

Distribution expansion planning is done to provide a distribution infrastructure in a smart grid manner that meets electricity customer requirements in terms of capacity, reliability, and quality in the most cost-effective manner. This entails the need to model, reinforce, and develop the existing infrastructure in a phase-wise manner so that infrastructure expansion is appropriate with the evolving electricity requirements. To fulfill the load estimates defined by the planning horizon, distribution planners must guarantee that there is acceptable substation capacity (i.e., distribution transformer) and feeder capacity [8,22–24]. The multiobjective particle swarm optimization (MOPSO) algorithm is used in this chapter to conduct the proposed optimization approach for optimal sizing and placement of the DG unit sectional switch in an active distribution network system.

2 Smart grid technology

Traditional and smart grid technologies are the two categories of grid technologies. Consumers are connected to an electric network, which is regulated by DS operators, in a conventional structure. Energy is provided by large, semicentralized utilities. The energy flow in a traditional grid technology is one-directional, more predictable, and controlled. The typical distribution is depicted in the diagram below.

3 Methodology

This chapter uses articles, theses, dissertations, reference works, textbooks, recognized and regulated documents, media, and software such as MATLAB, Microsoft Excel, and Visio for analysis and simulation to perform optimal expansion planning, implement optimization techniques, and the materials used in this work.

The methods exercised on optimal expansion planning in a smart grid environment of a radial distribution network are data collection, including ratings distribution transformers, cable size, voltage profile, and customer load. From the data collected in the distribution network, a single line diagram has been prepared with Visio.

Load flow analysis was performed using MATLAB2020 utilizing the backward/forward sweeping technique after gathering data for the system's above-mentioned parameters. The voltage profile at each node, the overall voltage deviation, and the active and reactive power loss for the present system are all analyzed as a result of this study.

The system under consideration has had its load forecasts done for the years 2027/28. Separated feeders, upgraded feeders, and upgraded feeders with DG and sectional switches are the three phases that the system should take to attain the offered solutions. In this study, all stages were subjected to a backward/forward power flow analysis to determine the optimum option. Using MOPSO techniques, we were able to achieve optimal DG and sectional switch sizing and location.

4 Overview of study area

Debre Markos is a town in the Amhara National Regional State, East Gojjam Zone, northwestern Ethiopia.

This work involves only one feeder, which is Debre Markos feeder 3. It is circled in red in Fig. 1 below. This feeder is supplied from Debre Markos substation by a 63 MVA three-winding transformer. There are three additional feeders with a 15 kV rating: D/M feeder 4, Amanuel feeder, and Lumame feeder, which are also supplied by the same three transformers. It feeds the towns of Amanuel and Lumame, respectively. There is another line with a 66 kV rating, i.e., Finote Selam town, which is supplied from D/M substation by the same 63 MVA three-winding transformer.

4.1 Existing Debre Markos distribution network

The system data of the study distribution network is described and the peak load on each feeder recorded from 2013/14 to 2017/18 at the Debre Markos distribution system is shown in Table 1.

For the last of 2017/18, there were 106 existing three-phase distribution transformers installed in this feeder. This distribution network is the radial-type distribution network in which the power is delivered from the source to the customer in a straightforward manner. This is cheap to

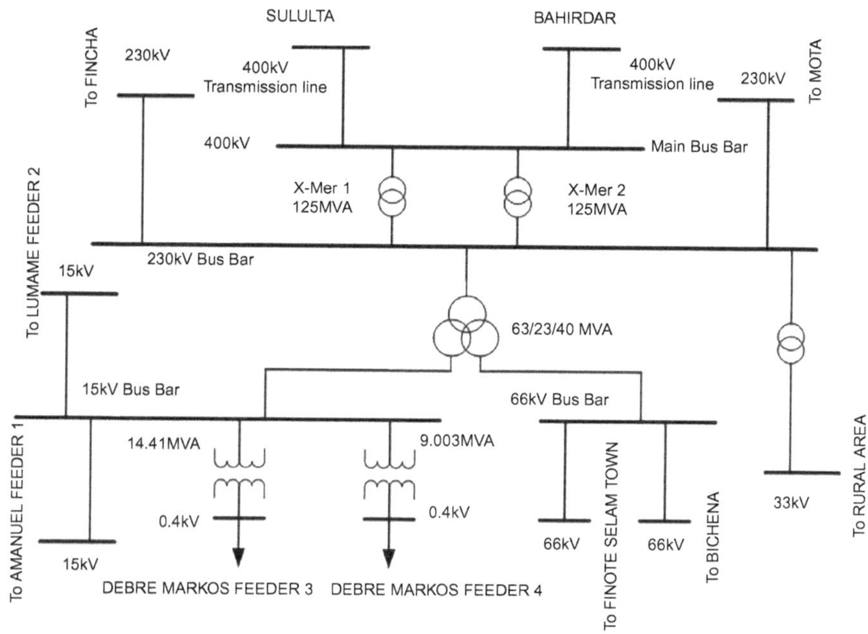

Fig. 1 Overview of D/M distribution substation.

design and protect but it is less reliable and subject to disturbances easily. In these medium-voltage 15 kV line feeders, aluminum conductors are used. The 15 kV Debre Markos feeder 3 existing distribution network feeder that is at 2017/18 is shown in Fig. 2.

4.2 Demand forecasting

The least square approximation approach is employed in this study to anticipate trends. The following equation [25] may be used to determine the current load demand for all load buses as well as the future load demand at the end of the year *n*th.

$$P_n = P_o(1+g)^n \tag{1}$$

where P_n is load demand at the end of the *n*th year, P_o is initial load, g is annual load growth rate, and n is number of years.

Based on the collected data, the forecasting of power demand in the period of 2017/18–2022/23–2027/28 in all the feeders are evaluated. The forecasting of the year (*n*th) can be computed based on the mathematical expression by using Eq. (1).

Table 1 The loading condition of the 15 kV feeders from 2013/14 to 2017/18.

Feeder name	Year	Yearly peak power demand (MW)	Voltage level (kV)	Current (A)	Total number of transformers	Length (km)	Total transformer capacity (kVA)
D/M feeder 3	2013/14	28.67	15	106.88	106	63.1601	23260
	2014/15	32.29	15	120.43			
	2015/16	37.15	15	138.42			
	2016/17	39.7	15	165.3			
	2017/18	49.68	15	188.2			
D/M feeder 4	2013/14	21.22	15	79	No details	No details	No details
	2014/15	24.2	15	90.22			
	2015/16	27.59	15	102.92			
	2016/17	26.4	15	120			
	2017/18	32.76	15	124.9			
Amanuel feeder	2013/14	0.24	15	112.71	No details	No details	No details
	2014/15	33.23	15	123.89			
	2015/16	36.29	15	135.33			
	2016/17	26.4	15	130.7			
	2017/18	36.6	15	140.5			
Lumame feeder	2013/14	30.97	15	115.42	No details	No details	No details
	2014/15	34	15	126.78			
	2015/16	27.78	15	103.58			
	2016/17	29.6	15	113.78			
	2017/18	35.88	15	136.9			

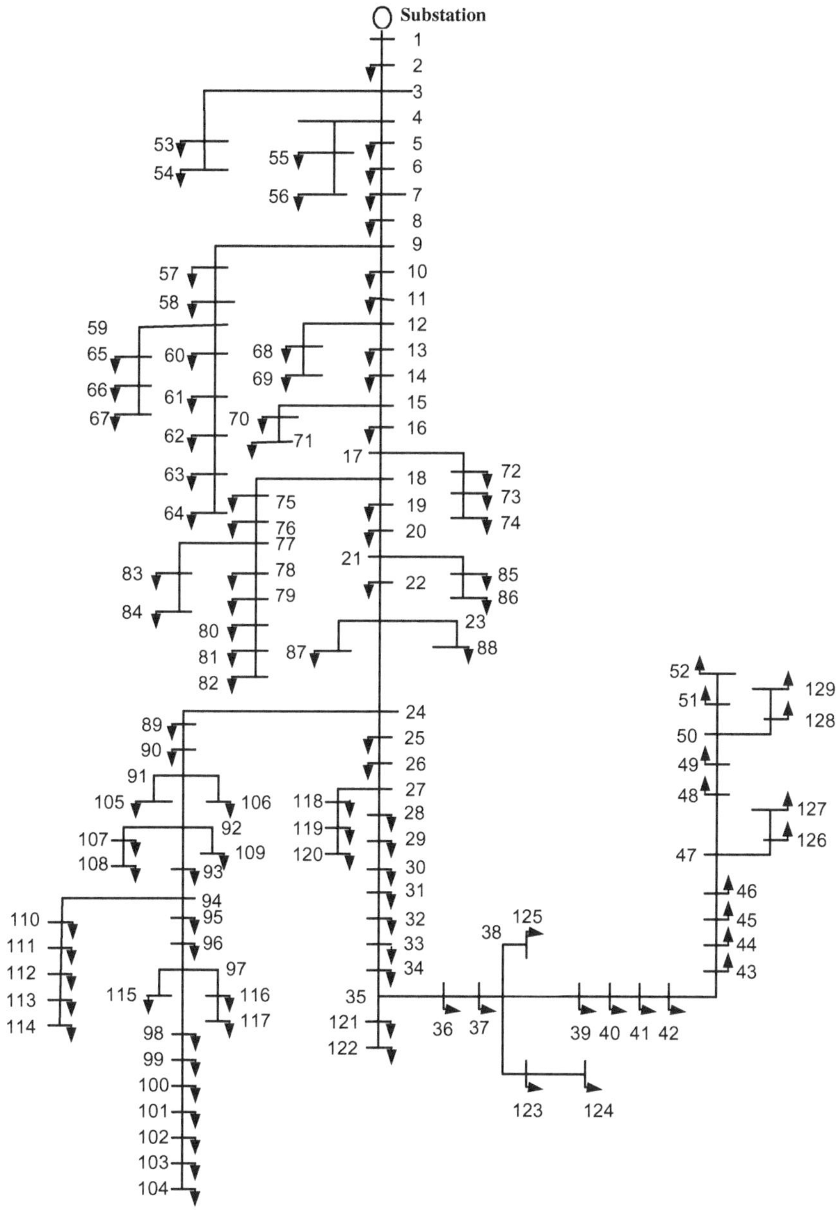

Fig. 2 Single line diagram of D/M feeder 3.

Using the forecasting technique stated above and Eq. (1), the forecasted electrical load of the distribution network from Table 2 for D/M feeder 3 is as follows:

- For the year 2017/18, the demand was 3.9 MW.
- For the year 2022/23, demand will be 7.66 MW.

Table 2 Forecasting for 2017/18–2027/28 (in MW).

	Existing demand@feeder				Predicted loading@feeder			
Year	**D/M 3**	**D/M 4**	**Amanuel**	**Lumame**	**D/M 3**	**D/M 3 growth rate (%)**	**D/M 4**	**D/M 4 growth rate (%)**
2013/14	2.39	1.77	2.52	2.58	2.364		1.801	
2014/15	2.69	2.02	2.77	2.83	2.694	13.95840677	1.982	10.04164739
2015/16	3.1	2.3	3.02	2.32	3.070	13.94989174	2.180	10.00841043
2016/17	3.31	2.2	2.20	2.47	3.498	13.95222584	2.398	10.01529052
2017/18	4.14	2.73	3.05	2.99	3.986	13.93520724	2.639	10.04169562
2018/19	–	–	–	–	4.543	13.96613004	2.904	10.04104831
2019/20	–	–	–	–	5.176	13.94239589	3.195	10.0143472
2020/21	–	–	–	–	5.898	13.95910481	3.515	10.01564945
2021/22	–	–	–	–	6.721	13.94461712	3.866	10.0284495
2022/23	–	–	–	–	7.659	13.96156231	4.256	10.04093945
2023/24	–	–	–	–	8.728	13.94842781	4.683	10.02545526
2024/25	–	–	–	–	9.945	13.95015755	5.152	10.01957644
2025/26	–	–	–	–	11.333	13.95173454	5.673	10.02911679
2026/27	–	–	–	–	12.914	13.9569086	6.237	10.02646281
2027/28	–	–	–	–	14.716	13.95108731	6.862	10.02137894

Where D/M is Debre Markos.

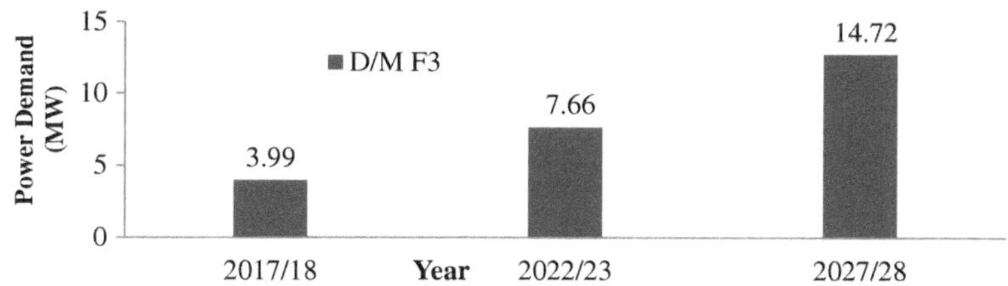

Fig. 3 Loading of D/M feeder 3 from 2016/17 to 2026/27.

- For the year 2027/28, loading will be 14.72 MW.

Forecasting for the D/M distribution network shows that the electrical demand grows dramatically, an increase shown in Fig. 3 below.

5 Result and discussion

The total peak demand taken from the existing networks and the results of load forecasting stated above in 2017/18, 2022/23, and 2027/28 will be 3.9 MW, 7.66 MW, and 14.72 MW for feeder 3, respectively. Use backward/forward MATLAB code to analyse the load flow on D/M distribution networks in cases of peak demand in 2017/18 and peak forecasted demands in 2022/23 and 2027/28 and determine the total voltage deviation, bus voltage, current, and power losses in each branch. The analysis results are shown in Table 3.

Results show that the present DS was not able to supply the future demand. Thus, appropriate expansion planning is the only solution, and it is done below. This expansion planning consists of separating and adding new feeders, with upgrading the separated and integrating optimal placing and sizing of DGs and sectional switches.

5.1 Divide present feeders and add new feeder

Due to covering the maximum demand loads of the town and the length of the conductors is also maximum, this is around 32 km from the distribution stations. This feeder has also maximum power losses and voltage deviation. Therefore, the feeder is separated into two: D/M F3A and D/M F3B.

As the feeder is separated, even if the losses and drop are reduced as shown in Table 4, there are transformers and feeders with overloading. Consequently, this separated feeder is upgraded to overcome the overloading problems.

Table 3 Analysis of the existing network (at peak load) in 2017/18, 2022/23, and 2027/28.

Feeder name	Total current (A) @2017/18	Demand (MW)	Line data	Demand@	Total power loss (kW)	Total reactive power loss (kvar)	Total voltage deviation (p.u)	Overloaded lines	Overloaded Trs
D/M feeder 3	188.2	4.14	2017/18	2017/18	519.3944	518.3686	15.6075	Yes	Yes
			2017/18	2022/23	2040.4	2033.9	31.0227	Yes	Yes
			2017/18	2027/28	7434.9	7391.8	58.6952	Yes	Yes

Table 4 Result analysis of the separated network in case of peak load 2017/18, 2022/23, and 2027/28.

Feeder	Line data	Load @	Total power loss (kW)	Total reactive power loss (kvar)	Total voltage deviation (p.u)	Overloaded lines	Overloaded Trs
D/M feeder 3 F3A	2017/18	2017/18	55.9532	57.81423	2.1528	No	No
	2017/18	2022/23	170.6984	175.9568	3.8814	Yes	Yes
	2017/18	2027/28	376.8642	388.5580	5.7412	Yes	Yes
D/M feeder 3 F3B	2017/18	2017/18	93.6031	96.7861	4.1405	No	Yes
	2017/18	2022/23	296.7958	307.8825	7.3219	Yes	Yes
	2017/18	2027/28	665.4087	690.7357	10.9415	Yes	Yes

Table 5 Upgraded conductors for future demand.

Feeder	Existing	Distance (km)	Upgraded wire type
F3A	AAC-95	7.458	AAC-200
		2.452	AAC-200
F3B	AAC-95	13.5281	AAC-200

5.2 Upgrade and installed new lines and distribution transformers

According to the current flow through the line resulted from the load flow on MATLAB software, the suitable wire size is selected [8,26] (Table 5).

The majority of the transformers are overburdened and required replacement. Additional transformer capability could be chosen based on the projected electricity demandin 2027/28 [8].

As shown in Table 6, the load flow analysis result, the power losses, and the voltage deviation are minimized radically while the line is also sufficient to conduct, but the distribution network faces transformer-overloading condition on the targeted year 2027/28. To overcome these challenges, DGs and sectional switches are incorporated with the appropriate places and capacity.

5.3 Integration of DGs and sectional switches in upgraded separated feeder

MOPSO techniques are executed on the two feeders (D/M F3A and D/M F3B) for the optimal place and size of DG and sectional switches, which are used to radically reduce the system voltage deviation and power losses. Fig. 4 represents the existing D/M feeder 3 with 129 buses.

As shown in Table 7, the load flow analysis result, the power losses, and the voltage deviation are minimized radically while the lines and transformers overloading conditions are sufficient to conduct future loads on the targeted year 2027/28. MOPSO is used to optimally place and size DGs for D/M F3A and D/M F3B, and these are at bus 15, 1.9128 MVA and at bus 18, 1.9999 MVA, respectively. The optimal numbers of sectional switches for D/M F3A and D/M F3B are nine and eight, respectively. In addition, the optimal placement of the sectional switches for D/M F3A and D/M F3B are at buses 3, 4, 9, 12, 15, 17, 18, 21, and 23 and at bus 5, 13, 16, 22, 25, 26, 36, and 37, respectively. The optimal locations of DGs and sectional switches for D/M F3A and D/M F3B are illustrated in Figs. 4 and 5.

Table 6 Upgraded separate and existing network load flow results (@ peak demand in 2027/28).

Feeder	Line data	Demand@	Total active power loss (kW)	Total reactive power loss (kW)	Total voltage deviation (p.u)	Overloaded lines	Overloaded Trs
Upgraded separated feeder D/M F3	Upgraded D/M F3A	2027/28	210.412	306.4271	3.7929	No	Yes
	Upgraded D/M F3B	2027/28	338.621	510.6102	6.9762	No	Yes
	Total	2027/28	545.238	814.4205	10.7691		

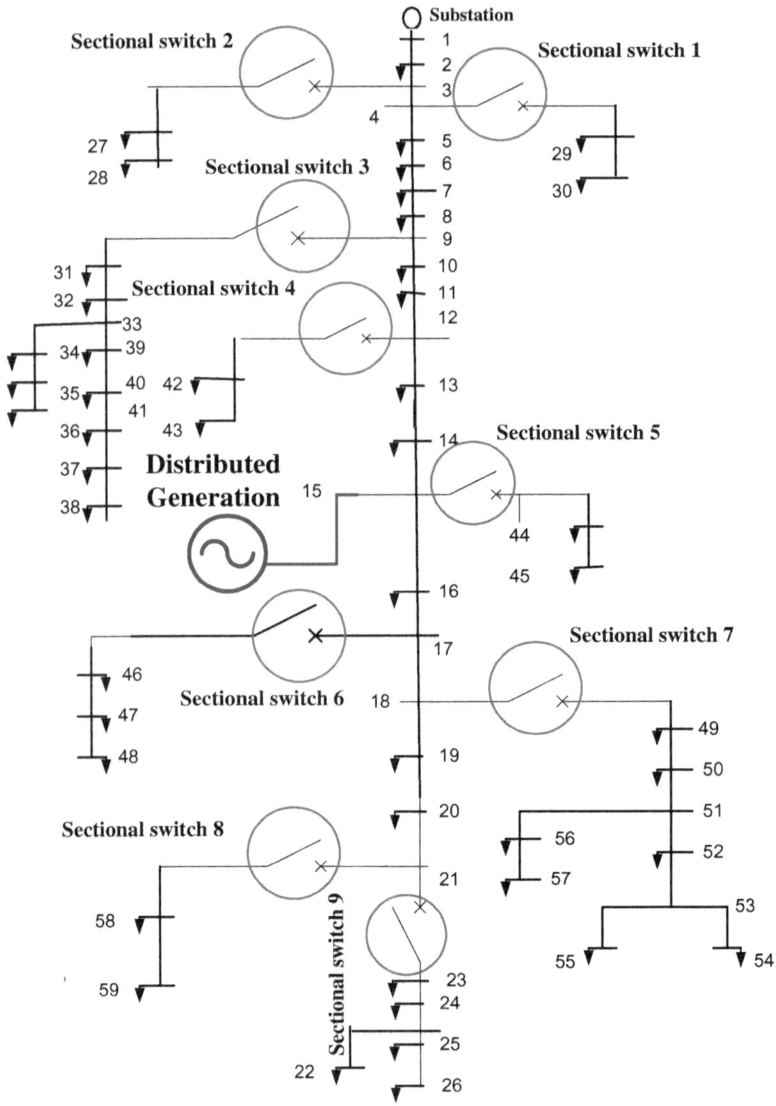

Fig. 4 Separated feeder D/M F3A with its DG sectional switch location by utilizing MOPSO.

Table 8 shows the optimum location, size, and real power losses before and after DG placement in the existing two separated feeders, D/M feeder 3 (Table 9).

Figs. 6 and 7 show the voltage profiles for the separated feeders F3A and F3B.

Table 7 Upgraded separated result analysis of feeders with DGs and sectional switches in case of peak load in 2027/28.

Feeder	Line data	Demand @	Total active power loss (kW)	Total reactive power loss (kvar)	Total voltage deviation (p.u)	Overloaded lines	Overloaded Trs
Upgraded separated feeder	Upgraded D/M F3A	2027/28	210.412	306.4271	3.7929	No	Yes
	Upgraded D/M F3B	2027/28	338.621	510.6102	6.9762	No	Yes
Upgraded separated feeder with DG	Upgraded D/M F3A with DG at bus 15 (1.9128 MVA)	2027/28	22.385	21.876	0.179	No	No
	Upgraded D/M F3B with DG at bus 18 (1.9997 MVA)	2027/28	75.017	105.103	0.313	No	No

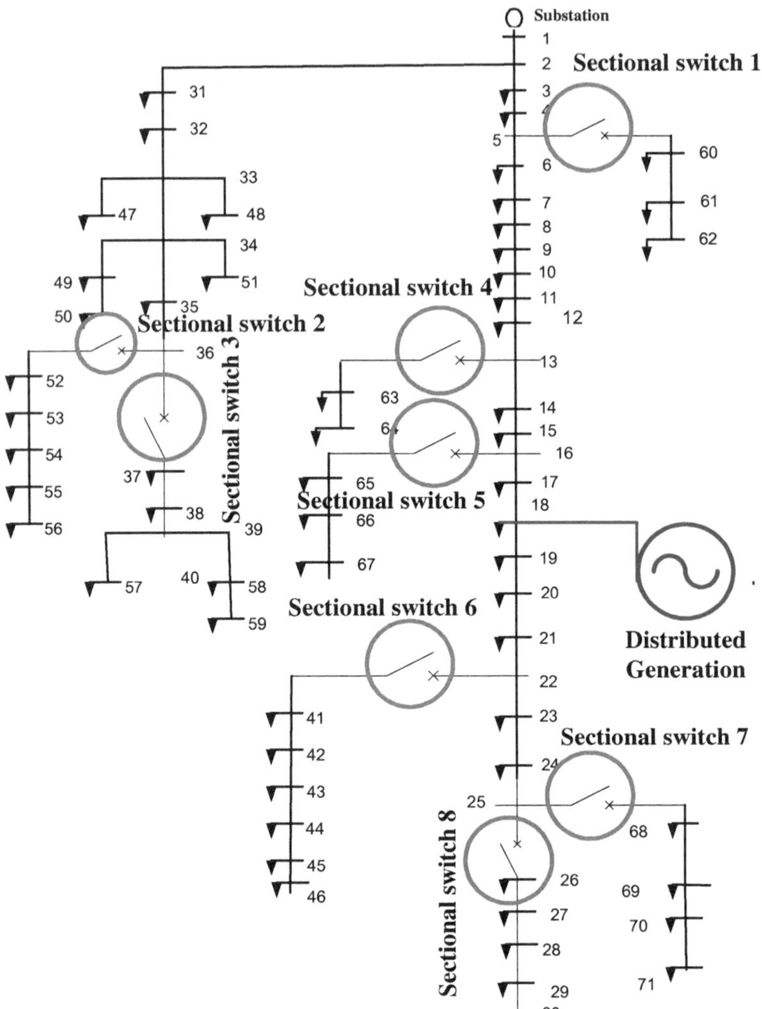

Fig. 5 Separated feeder D/M F3B with its DG and sectional switches utilizing MOPSO.

Table 8 Result of MOPSO on reduction in real and reactive power loss and voltage deviation with optimal placement of DG and sectional switches.

Feeder	Method	Bus	Optimum capacity (MVA)	Loss (active) (kW)	Loss (reactive) (kvar)	Variation (voltage) (p.u)
F3A	MOPSO	15	1.9128	22.385	21.876	0.179
F3B		18	1.9999	75.017	105.103	0.313

Table 9 Result of MOPSO with DGs and sectional switches.

	Loss (active) (kW)		Loss (reactive) (kvar)		Variation (voltage) (p.u)	
Different scenarios	**F3AU**	**F3BU**	**F3AU**	**F3BU**	**F3AU**	**F3BU**
Without DG	210.412	338.621	306.4271	510.6102	3.7929	6.9762
With DG by MOPSO	22.385	75.017	21.876	105.103	0.179	0.313
Reduction (%)	89.36	77.85	92.86	79.42	95.28	95.51

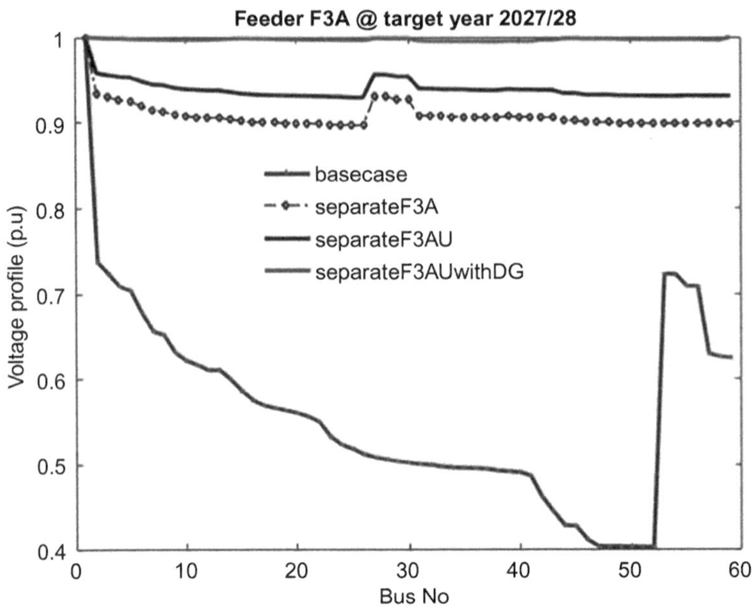

Fig. 6 For separated feeder F3A, voltage profile at various scenarios.

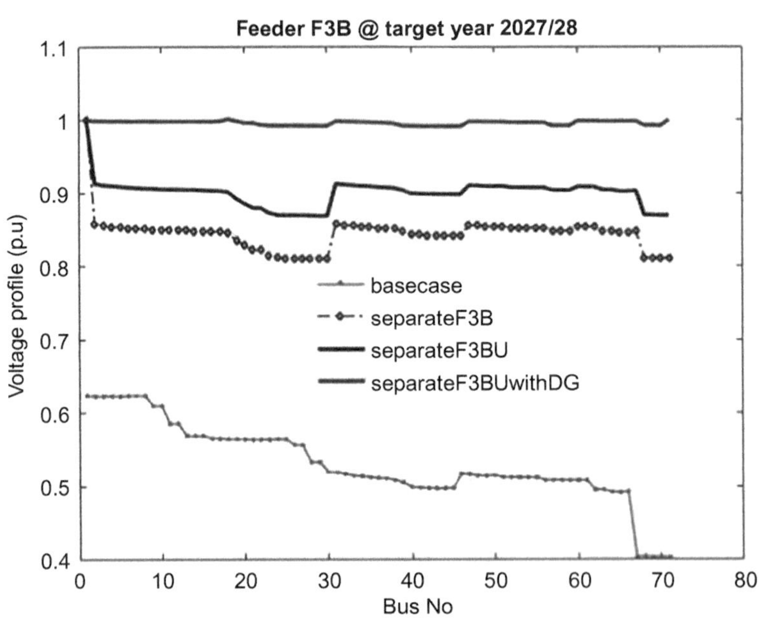

Fig. 7 For separated feeder F3B, voltage profile at various scenarios.

6 Conclusion

This chapter presented the optimal expansion planning of the smart DS. A practical case study of D/M DS was utilized. In expanding the distribution network, feeder separation, upgrading the components, and incorporating DG with a proper capacity level were done. The MOPSO technique was used to size and place DG in the distribution network to minimize losses and improve the voltage prolife.

References

[1] Y. Al-Mahroqi, I.A. Metwally, A. Al-Hinai, A. Al-Badi, Reduction of power losses in distribution systems, World Acad. Sci. Eng. Technol. 6 (2012) 498–505.
[2] T.F. Agajie, A.O. Salau, E.A. Hailu, M. Sood, S. Jain, Optimal sizing and siting of distributed generators for minimization of power losses and voltage deviation, in: 2019 5th International Conference on Signal Processing, Computing and Control (ISPCC), IEEE, 2019, October, pp. 292–297.
[3] A.T. Ferede, S.A. Olalekan, H.E. Abel, A.Y. Ayalew, Power loss mitigation and voltage profile improvement with distributed generation using grid-based multi-objective harmony search algorithm, J. Electr. Electron. Eng. 13 (2) (2020) 5–10.
[4] H. Farhangi, The path of the smart grid, IEEE Power Energy Mag. 8 (1) (2009) 18–28.
[5] H. Seifi, M.S. Sepasian, Electric Power System Planning: Issues, Algorithms and Solutions, Springer Science & Business Media, 2011.
[6] M.S. Sepasian, H. Seifi, A.A. Foroud, A.R. Hatami, A multiyear security constrained hybrid generation-transmission expansion planning algorithm including fuel supply costs, IEEE Trans. Power Syst. 24 (3) (2009) 1609–1618.
[7] B. Chen, J. Wang, L. Wang, Y. He, Z. Wang, Robust optimization for transmission expansion planning: minimax cost vs. minimax regret, IEEE Trans. Power Syst. 29 (6) (2014) 3069–3077.
[8] H. Falaghi, C. Singh, M.R. Haghifam, M. Ramezani, DG integrated multistage distribution system expansion planning, Int. J. Electr. Power Energy Syst. 33 (8) (2011) 1489–1497.
[9] J.F. Franco, M.J. Rider, R. Romero, A mixed-integer quadratically-constrained programming model for the distribution system expansion planning, Int. J. Electr. Power Energy Syst. 62 (2014) 265–272.
[10] P.S. Georgilakis, N.D. Hatziargyriou, A review of power distribution planning in the modern power systems era: models, methods and future research, Electr. Power Syst. Res. 121 (2015) 89–100.
[11] S. Bhowmik, S.K. Goswami, P.K. Bhattacherjee, A new power distribution system planning through reliability evaluation technique, Electr. Power Syst. Res. 54 (3) (2000) 169–179.
[12] S.K. Khator, L.C. Leung, Power distribution planning: a review of models and issues, IEEE Trans. Power Syst. 12 (3) (1997) 1151–1159.
[13] Y. Tang, Power distribution system planning with reliability modeling and optimization, IEEE Trans. Power Syst. 11 (1) (1996) 181–189.
[14] T.H. Fawzi, K.F. Ali, S.M. El-Sobki, A new planning model for distribution systems, IEEE Trans. Power Appar. Syst. 9 (1983) 3010–3017.
[15] A. Navarro, H. Rudnick, Large-scale distribution planning—part I: simultaneous network and transformer optimization, IEEE Trans. Power Syst. 24 (2) (2009) 744–751.

[16] E.G. Carrano, L.A. Soares, R.H. Takahashi, R.R. Saldanha, O.M. Neto, Electric distribution network multiobjective design using a problem-specific genetic algorithm, IEEE Trans. Power Deliv. 21 (2) (2006) 995–1005.
[17] R. Baghipour, S.M. Hosseini, A hybrid algorithm for optimal location and sizing of capacitors in the presence of different load models in distribution network, J. Oper. Autom. Power Eng. 2 (1) (2014) 10–21.
[18] T. Ackermann, G. Andersson, L. Söder, Distributed generation: a definition, Electr. Power Syst. Res. 57 (3) (2001) 195–204.
[19] C.L. Borges, D.M. Falcao, Optimal distributed generation allocation for reliability, losses, and voltage improvement, Int. J. Electr. Power Energy Syst. 28 (6) (2006) 413–420.
[20] C. Wang, M.H. Nehrir, Analytical approaches for optimal placement of distributed generation sources in power systems, IEEE Trans. Power Syst. 19 (4) (2004) 2068–2076.
[21] S.A.A. Kazmi, M.K. Shahzad, D.R. Shin, Multi-objective planning techniques in distribution networks: a composite review, Energies 10 (2) (2017) 208.
[22] E. H. S. and Q. (EHS & Q) and E. and Social, Ethiopia—Additional Financing for Energy Access Project Seven Towns Electricity Distribution Network Rehabilitation and Expansion Project, 2015.
[23] F. Kiessling, P. Nefzger, J.F. Nolasco, U. Kaintzyk, Overhead Power Lines: Planning, Design, Construction, 759, Springer, Berlin, 2003.
[24] London Power Associates Ltd in Association with Global Energy Consulting Engineers Pvt Ltd, Bahirdar Town Distribution Network Development Plan, 2013.
[25] T.F. Agajie, B. Khan, H.H. Alhelou, O.P. Mahela, Optimal expansion planning of distribution system using grid-based multi-objective harmony search algorithm, Comput. Electr. Eng. 87 (2020), 106823.
[26] V. Hosseinnezhad, M. Rafiee, M. Ahmadian, P. Siano, A comprehensive framework for optimal day-ahead operational planning of self-healing smart distribution systems, Int. J. Electr. Power Energy Syst. 99 (2018) 28–44.

CHAPTER 14.2

Parameter optimization of PV modules: An overview

Abdelhady Ramadan[a], Salah Kamel[a], Ragab A. El-Sehiemy[b], and Baseem Khan[c]
[a]Department of Electrical Engineering, Aswan University, Aswan, Egypt
[b]Kafrelsheikh University, Kafrelsheikh, Egypt
[c]Department of Electrical and Computer Engineering, Hawassa University, Hawassa, Ethiopia

Contents

1 Introduction

The development of industry and the rapid advancement of digital technology in combination with the increase in population have had a profound effect on increasing the demand for electricity around the world. Today, renewable energy (RE) resources play an important role in meeting the increasing demand for electricity. The most common sources of renewable power generation are solar, hydro, wind, geothermal, and tidal energies. Solar power is the most popular energy source because of its clean emissions and reliability that are not harmful to the environment [1].

Due to the poor efficiency of solar PV modules, many researchers have focused their efforts on improving the conversion efficiency using the

Active Electrical Distribution Network
https://doi.org/10.1016/B978-0-323-85169-5.00001-0

parameter extraction procedure, which is known as maximum power point tracking (MPPT). The presence of varying radiation levels as well as temperature changes in the surrounding environment has a significant impact on the output of solar panels, resulting in a loss of efficiency. As a result, parameter extraction and the MPPT algorithm are employed to boost overall efficiency. The cost of producing and installing PV panels is very high. As a result, solar panel simulation is critical for determining the properties of solar PV panels under various environmental variables, including temperature (k) and irradiance (w/2).

Several studies have offered numerical and analytical approaches for extracting parameters from PV panels using the manufacturer's data sheet in recent years. The values of these parameters, however, were not precise enough [2].

In this chapter, a brief introduction about optimization algorithms is presented and an overview of the most popular types applied for PV parameter estimation in the literature is also discussed.

2 Types of optimization algorithms

2.1 Optimization definition

The goal of optimizing an objective function is to identify a set of inputs that results in a maximum or minimal function evaluation. Optimization problems are already part of our lives, such as how we optimize our way to work, the way to finish our work tasks, and the way to educate our children.

Example 1: how much daily exercise is optimal?

Suppose every human has a fixed number of heartbeats during his life, so what is the optimal use of this number?

One idea is to stay in bed as the heartbeat is slow at rest.

Another idea is that a well-trained heart is slower than an untrained one in rest, but the heartbeats are very fast during training time. So how we can optimize the daily exercise time?

Example 2: Planning of transportation task

Create a system that instructs a logistics business on how to complete all transportation requests with the lowest possible cost. Locate routes on a map and try to get the shortest overall distance between numerous depots and pickup and delivery sites. It is given that vehicles (trucks and trains) have capacity restrictions, that pickup and delivery windows exist, and that there are constraints and rules, as shown in Fig. 1.

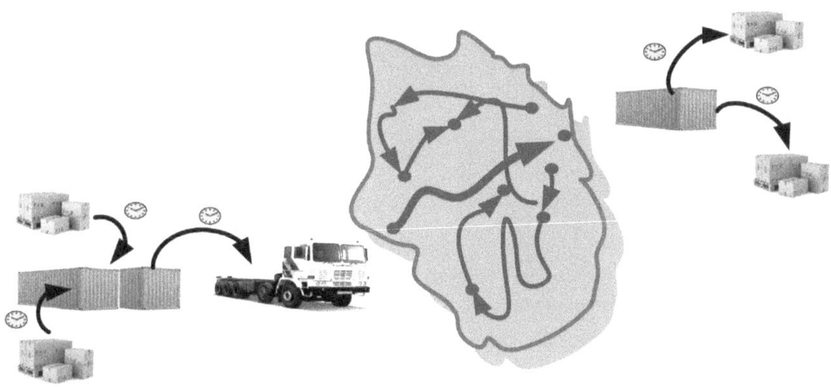

Fig. 1 Transportation planning optimization problem.

2.2 Optimization process

The main steps of the optimization process are presented in Fig. 2.

2.3 Design variables

The initial stage in the optimization process is the identification of the problem design variables. The design variables are extremely important to the design's correct operation and are changed during the optimization phase.

Design parameters usually remain fixed or vary in relation to the design variables [3].

2.4 Constraints

The lower and upper limit of the design variable and design parameters according to the optimization problem.

2.5 Objective function

The objective function is a mathematical expression in terms of design variable and design parameters. The main target of optimization is to minimize or maximize the objective function by finding the best values for the design variables and parameters within the defined constraints. For example, minimize the problem cost or time or maximize the equipment lifetime.

2.6 Variable bounds

This is the minimum and maximum number of design variables; this factor usually is not important for optimization algorithms.

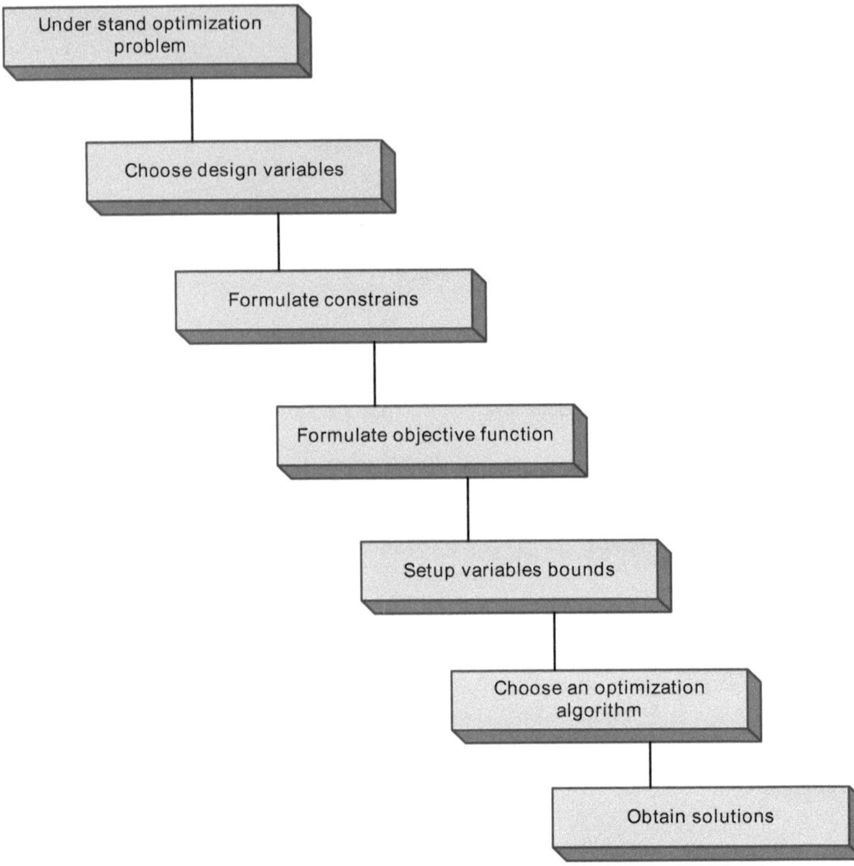

Fig. 2 Optimal design procedure main steps.

2.7 Optimization algorithm selection

There are many optimization algorithms in the literature. It is a challenging problem to choose one algorithm among hundreds of popular optimization algorithms. The selection of a suitable algorithm depends on the type of optimization problem. In the literature, different classifications of optimization problems have been proposed [4]:

2.7.1 Discrete optimization versus continuous optimization

Discrete optimization:

- The model sense for only discrete values of variable and parameters often a set of integers.

- Many discrete optimization algorithms generate a sequence of continuous sub problems.

Continuous optimization:

- The model sense for only continuous values of variable and parameters often a real value.
- Easier to use.

2.7.2 Unconstrained optimization versus constrained optimization

Variables constraint is an important factor that distinguishes between optimization problems.

Unconstrained optimization: *no constraints* for the variables

Constrained optimization: the variable values should be within the constraint's lower and upper limits. Based on the characteristics of constraints, this problem can be categorized as linear, nonlinear, quadratic, convex, etc. Further, based on the smoothness of the functions, it is categorized as differentiable or nondifferentiable.

2.7.3 None, one, or many objectives

The classification factor here depends on the number of objective functions.

No objective functions: this type of optimization problem has no objective function.

One objective function: There is just one objective function in this sort of optimization problem.

Multiobjective function: This type of objective problem has more than one objective in a multiobjective function.

2.7.4 Deterministic versus stochastic optimization

Deterministic optimization: It is commensurate with the problems that have accurate data. However, knowing the data for most of the real problems cannot be identified for more than one reason. Deterministic methods are better suited to linear than nonlinear optimization problems.

Stochastic optimization: These techniques produce and utilize random variables; therefore, to find the optimal solution, these techniques search the space in a global way. Metaheuristic algorithms are one of the most often used stochastic approaches. Metaheuristic algorithms are based on numerous natural concepts such as evolutionary and swarm intelligence algorithms as well as physical theory, as shown in Fig. 3. Evolutionary algorithms simulate the principals of evolution in nature [5].

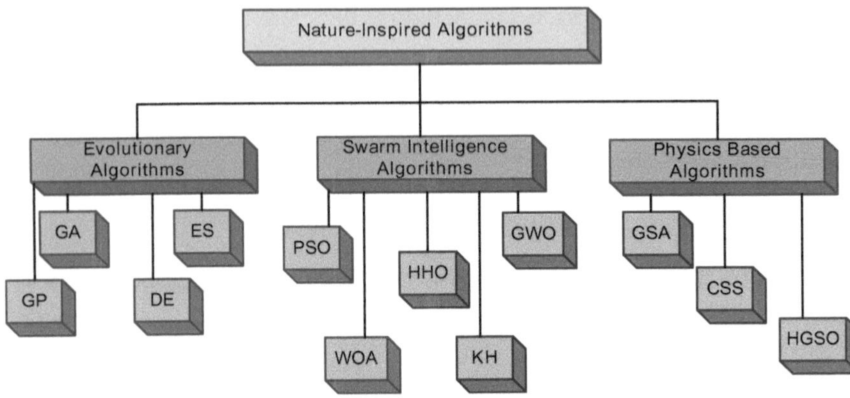

Fig. 3 Nature-inspired algorithm main classification.

3 Solar cell models

There are different mathematical models for photovoltaic (PV) solar cells [6]. Basically, solar cells can be classified by three different mathematical models, which are as follows:

1. Single diode (SD) model

   It is the simplest PV model. It has five unknown parameters, which are as follows: PV current (I_{pv}), saturation current of the diode (I_D), the factor of ideality a, series resistance (R_s), and parallel resistance (R_{sh}). This model has accuracy issues. Fig. 4 shows the single diode model.

2. Double diode (DD)

   This PV model is more elaborate. This model has the advantage of more precision but more complexity due to the addition of other unknown parameters of the second added diode. Other parameters are similar to the single diode model. Fig. 5 presents the double diode model of a PV cell.

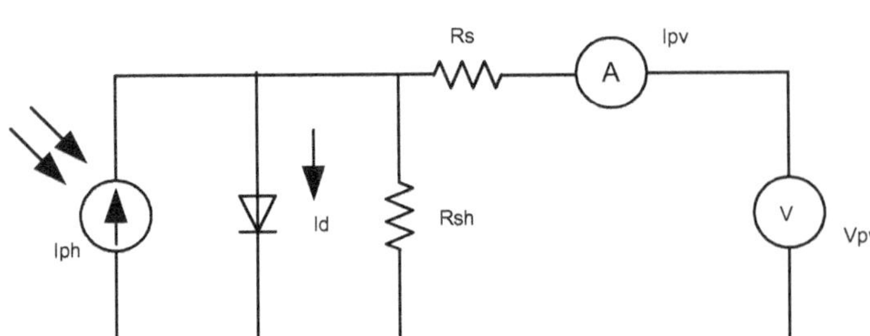

Fig. 4 Single diode model.

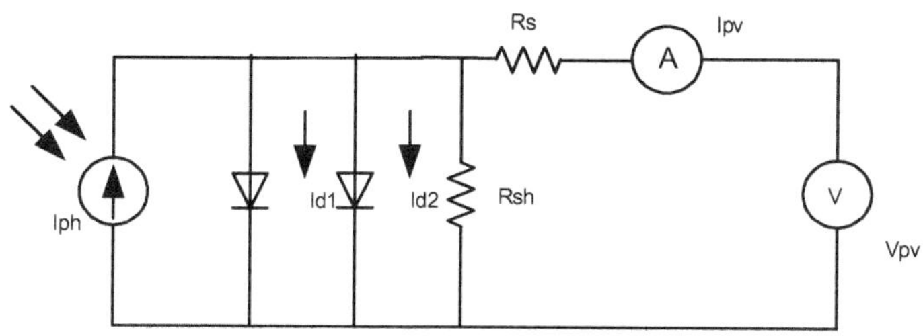

Fig. 5 Double diode model.

3. Three diode (TD) model

   It is a more detailed model to deal with low solar radiation in large PV plants. In this model, new parameters of leakage current are seen with gain limits. The model has nine parameters to compute.

4 Overview of the optimization parameters used in PV parameter extraction

PV parameter extraction is considered a big research area of recent years due to its direct effect on the cost and time of PV cell design. Different methods in the literature are applied to obtain the best parameters as well as an accurate PV model. The main classifications of this method are analytical or mathematical optimization and stochastic optimization [7,8]. Two subclassification methods are hybrid mathematical optimization and stochastic optimization and hybrid stochastic optimization; see Fig. 6.

Analytical methods use mathematical analysis to obtain the required model parameters. One of the famous analytical methods is curve fitting using the Newton-Raphson algorithm and gradient-based information methods [9]. This method has the advantage of fast and accurate performance, so it is suitable for simple models such as the SD model in PV models. But with more complex and nonlinear models such as the DD model and TDM, using these methods will be very complex and they may be trapped in local optima solutions.

Stochastic optimization techniques have been applied in the literature to overcome the drawbacks of analytical methods. As described in Section 2, stochastic optimization starts its search by random values. In each iteration change, this value is based on the learning rate and the required objective function; the popular types of stochastic techniques used for DD and TD models are metaheuristic algorithms [10]. The history of the application of this method is presented in [8,9]. The literature review in the previous

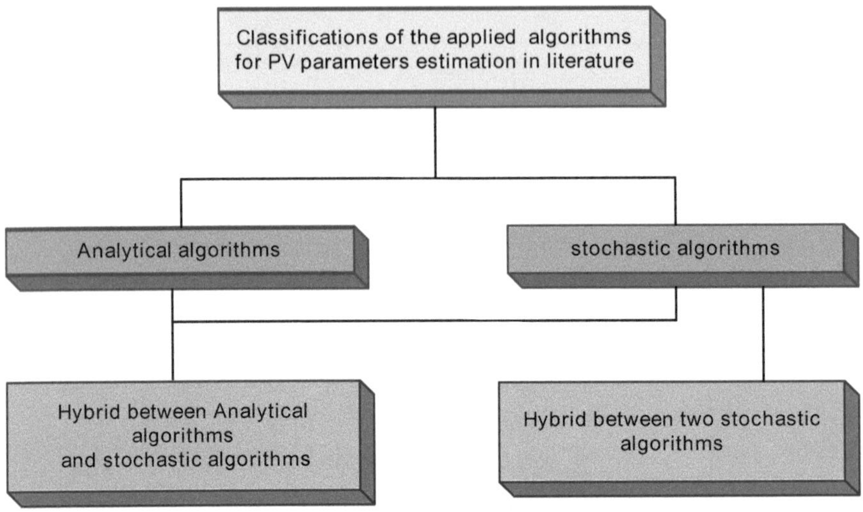

Fig. 6 Classifications of the applied algorithms for PV parameter estimation in the literature.

work focuses on SD and DD but the work on TD is missed. As this history is very large and to avoid repetition from other reviews [8,9], a brief notice about this work is presented in Table 1. The detailed study in this section focuses on the missed work in these references.

Table 1 A brief summary of the previous work in SD and DD.

Ref.	Technique	Type	Year
[11]	Memetic algorithm	Combines metaheuristic and gradient-based techniques	2015
[12]	Chaotic optimization approach (COA)	Metaheuristic	2019
[13]	Hybridized optimizer(HISA)		2019
[14]	Chaotic heterogeneous comprehensive learning particle swarm optimizer (HCLPSO)	Metaheuristic	2019
[15]	Improved opposition-based whale optimization algorithm (OBWOA)	Metaheuristic	2018
[16]	Particle swarm optimization with adaptive mutation strategy (MPSO)	Metaheuristic	2018

Hybrid mathematical optimization and stochastic optimization are used in the literature to employ the advantages of the two types in solving complex problems such as the DD model and TD model. The normal way is to simplify the optimization problem by calculating one or two of the problem parameters by an analytical method and estimate the rest of the parameters by stochastic techniques [11,17–19].

Hybrid stochastic optimization: this is implemented by make a hybridization between two stochastic techniques. The idea behind this way is to improve and enhance the performance of one algorithm to enhance and improve some drawbacks in one algorithm by the second one. This way is helpful for some special objective functions but it cannot be generalized for all problems. Table 1 presents a brief summary of the previous work in SD and DD.

In this section, a focus on recent optimization algorithms for PV parameter estimation is presented. The work is concentrated mainly on TD model application as it was missed in a different literature review. A summary about this work is presented in Table 2. Table 2 presents the algorithm in each reference, the type of algorithm based on the above classification, the PV model selected in each study, the proposed selected real PV systems, and finally the year of publication of each work. A detailed discussion is presented below:

Ref. [20] presents an application of the marine predators algorithm for the parameter estimation of the TD model. Ref. [21] presents an application of the improved wind-driven optimization algorithm. This improvement is based on hybridization between the differential evolution algorithm and the wind-driven optimization algorithm, to improve the exploration of the wind-driven optimization algorithm. In Ref. [22], an application of different metaheuristics for the parameter estimation of SD, DD, and TD is presented. These algorithms are genetic algorithm (GA), particle swarm optimization (PSO), Harmony Search (HS), flower pollination algorithm (FPA), simulated annealing (SA), and teaching-learning-based optimization (TLBO). In [17], the extraction of the SD model PV parameters is done through analytical methods using the three main points: short circuit current $(I_{sh}, 0)$, open circuit voltage $(0, V_{oc})$, and maximum power point (I_m, V_m). These three points are always presented in the datasheet of the PV cell by the manufacturer. The advantages of this method are fast implementation procedure and good accuracy. But the disadvantage here is the increasing complexity of calculation using two-dimensional and three-dimensional models. The PV parameters of the modified double diode model (MDDM) and modified triple diode model (MTDM) are estimated using two metaheuristic algorithms

Table 2 A summary of the previous work in SD, DD, and TD models.

Ref.	Technique	Type	Model	Application	Year
[20]	Marine predators algorithm	Metaheuristic	TD	KC200GT, MSX-60	2020
[21]	Improved wind driven optimization by differential evolution algorithm	Hybrid metaheuristic	TD	LG300N1C-A3, JAP6-60-250W/3BB(M2), Avancis Power Max smart 125W(M3)	2020
[22]	Different algorithms (GA), (PSO), (HS), (FPA), (SA), (TLBO)	Metaheuristic	SD, DD, TD	5 W polycrystalline	2020
[17]	Analytical method	Analytical Method	SD	BP3235T, KC200GT, BP-SX 150, and MSX60 PV modules	2020
[23]	Improved wind driven optimization	Metaheuristic	SD, DD	Kyocera—KC200GT 215 and RTC France solar cell	2017
[24]	Closed-loop particle swarm optimization and elephant herd optimization	Metaheuristic	MDD, MTD	MCSSC of area 7.7 cm^2 from Q6-1380 and CS6P-240P	2021
[25]	Particle swarm optimization	Metaheuristic	TD	30XLS	2019
[10]	Grasshopper optimization algorithm	Metaheuristic	TD	Kyocera KC200GT and Solarex MSX-60	2020
[18]	Sun flower optimization algorithm	Metaheuristic and analytical	TD	(Multicrystal KC200GT, polycrystalline MSX-60, and monocrystalline CS6K-280M)	2019

[19]	Harries-Hawks optimization algorithm	Metaheuristic and analytical	TD	(Multicrystal KC200GT, monocrystalline CS6K280M)	2020
[26]	Tree growth based optimization algorithm	Metaheuristic	SD, DD, TD	RTC France solar cells and PVM 752 GaAs thin-film cell Photowatt-PWP201	2020
[27]	Elephant herd algorithm	Metaheuristic	SD, DD, TD	Perovskite solar cell (PSCs)	2019
[28]	Elephant herd algorithm	Metaheuristic	SD	PSCs for energy efficiency improvement	2020
[29]	An interval branch and bound global optimizer	Metaheuristic and analytical	SD, DD, TDD	RTC France solar cells	2020
[29]	Turbulent flow of water optimizer	Metaheuristic	SD, DD, TDD	RTC France solar cells	2021
[30]	Elephant herd optimizer and closed loop PSO	Metaheuristic	Modified DD and Modified TD	MCS solar cells	2021
[31]	Forensic algorithm	Metaheuristic	SD, DD, TD	RTC France solar cells	2021
[32]	Improved bonobo optimizer	Metaheuristic	SD, DD, TD	RTC France solar cells	2021

in [24]. The change is accomplished by connecting an electric resistance in series with the second diode to simulate grain boundaries that limit cell conductivity. Closed-loop particle swarm optimization (CLPSO) and elephant herd optimization (EHO) are the two methods used. According to the findings of the paper, MTDM is more accurate than MDDM. Ref. [25] presents a simple application PSO for the parameter extraction of nine parameters of PV TDM. The PSO has an advantage of increasing the probability of reaching the global minimum solution for the problem in a short time. Ref. [10] proposes an application of the grasshopper optimization algorithm (GOA) for the parameter estimation of TDM using two examples (Kyocera KC200GT and Solarex MSX-60). In Ref. [18], a simplification for the complexity of the TDM nine parameters estimation has been proposed by calculating two parameters analytically (series resistance (R_s) and Photo-generated current (I_{ph})). The rest of parameters are estimated using the sunflower optimization algorithm. The proposed previous process is retained on the three main point measurements (short circuit point, open circuit point, and maximum power point). The authors of [18] presented a combination method between analytics and Harries Hawks optimization in [19]. The main advantages of this method are the complexity reduction for TDM parameter estimation and saving the time of real experiments by using the main three manufacturer measurements listed in solar cell datasheets. In [26], a metaheuristic algorithm called the tree growth-based optimization algorithm was used to estimate the parameters of the SD, DD, and TD models.

5 Conclusion

This chapter presented an overview of the various optimization techniques utilized for the purpose of solar photovoltaic parameter optimization. A detailed overview of the different optimization techniques based on the various characteristics was presented. Further, different types of PV models such as single diode, double diode, and three diode were discussed. A comparative analysis between the single diode and double diode models was also presented in tabular form.

Acknowledgement

The authors thank the support of the National Research and Development Agency of Chile (ANID), ANID/Fondap/15110019.

References

[1] A.A.K. Ismaeel, E.H. Houssein, D. Oliva, M. Said, Gradient-based optimizer for parameter extraction in photovoltaic models, IEEE Access 9 (2021).

[2] C. Saha, N. Agbu, R. Jinks, M.N. Huda, Review article of the solar PV parameters estimation using evolutionary algorithms, MOJ Solar Photoen. Syst. 2 (2) (2018) 63–75.

[3] A.J. Kulkarni, A. Abraham, G.K. Monash, Introduction to Optimization, September 2017.

[4] . www.neos-guide.org/.

[5] M.H. Nadimi-Shahraki, S. Taghian, S. Mirjalili, An improved grey wolf optimizer for solving engineering problems, Expert Syst. Appl. 166 (2021).

[6] K. Et-torabi, I. Nassar-eddine, A. Obbadi, Y. Errami, R. Rmaily, S. Sahnoun, A. El Fajri, M. Agunaou, Parameters estimation of the single and double diode photovoltaic models using a Gauss–Seidel algorithm and analytical method: a comparative study, Energy Convers. Manag. 148 (2017) 1041–1054.

[7] R. Tamrakar, A. Gupta, A Review: extraction of solar cell modelling parameters, Int. J. Innov. Res. Electr. Electron. Inst. Control Eng. 3 (1) (2015).

[8] B. Maniraj, A.P. Fathima, Parameter extraction of solar photovoltaic modules using various optimization techniques: a review, in: National Science, Engineering and Technology Conference (NCSET), 2020.

[9] F. Ghani, M. Duke, J. Carsona, Numerical calculation of series and shunt resistance of a photovoltaic cell using the Lambert W-function: experimental evaluation, Sol. Energy 87 (2013) 246–253.

[10] O.S. Elazab, H.M. Hasanien, I. Alsaidan, A.Y. Abdelaziz, S.M. Muyeen, Parameter estimation of three diode photovoltaic model using grasshopper optimization algorithm, Energies 13 (2020) 497.

[11] Y. Yoon, Z.W. Geem, Parameter optimization of single-diode model of photovoltaic cell using memetic algorithm, Int. J. Photoenergy 2015 (2015), 963562. 7 p.

[12] M. Ćalasan, D. Jovanović, V. Rubežić, S. Mujović, S. Đukanović, Estimation of single-diode and two-diode solar cell parameters by using a chaotic optimization approach, Energies 12 (21) (2019) 4209.

[13] K. Dhruv, Y. Goswami, R.V. Kumar, A novel approach to parameter estimation of photovoltaic systems using hybridized optimizer, Energy Convers. Manag. 187 (2019) 486–511.

[14] Y. Dalia, A. Dalia, M.B. Eteiba, N.S. Ponnuthurai, Static and dynamic photovoltaic models' parameters identification using chaotic heterogeneous comprehensive learning particle swarm optimizer variants, Energy Convers. Manag. 182 (2019) 546–563.

[15] E.M. Abd, O. Diego, Parameter estimation of solar cells diode models by an improved opposition-based whale optimization algorithm, Energy Convers. Manag. 171 (2018) 1843–1859.

[16] M. Manel, S. Anis, M.M. Faouzi, Particle swarm optimisation with adaptive mutation strategy for photovoltaic solar cell/module parameter extraction, Energy Convers. Manag. 175 (2018) 151–163.

[17] R. Ndegwa, J. Simiyu, E. Ayieta, N. Odero, A fast and accurate analytical method for parameter determination of a photovoltaic system based on manufacturer's data, Hindawi J. Renew. Energy 2020 (2020). 18 p.

[18] M.H. Qais, H.M. Hasanien, S. Alghuwainem, Identification of electrical parameters for three-diode photovoltaic model using analytical and sun flower optimization algorithm, Appl. Energy 250 (2019) 109–117.

[19] M.H. Qais, H.M. Hasanien, S. Alghuwainem, Parameters extraction of three-diode photovoltaic model using computation and Harries-Hawks optimization algorithm, Appl. Energy 195 (2020) 109–117.

[20] M.A. Soliman, H.M. Hasanien, A. Alkuhayli, Marine predators algorithm for parameters identification of triple-diode photovoltaic models, IEEE Access 8 (2020).
[21] I.A. Ibrahim, M.J. Hossain, B.C. Duck, M. Nadarajah, An improved wind b driven optimization algorithm for parameters identification of a triple-diode photovoltaic cell model, Energy Conver. Manag. 213 (2020).
[22] V. Obiora, C. Saha, A. Al Bazi, K. Guha, Optimisation of solar photovoltaic (PV) parameters using metaheuristics, Microsyst. Technol. (2020).
[23] M. Derick, C. Rani, M. Rajesh, M.E. Farrag, Y. Wang, K. Busawon, An improved optimization technique for estimation of solar photovoltaic parameters, Solar Energy 157 (15) (2017) 116–124.
[24] A.S. Bayoumi, R.A. El-Sehiemy, K. Mahmoud, M. Lehtonen, M.M.F. Darwish, Assessment of an improved three-diode against modified two-diode patterns of MCS solar cells associated with soft parameter estimation paradigms, Appl. Sci. 11 (2021).
[25] A. Harrag, Y. Daili, Three-diodes PV model parameters extraction using PSO algorithm, Revue des Energies Renouvelables 22 (2019) 85–91.
[26] A.A. Zaki Diab, H.M. Sultan, R. Aljendy, A.S. Al-Sumaiti, M. Shoyama, Z.M. Ali, Tree growth based optimization algorithm for parameter extraction of different models of photovoltaic cells and modules, IEEE Access 8 (2020).
[27] A.A. Zaky, et al., Energy efficiency improvement of water pumping system using synchronous reluctance motor fed by perovskite solar cells, Int. J. Energy Res. 44 (14) (2020) 11629–11642.
[28] R. Chenouard, R.A. El-Sehiemy, An interval branch and bound global optimization algorithm for parameter estimation of three photovoltaic models, Energy Conversion and Management 205 (2020).
[29] A.A. Zaky, et al., Optimal performance emulation of PSCs using the elephant herd algorithm associated with experimental validation, ECS J. Solid State Sci. Technol. 8 (12) (2019) Q249–Q255.
[30] M. Said, et al., Estimating parameters of photovoltaic models using accurate turbulent flow of water optimizer, Processes 9 (4) (2021) 627. https://www.mdpi.com/2227-9717/9/4/627.
[31] R.A. El-Sehiemy, A.M. Shaheen, A. Ginidi, S.S.M. Ghoneim, A forensic-based investigation algorithm for parameter extraction of solar cell models, IEEE Access (2020). https://ieeexplore.ieee.org/document/9303360.
[32] R.Y. Abdelghany, et al., Development of an improved bonobo optimizer and its application for solar cell parameter estimation, Sustainability 13 (7) (2021). https://www.mdpi.com/2071-1050/13/7/3863.

CHAPTER 14.3

Parametric optimization of a segmented thermoelectric generator in a concentrated photovoltaic system under fluctuating weather conditions

Chika Maduabuchi[a], Ravita Lamba[b], Chigbogu Ozoegwu[a], Howard O. Njoku[c,d], Mkpamdi Eke[a], and Emenike C. Ejiogu[e,f]

[a]Department of Mechanical Engineering, University of Nigeria, Nsukka, Nigeria
[b]Department of Electrical Engineering, Malaviya National Institute of Technology Jaipur, Jaipur, India
[c]Applied Renewable and Sustainable Energy Research Group, Department of Mechanical Engineering, University of Nigeria, Nsukka, Nigeria
[d]Department of Mechanical Engineering Science, University of Johannesburg, Johannesburg, South Africa
[e]Laboratory of Industrial Electronics, Power Devices and New Energy Systems, University of Nigeria, Nsukka, Nigeria
[f]Africa Centre of Excellence for Sustainable Power and Energy Development, University of Nigeria, Nsukka, Nigeria

Contents

Active Electrical Distribution Network
https://doi.org/10.1016/B978-0-323-85169-5.00020-4

1 Introduction

Photovoltaic (PV) systems are capable of generating electricity directly from the sun, thus making the PV a clean alternative energy generation technology [1]. Apart from electricity, the photovoltaic systems also produce a significant amount of heat during the photovoltaic process. Thus, a major portion of the incoming solar energy is lost in the form of heat. This waste heat can be utilized to generate electricity using a thermoelectric generator (TEG). This is made possible by the Seebeck effect, which is the generation of electricity at the junction of two dissimilar metals when exposed to a temperature gradient. A combination of PV and TEG systems can effectively utilize the incoming solar radiation because the PV mostly uses the visible and ultraviolet parts while the TEG relies on the infrared part. The photovoltaic-thermoelectric generator (PV-TEG) hybrid system can be developed using the direct coupling or spectrum splitting technique [2]. In the spectrum splitting method, beam splitters are used to discretize the incident solar irradiation at a precise cut-off wavelength for energy utilization using the PV and TEG, respectively [3]. On the other hand, the direct coupling method eliminates the requirement of the beam splitter because in a parallel arrangement, the TEG is attached to the PV backplate [4].

One of the effective means of improving the performance of stand-alone and hybrid systems is by implementing geometry optimization. Fan et al. [5] developed a numerical model to determine the optimal thermoelectric (TE) leg length and area for achieving TEG optimum power output. Their findings showed that increasing the TE leg length decreased the TEG maximum power output density.

Li et al. [6] optimized TE geometry in a hybrid PV-TEG system under steady conditions and found out that the optimal geometries in TEG and PV-TEG were different. Also, a comprehensive model was used by Hashim et al. [7] in investigating the optimal TE geometry in a PV-TEG system. Their findings postulated that the amount of TE material used must be taken into account when performing geometric optimization for maximum power. Lamba et al. [8] established that the temperature gradient varied inversely with TE leg length in the concentrated photovoltaic-thermoelectric generator (CPV-TEG) system hybrid. Also, Zhu et al. [9] developed a three-dimensional (3D) model to analyze a CPV-TEG system. It was shown that the CPV-TEG system provided an additional 4% increase in the power output due to the incorporation of the TEG. Shittu et al. [10] used COMSOL multiphysics software to investigate TE geometry in the CPV-TEG system under changing climate conditions.

From the literature reviewed, the feasibility of a CPV-TEG hybrid system and its ability to improve the performance of a stand-alone concentrated photovoltaic system have been demonstrated. Further, the merits of performing transient analysis instead of steady have been shown and the consequence of utilizing 3D models in place of two-dimensional models has been established. However, there is no study on the optimization of thermoelectric geometry in a CPV system using a segmented thermoelectric generator (STEG). In addition, the impact of TE leg geometry such as leg length, cross-section area, ceramic length, and fin length is yet to be studied in a hybrid concentrated photovoltaic system with a segmented thermoelectric generator.

Hence, this study presents a comprehensive 3D numerical analysis on optimum geometry in a CPV-segmented TEG (CPV-STEG) subject to fluctuating weather conditions. Also, the effect of TE geometric parameters such as leg length, area, and ceramic length on the overall performance of the CPV-STEG is studied. Apart from this, a real-time performance comparison of the CPV-STEG with and without optimum geometry is performed under dry season meteorological conditions peculiar to the University of Nigeria, Nsukka, Enugu, Nigeria, climate. Finally, the effect of concentration ratio on the CPV-STEG performance is analyzed.

The subsequent sections of this work are arranged thusly: an overview of the hybrid system arrangement and the thermoelectric materials adopted in this study is depicted in Section 2. The numerical model employed in the analysis of the CPV, STEG, and hybrid system is presented in Section 3 while Section 4 shows the results obtained accompanied by subsequent discussions. Finally, Section 5 presents the conclusions.

2 System description and materials

Fig. 1 depicts the exploded view of the hybrid CPV-TEG system containing the CPV, STEG, and heat sink systems. The CPV used in this study is a polycrystalline silicon PV module comprising different layers of top glass cover, ethylene vinyl acetate (EVA), PV cell, and tedlar polyester tedlar (TPT). The EVA is the encapsulation for binding the solar cell, cell top surface, and TPT. The STEG is made up of *p-type* DDFeCoSbS and *n-type* InCoSb at the hot regions while *p* and *n-type* BiTe are used at the cold regions. This arrangement is preferred due to the optimal performance of the materials at the given temperature range. The thermoelectric properties considered for the analysis of the hybrid system are temperature-dependent. Furthermore, the STEG is comprised of ceramic plates (specified as 96% alumina), copper electrodes, and solder paste. The cooling method used is water cooling and

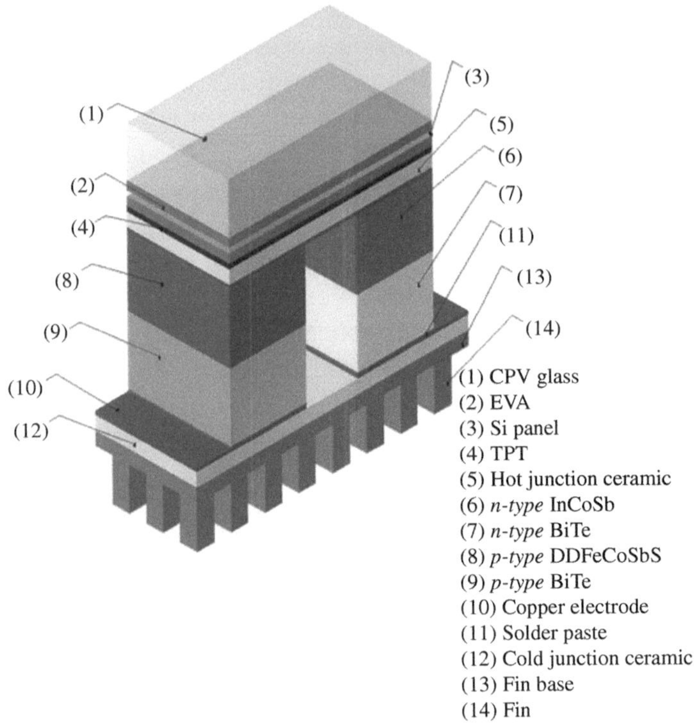

Fig. 1 Exploded view of the hybrid system.

the heat sink material is aluminum. Throughout this study, the convective heat transfer coefficient and concentration ratio of $500\,\mathrm{W/m^2K}$ and 30, respectively, are used.

3 Numerical model

The numerical model consists of CPV, TEG, and CPV-TEG hybrid system models. Initially, the CPV and STEG systems are modeled independently and then assembled as the hybrid system.

3.1 Concentrated photovoltaic system

The 3D energy equation for a homogenous and isotropic medium gives the temperature distribution in the photovoltaic layers as [11,12]

$$\rho_d C_p \frac{\partial T}{\partial t} - \vec{\nabla} \cdot (k \nabla T) = q_{abs} - P_{PV} \tag{1}$$

where C_p, k, ρ_d, T represent the specific heat capacity at constant pressure, thermal conductivity, and the density and temperature of each layer, respectively. In addition, q_{abs} and P_{PV} are the absorbed solar energy and polycrystalline silicon layer power output, respectively.

The solar energy absorbed by the nth layer is expressed as

$$q_{abs,n} = C \times G_{abs,n} \times A_n \times \alpha_n \tag{2}$$

$$G_{abs,n} = G_{abs,n-1} \times \left[(1-\alpha_{n-1}) - \rho_{n-1}\right] \tag{3}$$

where ρ_n, α_n, and A_n are the reflectivity, absorptivity, and area of the nth layer, respectively. Also, C and $G_{abs,n}$ are the concentration ratio and the incident solar radiation on the nth layer, respectively.

The CPV efficiency and power output are given as [13]

$$\eta_{CPV} = \eta_r[1 - \beta_{PV}(T_{PV} - T_r)] \tag{4}$$

$$P_{CPV} = q_{abs,3} \times \eta_{PV} \tag{5}$$

where T_r, η_r, and β_{PV} are the reference temperature (298.15 K), efficiency (17%), and PV thermal expansion coefficient (0.00451 K^{-1}), respectively. Also, η_{PV} is the PV efficiency and $q_{abs,3}$ is the absorbed solar energy by the PV cell layer. The thermal-electric coupled field equations are simultaneously solved with the finite element method.

Radiative heat loss at the PV glass top surface is given as

$$q_{rad} = \varepsilon\sigma\left(T_{PV}^2 + T_{sky}^2\right)\left(T_{PV} + T_{sky}\right) \tag{6}$$

where σ and ε are the Stefan-Boltzmann constant and emissivity, respectively. Likewise, T_{sky} is the sky temperature, expressed as [14]

$$T_{sky} = 0.0552\,T_a^{1.5} \tag{7}$$

where the ambient temperature is T_a.

At the PV top glass surface, the convective heat transfer is taken. The convection heat transfer coefficient is specified as [15]

$$h_{conv} = 5.7 + 3.8v \tag{8}$$

where v is wind speed in ms^{-1}.

3.2 Segmented thermoelectric generator

The STEG temperature distribution is obtained by solving the coupled field equations [14]

$$\vec{\nabla}\cdot\left(k\vec{\nabla}T\right)+J^2\rho-T\frac{dS}{dT}\vec{J}\cdot\vec{\nabla}T=0 \tag{9}$$

where k, ρ, S, $T\frac{dS}{dT}$, and $\vec{J}$ are the temperature-dependent thermal conductivity, electrical resistivity, Seebeck coefficient, Thomson coefficient, and current density vector, respectively.

The STEG power output and efficiency are given as

$$P_{STEG}=I^2R_L \tag{10}$$

$$\eta_{STEG}=\frac{P_{STEG}}{Q_h} \tag{11}$$

where I, R_L, and Q_h are the electric current, external load resistance, and STEG heat input, respectively.

3.3 Hybrid system

The CPV-STEG power output and efficiency are expressed as

$$P_{CPV-STEG}=P_{CPV}+P_{STEG} \tag{12}$$

$$\eta_{CPV-STEG}=\eta_{CPV}+\eta_{STEG} \tag{13}$$

4 Results and discussion

Transient real-time meteorological variations are considered in this study for dry season conditions at the University of Nigeria, Nsukka, Enugu, Nigeria on March 1, 2010. The simulations were performed for 24h to include day and night weather variations. Fig. 2 shows the hourly meteorological data taken from the Centre for Basic Space Science, University of Nigeria, Nsukka [16]. The transient solar radiation is depicted in Fig. 2A while atmospheric temperature and wind variations are captured in Fig. 2B.

4.1 Effect of thermoelement leg geometry

The variations of TE leg geometry parameter and leg length affect the STEG performance and thus, significantly alter the CPV-STEG system's overall performance. Hence, the effect of varying the TE leg length on the performance of the CPV, STEG, and CPV-STEG system is depicted in Fig. 3 using reference values at a maximum operation time of 3p.m. Fig. 3 demonstrates that the CPV cell average temperature increases with increasing leg length, thus decreasing the CPV and CPV-STEG power output and efficiency. This infers that using STEG with long legs will greatly reduce the

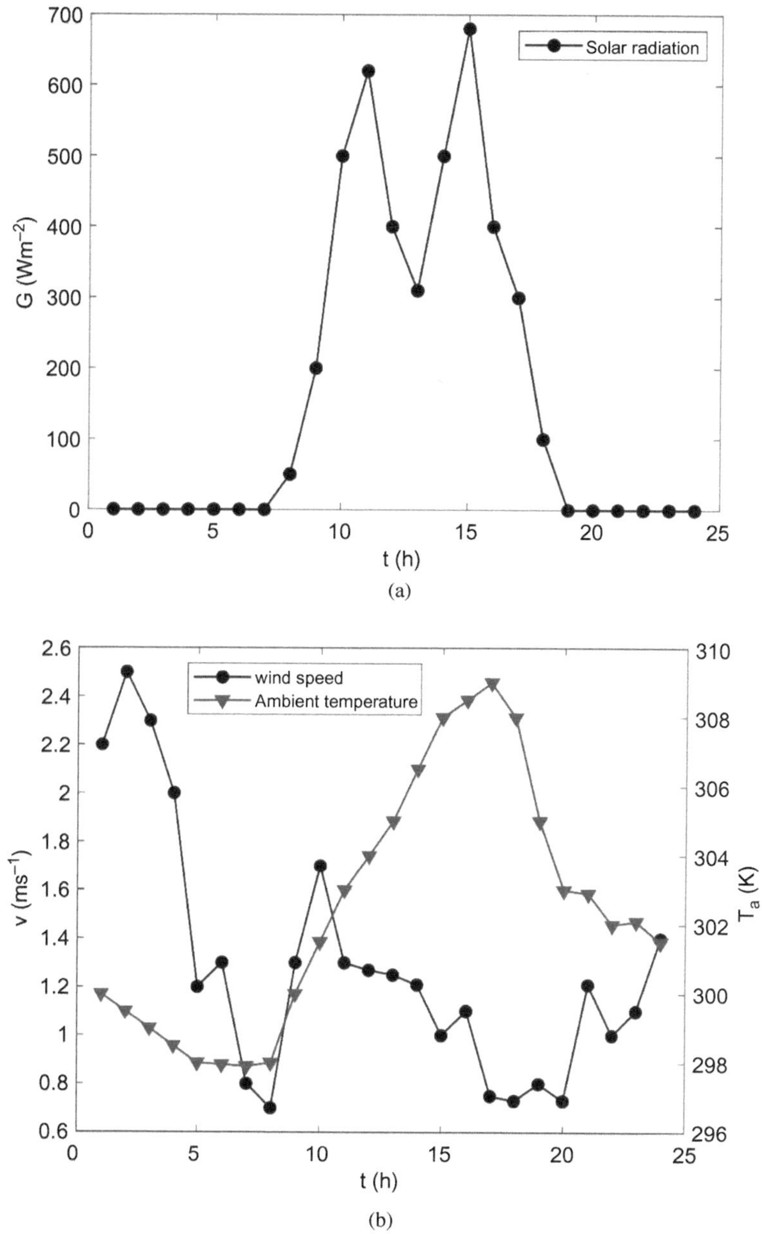

Fig. 2 Hourly meteorological data: (A) solar radiation, (B) wind speed and ambient temperature [16].

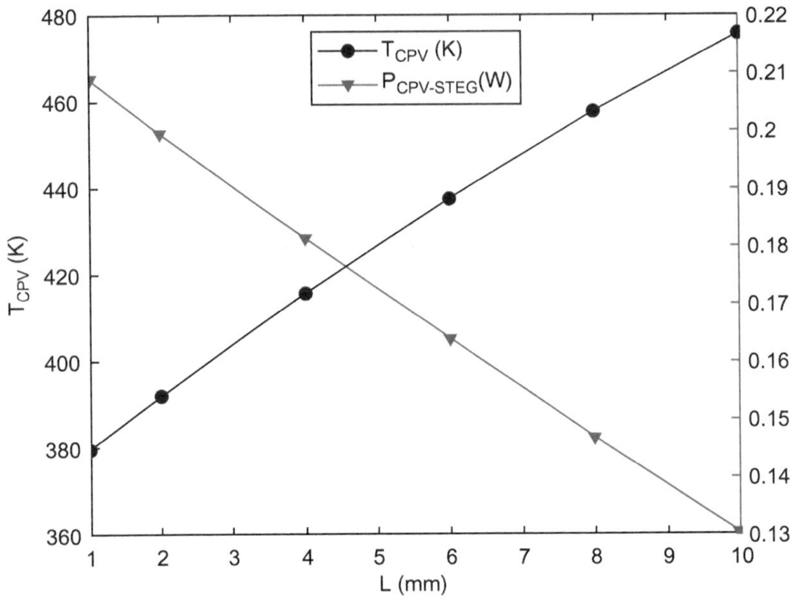

Fig. 3 Effect of length of TE leg variation on the temperature of the CPV cell and the power output of CPV-STEG.

thermal conductivity of the STEG. This in turn, will lead to a build-up of heat at the hot junction ceramic plate attached to the PV back plate, thus increasing the CPV cell temperature. Typical values show that using a leg length of 1 mm in comparison with 10 mm provided a 58% and 60% increase in the efficiency and power output, respectively.

The effect of increasing leg length on the STEG performance parameters including the temperature gradient, conversion efficiency, and power output has been captured in Fig. 4A–C. It is seen that the STEG temperature gradient, conversion efficiency, and power output increase linearly with the leg length. This is because increasing the leg length directly increases the thermal retention capability of the STEG, thus increasing its temperature gradient and performance parameters. This demonstrates the merits of performing a parametric optimization because the conditions for obtaining maximum system performance in a STEG differ from a CPV-STEG system.

4.2 Effect of fin length

Because the STEG cold junction is cooled using a heat sink, it becomes crucial to analyze the effect of the fin geometric parameters on the CPV-STEG system performance. Using an optimal leg length and area of 1 mm and

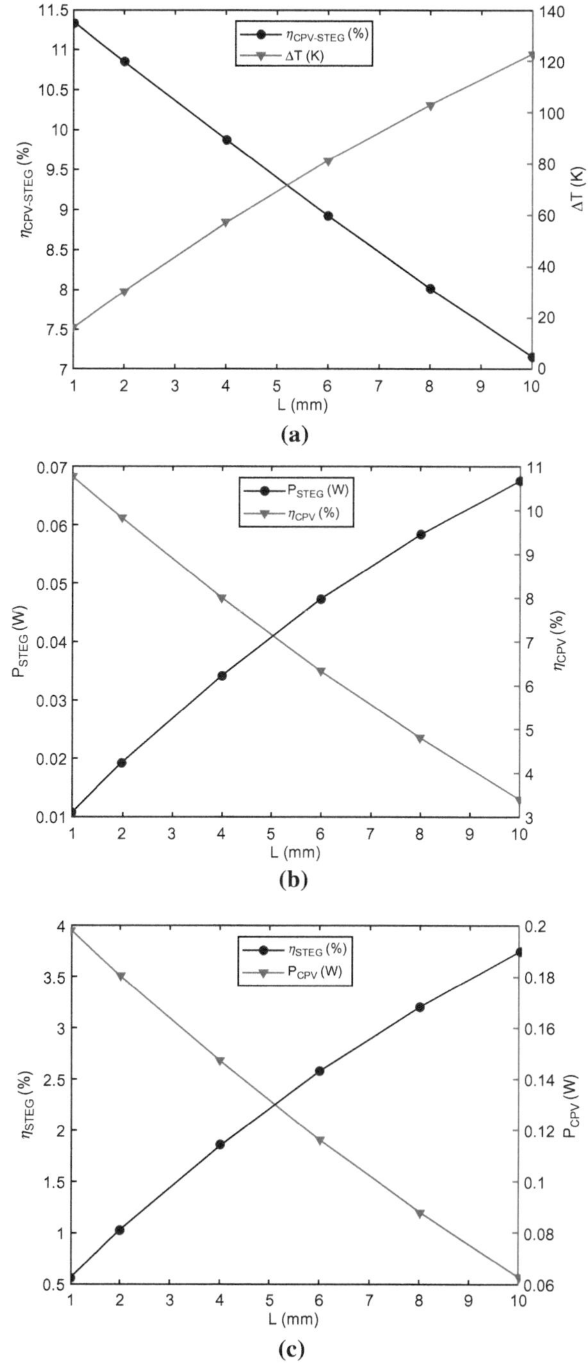

Fig. 4 Effect of length of TE leg variation on (A) CPV-STEG efficiency and STEG temperature gradient, (B) STEG power output and CPV efficiency, (C) STEG efficiency and CPV power output.

33 μm^2, respectively, as well as a ceramic length of 0.2 mm and a fin base length of 1.4 mm at a peak operating time of 3 p.m., the effect of increasing fin length on the hybrid CPV-STEG system performance is depicted in Fig. 5A and B. It is seen that increasing the fin length decreases the CPV cell average temperature, which in turn results in a corresponding increase in the CPV-STEG power output and efficiency. This is due to the enhanced rate of heat removal along the length of the longer fins, further showing that longer fin lengths are suitable for hybrid CPV-STEG systems.

4.3 Effect of concentration ratio

In concentrated PV systems, a very crucial operating parameter is the concentration ratio. Employing an optimum leg length and area of 1 mm and 33 μm^2, respectively, as well as a ceramic length of 0.2 mm, a fin base and fin length of 1.4 mm and 5 mm, respectively, and a heat transfer coefficient of 700 W/m^{2}K at a peak time of 3 p.m., The effect of concentration ratio on the performance of the CPV-STEG system is shown in Fig. 6A and B. It can be observed that an increase in the concentration ratio results in an increased CPV cell average temperature, which results in decreased CPV and CPV-STEG power output and efficiencies. This of course is a result of the increased STEG heat input. This implies that the concentration ratio in a CPV-STEG system should be the lowest possible to reduce overheating the CPV cell while generating substantial power output from the coupled STEG.

4.4 Effect of STEG geometry optimization

A very crucial result depicting the merits of optimizing a hybrid CPV-STEG system is shown in Fig. 7A–C. Utilizing reference values for heat transfer coefficient and concentration ratio of 500 W/m^{2}K and 30, respectively, which are maintained throughout this study, the performance of both systems is compared under fluctuating climate conditions as shown in Fig. 2, for a period of 24 h in order to capture day and night meteorological variations. Fig. 7A reveals that the CPV cell average temperatures obtained from the hybrid systems with optimized and original geometries were 378.53 K and 460.92 K, respectively. Hence, it is clearly demonstrated that the CPV cell average temperature can be significantly reduced by optimizing the STEG in a hybrid CPV-STEG system. Fig. 7B reveals that the CPV-STEG system with the optimum geometry generated a higher power output compared to the initial geometry. In addition, it is observed that the

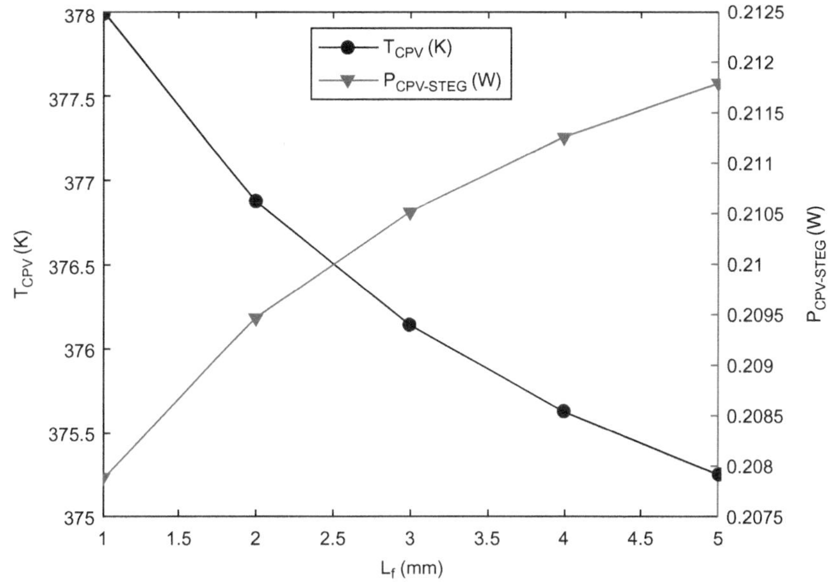

(a)

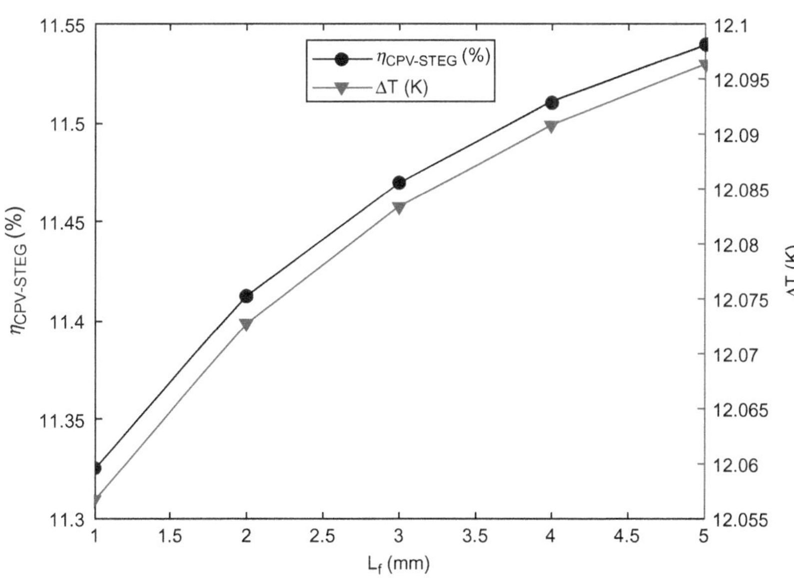

(b)

Fig. 5 Effect of fin length on (A) temperature of CPV cell and power output of CPV-STEG, (B) efficiency of CPV-STEG and temperature gradient of STEG.

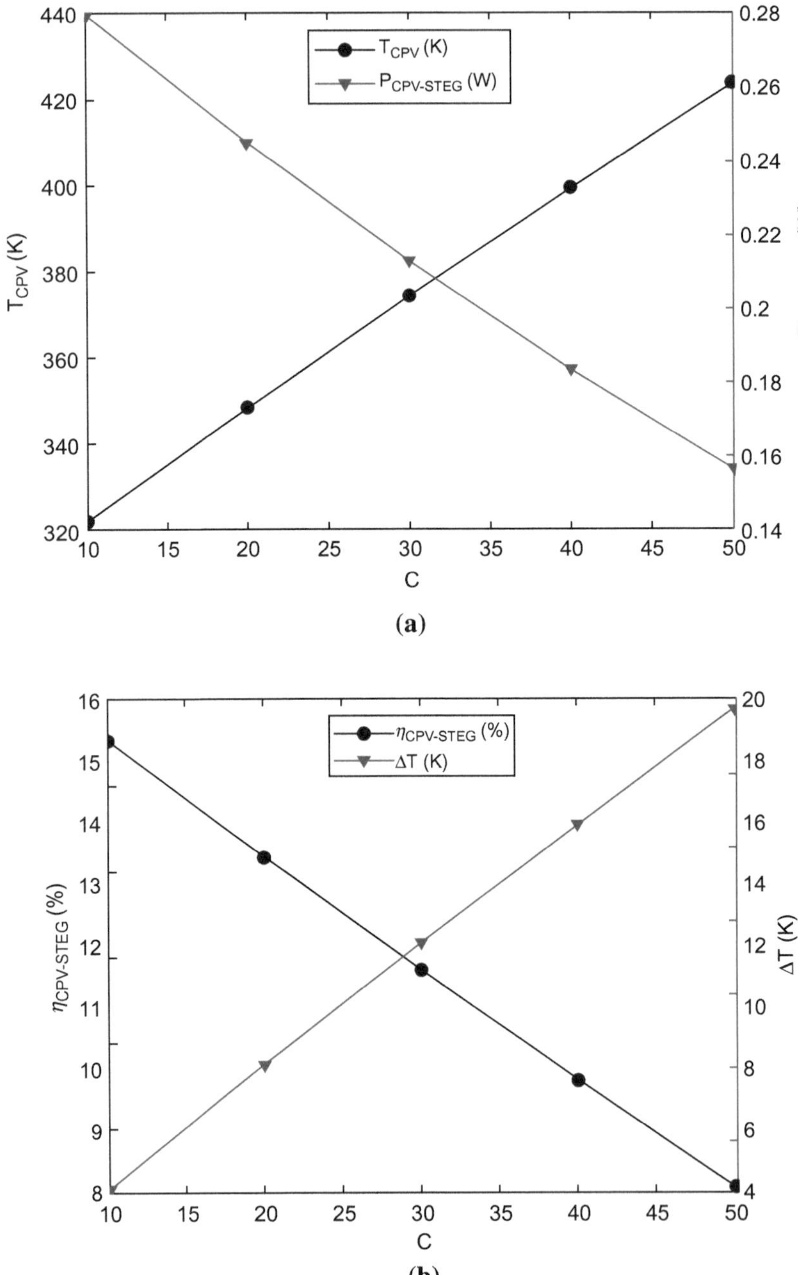

Fig. 6 Effect of concentration ratio on (A) temperature of CPV cell and power output of CPV-STEG, (B) efficiency of CPV-STEG and temperature gradient of STEG.

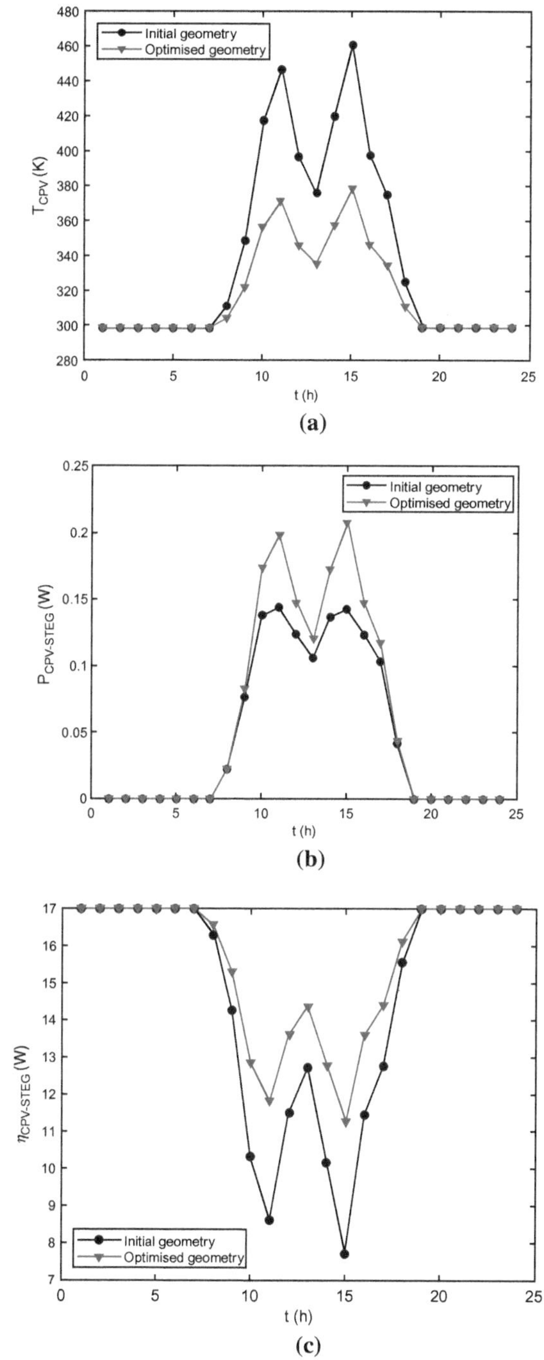

Fig. 7 Transient variation of CPV-STEG: (A) temperature of CPV cell, (B) power output, and (C) efficiency.

CPV-STEG with the optimized geometry was able to reduce the adverse effects of fluctuating weather conditions compared to initial geometry, which generated the lowest power output at very high solar radiation intensities. Thus, we postulate that STEG geometry optimization is key to alleviating the negative effects of fluctuating weather conditions. Under peak solar radiation value obtained at 3p.m., the CPV-STEG with optimized geometry increased the power output by 45.7% when the initial geometry was used. This significant increase illustrates the necessity of STEG geometry optimization in hybrid CPV-STEG systems. Fig. 7C further discloses that at 3p.m., the hybrid system with optimized geometry provided a 45.9% increase in efficiency when the hybrid system with original geometry was used.

5 Conclusions

The parametric optimization of TE geometry in a CPV-STEG hybrid system is carried out in this paper. A comprehensive 3D numerical analysis of the optimized segmented TEG geometry in a hybrid concentrated photovoltaic system under transient fluctuating climate conditions was implemented. The effects of TE leg length, concentration ratio, and fin length on the overall performance of the proposed system were parametrically investigated and the optimal values were determined. The numerical analysis was implemented for 24h to encompass day and night weather variations under dry season meteorological conditions at the University of Nigeria, Nsukka, Enugu, Nigeria. In addition, the effect of concentration ratio on the hybrid system performance was examined. The following conclusions were made:

- The adverse impact of fluctuating meteorological conditions can be significantly reduced by optimizing TE geometry in the CPV-TEG hybrid systems.
- At peak sunshine, this hybrid system with optimum geometry increased the maximum power output by 45.7% when initial geometry was used.
- Under very high solar radiation intensity, the CPV-STEG system with optimized geometry reduced and increased the CPV cell average temperature and efficiency by 21.8% and 45.9%, respectively, when the initial geometry was used.
- When the leg length is decreased from 8 to 1mm, the STEG and CPV-STEG system power outputs are decreased and increased by 83.7% and 60.3%, respectively.

Acknowledgment

The second, fifth, and sixth authors would like to acknowledge and appreciate the support offered by the Africa Centre of Excellence for Sustainable Power and Energy Development, University of Nigeria, Nsukka, toward the successful completion of this work.

Conflict of interest

The authors declare no known or outstanding personal or financial conflicts of interest.

References

[1] G. Li, T.M.O. Diallo, Y.G. Akhlaghi, S. Shittu, X. Zhao, X. Ma, et al., Simulation and experiment on thermal performance of a micro-channel heat pipe under different evaporator temperatures and tilt angles, Energy 179 (2019) 549–557, https://doi.org/10.1016/j.energy.2019.05.040.
[2] D. Narducci, P. Bermel, B. Lorenzi, N. Wang, K. Yazawa, Hybrid and Fully Thermoelectric Solar Harvesting, vol. 268, Springer International Publishing, Cham, 2018, https://doi.org/10.1007/978-3-319-76427-6.
[3] S. Shittu, G. Li, X. Zhao, X. Ma, Series of detail comparison and optimization of thermoelectric element geometry considering the PV effect, Renew. Energy 130 (2019) 930–942, https://doi.org/10.1016/j.renene.2018.07.002.
[4] S. Shittu, G. Li, Y.G. Akhlaghi, X. Ma, X. Zhao, E. Ayodele, Advancements in thermoelectric generators for enhanced hybrid photovoltaic system performance, Renew. Sustain. Energy Rev. 109 (2019) 24–54, https://doi.org/10.1016/j.rser.2019.04.023.
[5] L. Fan, G. Zhang, R. Wang, K. Jiao, A comprehensive and time-efficient model for determination of thermoelectric generator length and cross-section area, Energy Convers. Manag. 122 (2016) 85–94, https://doi.org/10.1016/j.enconman.2016.05.064.
[6] G. Li, S. Shittu, X. Ma, X. Zhao, Comparative analysis of thermoelectric elements optimum geometry between photovoltaic-thermoelectric and solar thermoelectric, Energy 171 (2019) 599–610, https://doi.org/10.1016/j.energy.2019.01.057.
[7] H. Hashim, J.J. Bomphrey, G. Min, Model for geometry optimisation of thermoelectric devices in a hybrid PV/TE system, Renew. Energy 87 (2016) 458–463, https://doi.org/10.1016/j.renene.2015.10.029.
[8] R. Lamba, S.C. Kaushik, Solar driven concentrated photovoltaic-thermoelectric hybrid system: numerical analysis and optimization, Energy Convers. Manag. 170 (2018) 34–49, https://doi.org/10.1016/j.enconman.2018.05.048.
[9] W. Zhu, Y. Deng, Y. Wang, S. Shen, R. Gulfam, High-performance photovoltaic-thermoelectric hybrid power generation system with optimized thermal management, Energy 100 (2016) 91–101, https://doi.org/10.1016/j.energy.2016.01.055.
[10] S. Shittu, G. Li, X. Tang, X. Zhao, X. Ma, A. Badiei, Analysis of thermoelectric geometry in a concentrated photovoltaic-thermoelectric under varying weather conditions, Energy 202 (2020) 1–13, https://doi.org/10.1016/j.energy.2020.117742.
[11] S. Shittu, G. Li, X. Zhao, Y.G. Akhlaghi, X. Ma, M. Yu, Comparative study of a concentrated photovoltaic-thermoelectric system with and without flat plate heat pipe, Energy Convers. Manag. 193 (2019) 1–14, https://doi.org/10.1016/j.enconman.2019.04.055.
[12] S. Shittu, G. Li, X. Zhao, X. Ma, Y. Golizadeh, Y. Fan, Comprehensive study and optimization of concentrated photovoltaic-thermoelectric considering all contact

resistances, Energy Convers. Manag. 205 (2020) 112422, https://doi.org/10.1016/j.enconman.2019.112422.
[13] D.L. Evans, Simplified method for predicting photovoltaic array output, Sol. Energy 27 (1981) 555–560.
[14] C.C. Maduabuchi, C.A. Mgbemene, Numerical study of a phase change material integrated solar thermoelectric generator, J. Electron. Mater. 49 (2020) 5917–5936, https://doi.org/10.1007/s11664-020-08331-3.
[15] S. Shittu, G. Li, Q. Xuan, X. Xiao, X. Zhao, X. Ma, Transient and non-uniform heat flux effect on solar thermoelectric generator with phase change material, Appl. Therm. Eng. 173 (2020) 1–15, https://doi.org/10.1016/j.applthermaleng.2020.115206.
[16] C. Nwosu, U.C. Uchenna, T. Madueme, Wind-solar hybrid power system for rural applications in the south eastern states of Nigeria, J. Electr. Syst. 8 (2012) 304–316.

Index

Note: Page numbers followed by *f* indicate figures, *t* indicate tables, and *b* indicate boxes.

Kolibri 360 GmbH
Ferdinand-Jühlke-Straße 7
99095 Erfurt, Deutschland
produktsicherheit@kolibri360.de